Introducing the Synergy Between Forensic Science and Environmental Analysis

Mahipal Singh Sankhla
Kumud Kant Awasthi • Tina Sharma
Rajeev Kumar
Editors

Introducing the Synergy Between Forensic Science and Environmental Analysis

Springer

Editors
Mahipal Singh Sankhla
School of Basic and Applied Sciences
K.R. Mangalam University
Gurugram, Haryana, India

Kumud Kant Awasthi
School of Basic and Applied Sciences
K. R. Mangalam University
Gurugram, Haryana, India

Tina Sharma
Department of Forensic Science
Bahra University
Shimla, Himachal Pradesh, India

Rajeev Kumar
Department of Forensic Science
IILM University
Greater Noida, UP, India

ISBN 978-981-95-0198-4 ISBN 978-981-95-0199-1 (eBook)
https://doi.org/10.1007/978-981-95-0199-1

This Springer imprint is published by the registered company Springer Nature Singapore Pte Ltd.
The registered company address is: 152 Beach Road, #21-01/04 Gateway East, Singapore 189721, Singapore

This book, Introducing the Synergy Between Forensic Science and Environmental Analysis, is lovingly dedicated to Lord Krishna & Radha Rani, whose endless grace and wisdom light the way through every challenge. With a heart full of devotion, we seek His guidance in our journey to uncover truths and bring healing, just as His teachings guide us toward compassion and understanding in all that we do.

Preface

During an age of increasing environmental damage and sophisticated environmental crimes, the necessity for an interdisciplinarity in learning about and limiting environmental damage has never been more relevant. Publishing *Introducing the Synergy Between Forensic Science and Environmental Analysis* was born out of a need for a timely and much-needed addition to the emerging science of environmental forensics—a field that unites the investigative science of forensic science with the ecological knowledge of environmental studies. This book seeks to fill the knowledge gap between these fields with a comprehensive framework that unites scientific theory, advanced methodologies, and practical case applications.

The impetus for this volume arose from a mutual recognition among the editors and authors that environmental forensics is increasingly becoming important in academic, legal, industrial, and policy communities. However, rich scholarly resources that cover its multi-dimensional aspects in an integrated and coherent manner are still in short supply. This book meets that requirement by presenting a strong collection of core principles, practical methods, and directions for the future, all framed in terms of the real-world demands of contemporary environmental investigations.

This book caters to a broad range of readers. It is a core textbook for undergraduate and postgraduate students in forensic science, environmental science, ecology, and related disciplines. In addition, it gives applied insights and methodological guidance to researchers, forensic experts, environmental consultants, and policymakers who work in environmental regulation, crime investigation, and restoration planning. Every chapter is composed to be understandable but precise, making it possible for both initial students and experienced professionals to interact with the content meaningfully.

The structure of the book follows a rational sequence—from basic ideas and ecological principles to sophisticated analytical instruments and targeted applications. The initial chapters present the theoretical basis of environmental forensics and forensic ecology. Later sections go into detail about particular investigative methods, including source identification of pollution, ecological crime scene investigation, and forensic responses to wildlife crime. The later chapters discuss soil and sediment examination, microplastic contamination, ecosystem recovery, and future-looking ideas for the merging of environmental forensics with sustainable development and environmental justice.

This book is a product of collective scholarship from a diverse pool of scholars, practitioners, and young researchers from all over India. The authors were chosen not just for their specialization in their respective fields but for their passion for interdisciplinary and applied research. All the chapters embody the collective power of collaborative scholarship and are a reflection of a common vision to make science more applicable in solving environmental problems.

We are immensely thankful to all the research scholars, Ph.D. researchers, and postgraduate students who offered critical feedback, data, drafting, and editorial assistance along the way in the formulation of this book. Their intellectual curiosity, high standards, and dedication to scientific excellence have been instrumental in crafting the content and organization of this book. Without their guidance, the multi-authored and comprehensive nature of this volume would not have been achievable.

We would also like to express our sincere appreciation to all the contributing authors who provided their precious experience and subject expertise. Their prompt contributions, cooperation throughout the editing process, and commitment to improving the quality of academic research have made the book an important and lasting academic publication.

We wish that this book stimulates additional research and innovation at the nexus of environmental studies and forensic science. Let it act as an impetus to research, learning, and action toward achieving ecological integrity and environmental justice.

Gurugram, Haryana, India — Mahipal Singh Sankhla
Gurugram, Haryana, India — Kumud Kant Awasthi
Shimla, Himachal Pradesh, India — Tina Sharma
Greater Noida, UP, India — Rajeev Kumar

Contents

Editors and Contributors

About the Editors

Mahipal Singh Sankhla is an Assistant Professor in Forensic Science at the School of Basic and Applied Sciences, K.R. Mangalam University, Gurugram, Haryana. He holds B.Sc. (Hons.), M.Sc., and Ph.D. in Forensic Science from Galgotias University, Greater Noida. He specializes in forensic toxicology, heavy metal toxicity, environmental forensics, nanotechnology, and fingerprint analysis. His professional training includes tenures at Forensic Science Laboratory (FSL) Lucknow, Central Bureau of Investigation- Central Forensic Science Laboratory (CBI-CFSL) New Delhi, Codon Institute of Biotechnology, and Rajasthan State Mines & Minerals Limited (RSMML) Udaipur. He previously served at Chandigarh University and Vivekananda Global University, Jaipur. He has been awarded a Department of Science & Technology (DST) funded Junior Research Fellowship at Malaviya national Institute of Technology (MNIT) Jaipur, along with the Young Scientist, Young Researcher, and Forensic Researcher Awards. As a mentor for NITI Aayog's Atal Innovation Mission, he supports STEM development in schools. He has published over 170 peer-reviewed articles, 30+ book chapters, edited 8 books, and holds 2 copyrights, 1 German patent, 3 design grants, and 70+ Indian patents. He is an active member of editorial boards, review panels, and organizing committees in national and international scientific events.

Kumud Kant Awasthi is currently working as a Professor at the School of Basic and Applied Sciences, K. R. Mangalam University, Gurugram, Haryana, India. Prior to this position, he worked at the Department of Life Sciences, Vivekananda Global University (VGU), Jaipur, India. He earned his B.Sc. from Chhatrapati Shahu Ji Maharaj University (CSJM) University, and M.Sc. and Ph.D. in Zoology from the University of Rajasthan. An expert in nanotoxicology and environmental toxicology, he has substantial experience in nanoparticle toxicity using in vivo and in vitro models. He has secured Indian and German patents and authored several books and copyrights. He is an active Institute innovation Council (IIC) member and Intellectual Property Right (IPR) Cell Nodal Officer, and serves as a coordinator for Global Initiative of Academic Networks (GIAN) courses and JNU's Animal Welfare programs. He was invited by the Oxford Centre for Animal Ethics and serves on Committee for the Purpose of Control and supervision of Experimentation on Animals (CPCSEA's) animal ethics committees. A life member of various scientific societies, he regularly reviews for reputed journals and is involved in organizing national and international academic events.

Tina Sharma is Professor of Forensic Science at Bahra University, Shimla. She specializes in biodiversity crime, fingerprint examination, forensic biology, DNA profiling, and toxicology. She holds a B.Sc. in Biotechnology and an M.Sc. in Forensic Science from Punjabi University, Patiala, and is University Grants Commission-National Eligibility Test (UGC-NET) qualified. Her training includes CFSL Chandigarh, CBI-CFSL New Delhi, and State Forensic Science Laboratory (SFSL) Himachal Pradesh. She has published 98 DNA sequences in NCBI, 30+ research papers, 20+ book chapters, and edited 2 books. She holds 60+ patents and 2 design registrations. A UGC-JRF and SRF awardee, she has guided 4 Ph.D. and 40+ PG students.

Rajeev Kumar is currently working as a Professor and Head in the Department of Forensic Science, at IILM University, Greater Noida, India. He had worked as a Senior Scientific Assistant (Chemistry) in Govt. Forensic Science Laboratory, Delhi for 18 months. He completed his B.Sc. and M.Sc. in Forensic Science from Agra University and Ph.D. in Forensic Toxicology from Sam Higginbottom University of Agriculture, Technology and Sciences (SHUATS), Prayagraj, India. He has been awarded UGC-JRF and SRF and recognized for excellence in reviewing by International Journal for Innovative Research in Science and Technology (IJIRST). He has published 125+ peer-reviewed articles, 5 patents, and numerous book chapters. He serves on the editorial boards of 28 international journals and reviews for 32 journals globally. He has contributed extensively to conferences as a speaker, organizer, and session chair. He has supervised 6 Ph.D. scholars (awarded) and 8 ongoing, with more than 80 M.Sc. dissertations completed under his guidance. His prolific academic and research contributions have established him as a key figure in forensic science.

Contributors

Shobha Ahi Kings College London, London, UK

Shivam Arya Department of Forensic Science, ICFAI University, Jaipur, India

Department of Forensic Science, Galgotias University, Greater Noida, India

Garima Awasthi School of Basic and Applied Sciences, K. R. Mangalam University, Gurugram, Haryana, India

Kumud Kant Awasthi School of Basic and Applied Sciences, K.R. Mangalam University, Gurugram, Haryana, India

Benex K. Babu Department of Forensic Science, Royal School of Life Sciences, The Assam Royal Global University, Betkuchi, Guwahati, Assam, India

Meghna Bordoloi Department of Forensic Science, Royal School of Life Sciences, The Assam Royal Global University, Betkuchi, Guwahati, Assam, India

Shivam Chourasiya Faculty of Forensic Science, Galgotias University, Greater Noida, Uttar Pradesh, India

Jyoti Dalal Department of Forensic Science, Lovely Professional University, Phagwara, Punjab, India

Amity School of Forensic Sciences, Amity University, Punjab, Mohali, India

Pundarikaksha Das Department of Forensic Science, Royal School of Life Sciences, The Assam Royal Global University, Betkuchi, Guwahati, Assam, India

Anoop M. Divakaran Department of Science and Technology, Toyota Technological Institute, Nagoya, Japan

Anjali Dubey Faculty of Forensic Science, Medi-Caps University, Indore, Madhya Pradesh

Archana Gautam Department of Forensic Science, SOS, ITM University, Gwalior, Madhya Pradesh, India

Nilesh Gupta Faculty of Pharmacy, Medi-Caps University, Rau, Indore, Madhya Pradesh, India

Chitra Jangid Department of Forensic Science, Lovely Professional University, Phagwara, Punjab, India

Lokesh Kumar Jangir Department of Physics, National Sun Yat-sen University, Kaohsiung, Taiwan

Priyadharshini Kalyanasundaram Department of Social Sciences, School of Social Sciences and Languages, Vellore Institute of Technology, Vellore, Tamil Nadu, India

Rajdip Khaniya Department of Biochemistry and Forensic Science, Gujarat University, Ahmedabad, Gujarat, India

Akash Kumar Dr. A.P.J. Abdul Kalam Institute of Forensic Science and Criminology, Bundelkhand University, Jhansi, Uttar Pradesh, India

Chandra Kumar School of Engineering, Faculty of Science, Engineering and Technology, Universidad Mayor, Santiago, Chile

Rajeev Kumar Department of Forensic Science, IILM University, Greater Noida, U.P., India

Sneha Lohar Department of Forensic Science, School of Bioscience and Technologies, Galgotias University, Greater Noida, Uttar Pradesh, India

Sneha Madhavan Department of Forensic Science, University of Calicut, Kerala Police Academy, Thrissur, Kerala, India

Dipak Kumar Mahida Department of Biochemistry and Forensic Science, Gujarat University, Ahmedabad, Gujarat, India

Vishal M. Makwana Department of Biosciences, Veer Narmad South Gujarat University, Surat, Gujarat, India

Badal Mavry Department of Forensic Science, School of Bioscience and Technologies, Galgotias University, Greater Noida, Uttar Pradesh, India

Annwesha Mazumdar Regional Forensic Science Laboratory, Durgapur, West Bengal, India

Ankita Patel Department of Biochemistry and Forensic Science, Gujarat University, Ahmedabad, Gujarat, India

Varsha Rani Patel Faculty of Forensic Science, MATS University, Raipur, Chhattisgarh, India

Chhote Raja Patle Department of Forensic Science, Medi-Caps University, Indore, Madhya Pradesh, India

Shakti Prakash School of Behavioural Sciences and Forensic Investigation, Rashtriya Raksha University, Gandhinagar, India

Saba Rashid School of Basic and Applied Sciences, K. R. Mangalam University, Gurugram, Haryana, India

Mansha Pragyan Ray School of Forensic Biotechnology, Department of Biotechnology, National Forensic Science University, Gandhinagar, Gujarat, India

Suryapratap Ray Department of Biosciences, School of Biosciences and Technology, Vellore Institute of Technology, Vellore, Tamil Nadu, India

Mahipal Singh Sankhla School of Basic and Applied Sciences, K. R. Mangalam University, Gurugram, Haryana, India

Vaibhav Saran Department of Forensic Science, Amity University, Mohali, Punjab, India

Anuj Sharma Department of Forensic Science, SOS, ITM University, Gwalior, Madhya Pradesh, India

Girraj Sharma Jindal Institute of Behavioral Sciences, O.P. Jindal Global University, Sonipat, Haryana, India

Kavita Sharma Shri Vaishnav Vidyapeeth Vishwavidyalaya, Indore, Madhya Pradesh, India

Tina Sharma Department of Forensic Science, Bahra University, Shimla, Himachal Pradesh, India

Vaibhav Sharma Department of Medical Elementology and Toxicology, Jamia Hamdard University, New Delhi, India

Harshita Shekhawat Department of Forensic Science, Royal School of Life Sciences, The Assam Royal Global University, Betkuchi, Guwahati, Assam, India

Anu Singla APJ Abdul Kalam Institute of Forensic Science & Criminology, Bundelkhand University, Jhansi, Uttar Pradesh, India

U. Sooraj Krishna Department of Forensic Science, M. E. S. Kalladi College, Mannarkkad, Palakkad, Kerala, India

Neeharika Srivastava Department of Forensic Science, Galgotias University, Greater Noida, Uttar Pradesh, India

Janmejay Singh Tomar Department of Pharmacy, Institute of Professional Studies, Gwalior, Madhya Pradesh, India

Supriya Tomar Department of Forensic Science, Medi-Caps University, Indore, Madhya Pradesh, India

Baljeet Yadav School of Basic and Applied Sciences, K.R. Mangalam University, Gurugram, Haryana, India

Murali Manohar Yadav Dr. A.P.J. Abdul Kalam Institute of Forensic Science and Criminology, Bundelkhand University, Jhansi, Uttar Pradesh, India

Abbreviations

AFIS	Automated Fingerprint Identification System
BTEX	Benzene, Toluene, Ethylbenzene, Xylene
CERCLA	Comprehensive Environmental Response, Compensation, and Liability Act
CODIS	Combined DNA Index System
CPCB	Central Pollution Control Board
CSIA	Compound Specific Isotope Analysis
DNA	Deoxyribonucleic Acid
DRO/TPH	Diesel Range Organics/Total Petroleum Hydrocarbons
EIA	Environmental Impact Assessment
EPA	Environmental Protection Agency
FAO	Food and Agriculture Organization
FSL	Forensic Science Laboratory
GC-MS	Gas Chromatography–Mass Spectrometry
GHG	Greenhouse Gas
GIS	Geographic Information System
GSR	Gunshot Residue
ICP-MS	Inductively Coupled Plasma Mass Spectrometry
IEA	International Energy Agency
IPCC	Intergovernmental Panel on Climate Change
LC-MS	Liquid Chromatography–Mass Spectrometry
LOP	Limit of Prosecution
MOEFCC	Ministry of Environment, Forest and Climate Change (India)
NGO	Non-Governmental Organization
NIBIN	National Integrated Ballistic Information Network
NIH	National Institutes of Health
NIST	National Institute of Standards and Technology
NOx	Nitrogen Oxides
OSHA	Occupational Safety and Health Administration
PAH	Polycyclic Aromatic Hydrocarbons
PCB	Polychlorinated Biphenyls
PCR	Polymerase Chain Reaction
PM	Particulate Matter
RCRA	Resource Conservation and Recovery Act

SOP	Standard Operating Procedure
SOx	Sulfur Oxides
STR	Short Tandem Repeat
UNEP	United Nations Environment Programme
USGS	United States Geological Survey
VOC	Volatile Organic Compounds
WHO	World Health Organization

Introduction to Environmental Forensic

1

Sneha Lohar, Badal Mavry, Anuj Sharma, Vaibhav Sharma,
Mahipal Singh Sankhla, Rajeev Kumar, Tina Sharma,
Kumud Kant Awasthi, and Lokesh Kumar Jangir

Abstract

Environmental forensics is a distinct area of forensic science that refers to the methodical examination of contamination events. It integrates the scientific principles of chemistry, biology, geology, environmental science, and forensic techniques for the purpose of resolving cases involving the release of contaminants. The origins and history of contemporary environmental forensics date back to the contamination of the Love Canal in New York in 1978. Formulated for the purpose of dealing with environmental liability legislation during the 1980s,

S. Lohar · B. Mavry
Department of Forensic Science, School of Bioscience and Technologies, Galgotias University, Greater Noida, Uttar Pradesh, India

A. Sharma
Department of Forensic Science, SOS, ITM University, Gwalior, Madhya Pradesh, India

V. Sharma
Department of Medical Elementology and Toxicology, Jamia Hamdard University, New Delhi, India

M. S. Sankhla (✉)
School of Basic and Applied Sciences, K.R. Mangalam University, Gurugram, Haryana, India

R. Kumar
Department of Forensic Science, IILM University, Greater Noida, UP, India

T. Sharma
Department of Forensic Science, Bahra University, Shimla, Himachal Pradesh, India

K. K. Awasthi
School of Basic and Applied Sciences, K. R. Mangalam University, Gurugram, Haryana, India

L. K. Jangir
Department of Physics, National Sun Yat-sen University, Kaohsiung, Taiwan

M. S. Sankhla et al. (eds.), *Introducing the Synergy Between Forensic Science and Environmental Analysis*, https://doi.org/10.1007/978-981-95-0199-1_1

especially in the USA, this scientific field is frequently applied for use within a legal, regulatory framework and to a lesser degree for the purpose of formulating policy and establishing cause. Environmental forensic analyses typically include a set of environment-related samples, e.g., soil, water, and air, and subsequent in-depth analysis by advanced chemical and biological techniques to reconstruct the events of contamination and differentiate between anthropogenic and natural impurities. The findings are applied to guide lawsuits, to guide regulation, and to remediate toxic or hazardous waste sites and may be crucial to safeguard public health when there are contaminants that influence air quality or tap water. As the globe is confronted with more daunting global environmental problems, environmental forensics is surfacing as an asset for identifying responsibility, fostering responsible management, and safeguarding health and safety.

Keywords

Environmental forensics · Forensic science · Contaminations · Environmental problems

1.1 Introduction

Environmental forensics is a branch of forensic science devoted to the rational and scientific investigation of environmental contamination events. Environmental forensics attempts to objectively evaluate physical, chemical, and historical facts in order to establish the originality, timeline, and extent of environmental contaminant releases on a defendable basis (Gałuszka et al. 2015). The environmental forensics aims essential are the identification of pollutants nature, the point of discharge, and the potential risk of harm in relation to the pollutants, which is often related to legal or regulatory investigations. Environmental forensics dates back to the 1980s and was in major part in response to the establishment of the environmental liability laws in the United States (Kaljurand 2014). There ensued the need to settle environmental responsibilities on contaminant sources and, therefore, the need to formulate accurate scientific approaches to apportion liability. Although different areas, including the European Union, had set up similar legal frameworks, environmental forensic approaches received a higher profile and were utilized in a larger global region (Ljungberg 2002).

This is crucial in terms of supporting investigations, regulatory decisions on a large scale. Critical functions include identifying culpable parties for established contamination sites, distributing liability in mixed waste incidents, identifying former owners associated with contaminant discharges, and evaluating the probability that chemical exposure is either related to present or future health problems (Morrison 1999). Environmental forensics is also critical in terms of x-raying naturally occurring contaminants from those which are human activity related, and in that vein pin down the origin of major marine oil spills and larger-scale environmental events. Environmental forensic investigation mainly involves the extraction and interpretation of relevant evidence. For example, contaminant samples of interest are available within physical materials such as soil, water, or air. Given that environmental variables have the potential to alter the samples, a detailed analytical process is an essential tool in proving when and at what place the contamination took place (Murphy and Morrison 2014). Chemical and biological methods are commonly employed to identify negligible or unrecorded contaminants with unclear evidence in historical records. These investigation findings tend to give critical evidence that can be used in legal and regulatory procedures and help determine who and guide the implementation of appropriate remedial measures. This is an area that combines the scientific fundamentals of chemistry, biology, environmental science, and geology with disciplined approaches of forensic science for managing contamination problems (Spikmans 2015). Parts of this discipline typically include environmental chemists, hydrologists, botanists, and atmospheric scientists, who pool their expertise to deal with difficult contamination scenarios. It is important to observe scientific protocols during investigations as the integrity of evidence is upheld and thus increases reliability and acceptance of investigative savings in legal situations (Stout and Wang 2016). In the end, though, environmental forensics goes further than the measurement of the environmental damage. Environmental forensics is central to the protection of public health, especially in situations where contaminants infiltrate the drinking water supply or are distributed through the air. With worldwide environmental problems continuing to exist, environmental forensics will prove more important for the support and maintenance of accountability, as well as environmental stewardship and public health protection (Wang et al. 1964).

This chapter presents a comprehensive and systematic description of environmental forensics as an aspect of forensic science, which incorporates both introductory and sophisticated techniques for the evaluation of contaminant age and tracking the source of contamination.

1.2 Historical Perspective of Environmental Forensic

Historically, the majority of forensic environmental investigations have relied heavily on laboratory interpretation of contamination events. This reliance was because there was little trustworthy and field-capable technology available for sensing pollutants in different environmental settings. At first, field researchers were limited to making simple site measurements that involved PH, conductivity, and dissolved oxygen. Such measurements were very critical in general environmental tracking but did not offer sufficient evidence for specific origin or type of contamination identification through scientific analysis. Thus, institutions routinely had a procedure that involved collecting environmental substances at the scene and dispatching them to advanced labs for fine chemical analysis. The focal analytical method at this stage was chemical fingerprinting, which facilitated police forensic scientists' ability to determine the similarity of the chemical fingerprints of contaminated samples with potentially source substances. Petrisor showed in his 2014 publication how this technique could help create a source by using the peculiarity of the chemical fingerprints of the samples from the environment to make this relationship (Petrisor 2014). But the analysis accounted for trends in chemical signatures amidst natural processes such as evaporation, biodegradation, and photooxidation. In the 2000s, the rapid development in analytical chemistry expanded the realm of environmental forensics to reduce a great deal of conventional pollutants to a variety of causes such as heavy metals (e.g., mercury and arsenic), pesticides, PAHs, crude oil, sewage, and asbestos. The literature has discussed these pollutants extensively (Mudge 2008). High levels of consistency in environmental forensic studies conducted in various laboratories were achieved with the help of Stout and Wang (2016) and the adoption of consistent analytical procedures by the US Environmental Protection Agency (USEPA). In the last decade, the science of environmental forensics has extended to new substrates, including emerging contaminants such as pharmaceuticals, personal care products, and illicit drugs, coinciding with their recent importance in environmental public health (Ramsey and Hewitt 2005). There is also increasing overlap of forensic and environmental sciences in epidemiological investigations, where wastewater analysis is a central approach in the measurement of drug use (Béen 2015). Such inter-disciplinary work shows the increased application of environmental forensic techniques for public health and urban surveillance. The current field-portable instruments have brought a revolutionary era to this field. Lack of prior use has not prevented the recent attention to shift to these tools for their fast-screening techniques, with evident great potential for more streamlined and prompt forensic investigations (Valcárcel and Cárdenas 2005). Ultimately, the development of environmental forensics presents us with an evolutionary chain, starting from isolated laboratory labs toward a more interactive, mobile, and field-focused approach supported by the eternal evolution of scientific knowledge as well as the implication of technical innovations (Buchberger 2011).

1.3 Forensic Techniques for Contaminant Age Dating and Source Identification

There are many different forensic methods for determining age and source, such as corrosion models for underground storage tanks, a compound's commercial availability, chemical associations with specific manufacturing processes, stable isotope analysis, degradation models, models of pollutant movement, and biomarkers (Morrison 2000). For these methods to be useful as forensic evidence in environmental lawsuits, they must be carefully considered and used. The underlying presumptions and the caliber of the data are scrutinized and often successfully contested when presented as scientific proof. One can determine which methodology, or combination of procedures, can best suit to protect the environment from such contaminants by identifying the source and limiting the use (Li et al. 2025).

1.3.1 Underground Storage Tank Corrosion Models

A system of evaluation metrics must be established in order to provide an accurate, reliable, and objective evaluation result, which is also a guarantee for the long-term effective operations of tanks, as the preliminary health evaluation reflects the current operational status of tanks (Feng et al. 2019). It's commonly stated that using tank corrosion models to determine when a liquid leak from an underground storage tank (UST) occurred is a good way to age-date the discharge (Morrison 2000). Numerous models for assessing the harm caused by atmospheric corrosion have been developed using various methodologies that vary depending on the goals of the model. These models can be divided into two categories: first level and second level. The first-level models are founded on physics and chemistry laws (Landolfo et al. 2010). In this instance, the development of corrosion products and the dissolution of metal are assessed at the microscopic level in terms of current. The second-level models, which are derived from the observation and interpolation of experimental data, are helpful for engineering applications and enable the evaluation of the corrosion rate as a function of mass and/or thickness loss with time (Roberge 2000). The Mean Time to Corrosion Failure (MTCF®), Tank Environmental Profiling (TEP®), and Tank Suitability Study (TSS) models are among the corrosion models used for subterranean storage tanks (Morrison 2000). Typically, the mass loss per unit area per unit time or the rate of penetration through thickness loss is used to represent the corrosion rate. Note that the thickness loss in corrosion investigations is often the average of the thickness losses of a specimen's groundward and exposed surfaces (Albrecht and Hall Jr 2003). The Mean Time to Corrosion Failure (MTCF1) model is a noninvasive process that employs statistics to estimate the tank's anticipated leak-free life after analyzing the corrosion-inducing features of a UST site (excavation) (Morrison 2000). To assess whether an underground storage tank is appropriate for cathodic protection, the Tank Environmental Pro®ling (TEP1) technique examines hydrocarbon concentrations and other physical and chemical features near the tank. The TSS model is predictive within a particular range of variables, such as soil resistivity (3200–1,100,000 ohm/cm^3), total chloride (10–343 mg/kg), soil pH (3.6–11.4), total sulfide (0–48 mg/kg), and total solids (78–99.8%) (Morrison 2000).

1.3.2 Commercial Availability of a Chemical as an Indicator

The commercial availability of a chemical can occasionally be used by a forensic investigator to date the discharge of a pollutant and to identify possible culprits. Sometimes, the only way to bracket ownership or site operating timelines with potential chemical usage and release is to indirectly link to a specific party (Murphy and Morrison 2014). This strategy is not limited to substances with a comparatively limited supply. The presence and/or concentration of additives (reaction inhibitors, antioxidants, light inhibitors, oxygenates, stabilizers, etc.) can frequently further distinguish chemicals with a wide historical availability. These additives can occasionally be linked to relatively short timeframes that correspond with changing industrial practices or regulations. Analysis of commercial availability is especially helpful for fuel additives, insecticides, herbicides, fungicides, chlorinated solvents, and chlorofluorocarbons (Mudge 2008).

1.3.3 Chemicals Unique to Manufacturing Activities

1.3.3.1 Volatile Organic Compounds (VOCs)

VOCs are volatile organic chemicals that readily evaporate at room temperature and are extensively used in manufacturing as solvents, thinners, and cleaning agents. Substances such as toluene, xylene, acetone, and methyl ethyl ketone are frequently used in paints, coatings, adhesives, and pharmaceuticals. Industrial processes release these chemicals, which tend to accumulate in air and water, with associated health hazards such as toxicity, carcinogenicity, and irritation. They also form ground-level ozone and smog through photochemical reactions, causing damage to ecosystems and human health. VOCs pose a significant problem both indoors and outdoors because they are widely used and last for a long time (David and Niculescu 2021).

1.3.3.2 Heavy Metals

Heavy metals are naturally occurring substances that are emitted into the environment from several manufacturing activities, such as metal fabrication, the manufacture of electronics, and electroplating (Oladimeji et al. 2024). Lead, mercury, cadmium, chromium, and nickel are of special concern because of their toxicity, persistence in the environment, and bio-accumulatability in living organisms. Heavy metals may contaminate air, water, and soil and cause long-term environmental and health effects, such as neurological and organ damage, cancer, and disruption of ecosystems (Balali-Mood et al. 2021).

1.3.3.3 Acid Gases

Acid gases are formed through combustion, chemical processing, and metal refining. Sulfur dioxide (SO_2) and nitrogen oxides (NO_x) mainly result from fossil fuel burning and are significant acid rain contributors that impact ecosystems, structures, and bodies of water. HCl is emitted during chemical synthesis and waste burning. Acid gases can lead to respiratory illnesses in humans, damage infrastructure, and injure aquatic and terrestrial organisms (Sedyaaw et al. 2024).

1.3.3.4 Polyaromatic Hydrocarbons (PAHs)

PAHs constitute a series of chemicals resulting from the incomplete combustion of organic products like coal, oil, and wood. Benzo(a)pyrene, naphthalene, and phenol are some that occur in emissions from industrial processes, power generation, and waste incineration. PAHs resist breakdown in the environment and are harmful, and some of them have been proved to be carcinogenic. They have the potential for accumulating in water and soil and causing harm to wildlife and human health (Abdel-Shafy and Mansour 2016).

1.3.3.5 Chlorinated Solvents

Chlorinated solvents have extensive applications in manufacturing as degreasers, cleaners, and chemical intermediates. 1,1,1-trichloroethane, trichloroethylene (TCE), and perchloroethylene (PCE) are typical ones. They are volatile and can thus pollute air, land, and groundwater. They are persistent and toxic, and some of them are known to be carcinogens. Their extensive use and biodegradation resistance make them potent environmental pollutants (Moran et al. 2007).

1.3.3.6 Wastewater Pollutants

Industrial manufacturing wastewater commonly includes chemical oxygen demand (COD), oil residues, and suspended solids. COD quantifies oxygen consumed in the oxidation of organic compounds, a measure of the load of pollution. Oil residues deposit on surfaces, are toxic to aquatic life, and destroy ecosystems. Suspended solids lower water transparency, foul filters, and transport other contaminants. Such wastes are controlled in industrial effluent to safeguard water quality and ecosystem health (Fleyfel et al. 2024).

1.3.4 Compound Specific Isotopic Analysis (CSIA)

Compound Specific Isotopic Analysis (CSIA) is an advanced analytical method applied in environmental forensics to track the origin, pathway, and fate of pollutants in the environment (Kuntze et al. 2020). CSIA measures the concentration of stable isotopes of different chemical elements (like carbon, hydrogen, oxygen, nitrogen, sulfur, and chlorine) in an individual organic compound that exists in environmental materials (such as groundwater, soil, or air) (Frantz 2017).

1.3.5 Contaminant Transport Models

Contaminant transport models are complex computational models that simulate and predict the movement, alteration, and fate of chemicals during their passage through different environmental media. These include soil, water (surface as well as ground), and air (Tong et al. 2022). The models are founded on physics, chemistry, and mathematics principles and encompass a wide variety of natural processes governing contaminant fate, such as advection (mass transport with moving water or air), dispersion (turbulence- and mixing-induced spreading), diffusion (molecular movement from high to low concentration), sorption (attachment onto soil or sediment grains), chemical and biological breakdown, and volatilization (transfer from liquid or solid to gas phase) (Zheng and Bennett 2002). Apart from forensic applications, the models are required for risk assessment, whereby scientists and engineers can study potential routes of exposure, predict likely future scenarios of contamination, and establish impacts on human health and ecosystems (Bear and Cheng 2010). The models also aid in remediation planning by simulating how different cleanup methods—such as pump-and-treat systems, in situ chemical oxidation, or monitored natural attenuation—will work and determine the most cost-efficient and sustainable solutions (Duffy et al. 2000).

1.4 Advanced Analytical Approaches in Environmental Forensics

Investigating pollution occurrences in an effort to determine their cause is the goal of the field of environmental forensics. These investigations' primary goal is to gather evidence (Mudge 2008). This proof often takes the shape of tangible evidence, such as the pollutant, while accounting for changes in the contamination brought about by interactions with the environment (Mach and Verbeck 2018). Numerous possible contaminants exist, each having a potential forensic technique of identification as well as intrinsic dangers. Although there are several methods documented in the scholarly literature.

1.4.1 The Historical Laboratory-Based Approach to Analysis

Laboratory-based findings have long been used in environmental forensic investigations to aid in scene interpretation. This is mainly due to the lack of appropriate and trustworthy field-portable technology that could detect contaminants in environmental matrices until recently. While some relatively basic measurements, like pH, conductivity, and dissolved oxygen, could be made in the field, they have limited use in environmental forensic investigations, even though they might provide some information on environmental harm (Spikmans 2019). Therefore, sending samples to an appropriate laboratory for in-depth examination is the standard procedure. In this procedure, the samples undergo chemical fingerprinting in addition to analysis to identify the kind of contamination (Wang et al. 1964; Petrisor 2014). Analytical techniques have been created for a wide variety of typical contaminants. Heavy metal pollution, including that caused by mercury and arsenic, sewage, asbestos, pesticides, herbicides, polycyclic aromatic hydrocarbons (PAHs), crude oil, and associated refined products, is a few examples of prevalent contaminants (Murphy and Morrison 2006). For a long time, several methods have been employed to study environmental forensics, some of which include the following:

1.4.1.1 Spectroscopy

In spectroscopy, a sample of interest is exposed to incoming radiation, sometimes from a laser, and the emission or absorption that results is observed (Hollas 2004). Like fingerprints, each compound has a distinct resulting spectrum that may be used to identify a particular chemistry. The existence is demonstrated by these data, which are almost qualitative in character (Mach and Verbeck 2018). Water samples are frequently exposed to a wide range of contaminants. Aquifers and other underground sources are contaminated by industrial runoff and other sources in addition to surface water. Because pollutants must flow into subterranean sources, contamination may not usually appear right away. Hydrocarbon pollution from these underground sources can be detected using in-line infrared equipment (Lay-Ekuakille et al. 2013).

1.4.1.2 Mass Spectrometry and Chromatography

The mass-to-charge ratio (m/z) of the chemical is used by mass spectrometry to identify added ions. Ionization techniques come in a variety of forms and may be divided into two categories: hard and soft. High-energy ionization events, such as electron ionization, are used in hard ionization techniques, which frequently cause the target molecule to fragment (Gałuszka et al. 2015).

1.4.2 Field-Portable Equipment in Environmental Forensics

A significant amount of field-portable equipment has already been used in environmental forensic investigations. An outstanding summary of field-portable tools for environmental study was given by Gałuszka et al. (2015). There have been advances in using this equipment for environmental monitoring since they published their review paper (Gałuszka et al. 2015). The conditions greatly determine the use and suitability of field-portable equipment. This technology is best applicable to rapidly acquire information in the field, and it is best applied when there's an emergency situation and it's important to immediately identify usually unknown poisons. Emergency response relies significantly on the ability of the site investigator's visual impression of the scene without field-portable equipment (Kalnicky and Singhvi 2001; Mulligan et al. 2006; Lam et al. 2018). It is still required that some of the equipment addressed below has this ability in an environmental context; however, it is viable.

1.4.2.1 Portable X-Ray Fluorescence

Because it has been used so extensively in environmental research, portable X-ray fluorescence (XRF) can be considered a routine technique. It has gained popularity due to its cost-effectiveness and fast turnaround times (Kalnicky and Singhvi 2001; Gałuszka et al. 2015). A chemical profile that can be used for source identification can be generated by the method, as well as identifying a range of contaminating elements, like lead, copper, zinc, tin, arsenic, chromium, and mercury. The portable XRF is a very appropriate method for environmental forensic investigations because of its nondestructive characteristics (Mulligan et al. 2006; Bartley et al. 2007; Bosco 2013).

1.4.2.2 Portable Gas Chromatography-Mass Spectrometry

Gas chromatography-mass spectrometry (GC-MS) equipment is often used in laboratories for the screening and analysis of organic chemicals; hence, the field application of portable GC-MS instruments offers a real translation of laboratory-based capabilities in the field (Mulligan et al. 2006; Jjunju et al. 2014). It has been shown by several studies that using MS without GC yields satisfactory findings on a variety of sample types, including air toxics. Nonetheless, hyphenating GC to MS yields extra information that is frequently preferred, especially for complicated mixes. In addition to providing chemical profiles for source determination, portable GC-MS has the capacity to identify organic compounds with a high degree of confidence (Eckenrode 2001; Contreras et al. 2008). Despite significant recent improvements, portable GC-MS devices still fall short of their laboratory counterparts in terms of capability, mostly because of the requirement for quick findings. More compounds can be discovered thanks to the continued superiority of laboratory-based GC-MS technologies in compound separation. In environmental studies, where matrix impacts must be taken into consideration, this is especially crucial (Spikmans 2019).

1.4.2.3 Portable Fourier Transform Infrared and Portable Raman

Since Raman and mid-infrared spectroscopy are thought to be complementary, they are frequently employed in tandem. Raman offers the benefit of providing spectra through plastic or glass, enabling sample analysis within a container. The benefit of this is that the evidence is not tampered with and the first responder does not have to come into touch with the substance, creating a safer working environment (Carron and Cox 2010; Bartick 2012). Additionally, because the water signal is so weak, Raman can evaluate water samples. Even if Fourier transform infrared (FTIR) cannot do this, it is not affected by fluorescence interference, a problem that frequently arises in Raman analysis (Matthews et al. 2006; Bartick 2012). Infrared-based devices are frequently used in air monitoring to detect the presence of air pollutants, especially CO2 (Ashley 2003).

1.4.3 Environmental Applications of Field-Portable Equipment

A variety of field-portable devices that enable the screening of various pollutants across a range of environmental sample matrices are commercially available. More information about a few of the field-portable tools that mimic lab capabilities and are thus appropriate for environmental forensic investigations is provided below. Field-portable equipment should ideally be able to detect the pollution as well as provide a chemical fingerprint of the pollutant, so that source tracking may be carried out (Spikmans 2019).

1.4.3.1 On-Site Screening and Fast Response

With portable instruments, there is real-time, immediate analysis of environmental samples like air, water, and soil, which is critical for emergency response and monitoring processes. It provides fast screening to detect possible hazards and make prioritization of the samples for further laboratory analysis, which means much less sample transportation and handling (Narayana et al. 2024).

1.4.3.2 Pollutant Detection and Monitoring

Field-portable instruments are utilized to detect contaminants of a broad spectrum, including VOCs, heavy metals, pesticides, and pathogens. Gas chromatographs (GC), mass spectrometers (GC-MS), X-ray fluorescence (XRF) analyzers, and optical spectrometers are some of the devices utilized for detecting contaminants in air, water, and soil (Bednar et al. 2011).

1.4.3.3 Green Analytical Chemistry

Many handheld instruments are nondestructive, consume minimal or no reagents, and generate minimal or no analytical waste, and thus enable more environmentally friendly analytical methods. This is particularly beneficial in reducing the environmental impact in repeated or large-scale monitoring campaigns (Gałuszka et al. 2015).

1.4.3.4 Applications in Low-Resource and Remote Regions

Portable devices, especially microfluidic and biosensor-type tools, are specifically suited for use in low-resource settings, where standard laboratory equipment is not feasible due to cost, size, or complexity. The devices are constructed for simplicity of design, affordable prices, and high automation, and find applications for field use in identifying contaminants such as metal ions, pathogens, pesticides, and antibiotic residues (Sekhwama et al. 2024).

1.5 Age Dating and Source Identification of Key Environmental Contaminants

1.5.1 Petroleum Hydrocarbons

Petroleum hydrocarbons are petroleum molecules made up almost exclusively of the elements C and H. Based on the molecular weight of individual hydrocarbons, they constitute a continuum over a wide range rather than being separate entities. Diesel fuel, gasoline, and associated goods have hundreds, maybe thousands, of distinct PHs. There are four main structural groupings into which the PHs may be separated (Kuppusamy et al. 2020).

(a) Alkanes, also known as paraffins, are saturated hydrocarbons in which every carbon atom makes four single bonds with the other C and H atoms that comprise the chemical. Furthermore, these hydrocarbons are aliphatic, i.e., their carbon atoms are joined by straight or branched chains. Hexane, heptane, octane, and decane are some of the chemicals included in this category.
(b) Isoparaffins are branched-chain hydrocarbons, specifically alkanes, meaning they are saturated and have a carbon backbone.
(c) Cycloalkanes (also known as naphthalenes): These saturated hydrocarbons are distinguished by their ring-like structure. Hydrocarbons in this category include methylcyclopentane and ethylcyclo-p-hexane.

(d) Alkenes (or olefins): These hydrocarbons are unsaturated and have aliphatic as well as at least two carbon atoms joined by various covalent bonds. Propene and ethylene fall under this category.
(e) Arenes (also aromatics) group contains one benzene ring. These compounds are termed as BTEX (benzene, toluene, ethylene, and xylene). These types of compounds are described as polynuclear or polycyclic aromatic hydrocarbons (PAHs) since they consist of three or more closed-ring structures. Pyrene and phenanthrene are some of these types of compounds.

The primary hydrocarbon classes encountered in petroleum products are paraffins, iso-paraffins, aromatics, naphthenes, and olefins, or a PIANO for short. PIANO analysis provides detailed compositional data important for environmental assessments, fuel characterization, and refining operations. Gas chromatography (GC) is the most prevalent method for fractionating and quantifying these fractions, particularly when used in conjunction with mass spectrometry (MS) or flame ionization detection (FID), providing exceptional sensitivity and resolution for complex mixtures.

1.5.1.1 Analysis of Biomarkers in Petroleum

Biomarkers are intricate organic compounds constituted by hydrogen, carbon, and other elements. Biomarkers often possess minimal structural changes from their parent organic compounds in living things and occur in rocks, sediments, oil, and bitumen (He et al. 2018). Petroleum can stay constant for billions of years down and is largely made up of lipid-derived units (Peters et al. 2005). Because different geological circumstances may lead to the formation of petroleum, each petroleum sample often has a distinct biomarker fingerprint. Different temperature regimes during burial, maturation, and oil formation: (1) different source rocks that were deposited in different depositional settings; and (2) variable migration paths and in-reservoir modifications are the primary contributors of the wide variations in petroleum composition.

The main methods for analyzing the biomarkers in petroleum are gas chromatography-mass spectrometry (GC-MS) or gas chromatography-mass spectrometry-mass spectrometry (GC-MS-MS). The most prevalent hydrocarbon fractions in petroleum, the saturated and aromatic fractions, are usually the subjects of analyses.

Because they convey information about the organic matter input, depositional environment, age, and thermal history of the petroleum source rock, as well as the degree of petroleum alteration in the reservoir rocks, biomarkers are helpful in the geochemistry of petroleum exploration and production (Peters and Moldowan 1993). For instance, porphyrins and the acyclic diterpenes pristane and phytane are three frequent components in petroleum that are produced by the diagenesis and catagenesis of chlorophyll.

The age of the source rock can be determined using biomarkers (Peters and Moldowan 1993). Older Precambrian rocks and considerably younger Tertiary rocks both include stanes and hopanes. As a result, their existence in petroleum provides less insight into the source rock's age.

1.5.1.2 Polycyclic Aromatic Hydrocarbons (PHAs)

Heavy petroleum fractions and residues include unsaturated multi-rings, which are composed of many benzene and naphthalene rings connected to one another. These aromatics exist as solids in isolation and are known as polynuclear aromatics (PNAs) or polycyclic aromatic hydrocarbons (PAHs) (Hewstone 1994). The aromatic-ring configuration, or the variation in size and structure, has a substantial impact on the physical and chemical properties of any specific PAH molecule. When wood, coal, oil, gas, garbage, and other organic materials like tobacco and charbroiled meat are not entirely burned, PAHs are produced. Some PAHs are used to create colors, plastics, insecticides, and medications (Kuppusamy et al. 2016).

The majority of the more than 100 distinct PAHs are found in complicated mixes. Several significant low-molecular-weight (LMW) and high-molecular-weight (HMW) compounds feature three or fewer aromatic rings. Benzo[a]anthracene, benzo[a]pyrene, benzo[e]pyrene, benzo[b]fluoranthene, benzo[g,h,i]perylene, benzo[j]fluoranthene, benzo[k]fluoranthene, chrysene, dibenzo[a,h]anthracene, fluoranthene, fluorene, and indeno[1,2,3-c,d]pyrene are among the compounds classified as PAHs (Kuppusamy et al. 2017).

1.5.1.3 Pristane/ Phytane Ratios

Assuming that phytane is derived from reduction (anaerobic) source degradation of planktons by microorganisms and pristane is derived from oxygenated (aerobic) source degradation of planktons, it is the ratio of phytane to pristane abundance given redox conditions in the aquatic environment. Therefore, an toxic state (pyrogenic source) is indicated by a pristane/phytane ratio less than unity (1), and an anoxic condition (biogenic source) is indicated by a ratio larger than 1 (Fuqi et al. 2019; Omokpariola et al. 2022).

This ratio aids in differentiating between pyrogenic (combustion or aerobic breakdown) and biogenic (natural organic matter under anoxic circumstances) sources. It is extensively employed in environmental forensics and petroleum geochemistry to determine the source and modification of hydrocarbons in soils, sediments, and polluted waterways.

Pr/Ph ratios also support other geochemical markers and isotopic techniques used for contaminant age dating, even if their primary function is to show source and depositional environment. For instance, Pr/Ph ratios can determine the time and origins of contamination episodes when paired with radiometric dating or tracer investigations (such as radium isotopes or chlorofluorocarbons) (Lauer and Vengosh 2016).

1.5.1.4 BTEX Ratios and Their Significance

Benzene, toluene, ethylbenzene, and xylene are the volatile organic chemicals that make up BTEX (Littlejohns and Daugulis 2008). At room temperature, these are the liquid monocyclic aromatic hydrocarbons. They naturally occur when crude oil is refined to create a range of petroleum products, such as solvents, lubricants, gasoline, and diesel fuels, many of which contain trace amounts of BTEX. They are also naturally present in gas emissions from forest fires and volcanoes. BTEX enter the environment through cigarette smoke, emissions from cars and airplanes, and the production or use of materials or goods that contain them (Rad et al. 2014).

The simplest aromatic hydrocarbon is benzene, which has a weight percentage of 92.3% carbon and 7.7% hydrogen. It is classified as a category one carcinogen and is arguably the most resistant of the BTEX chemicals (Da Silva and Alvarez 2002). Because of this, several nations have laws that restrict the quantity of benzene that may be found in petroleum products like gasoline and fuel oil. Styrene, the substrate for the production of polystyrene, is made using ethylbenzene in large quantities. Additionally, it functions as a solvent in gasoline, dyes, and inks. Methylbenzene, another name for toluene, is a solvent used in paints, coatings, gums, oils, and resins. In addition to being a solvent with a variety of uses, xylene, which contains the three isomers of dimethylbenzene (m-xylene, p-xylene, and o-xylene), is commonly employed in the manufacture of polyester garments and plastic bottles.

1.5.2 Chlorinated Solvents

Chlorinated solvents such as tetrachloroethene (perchloroethylene or PCE), trichloroethylene (TCE), and methyl chloroform (1,1,1-trichloroethane or TCA) are among the most common volatile organic compounds (VOCs) found in soil and groundwater (Aeppli et al. 2009).

Perchloroethene (PCE) and trichloroethene (TCE) are among the most common pollutants at industrial sites. This is due to the fact that they find extensive application in metal degreasing, dry cleaning, and chemical production. Both of these chemicals have long environmental residence times, and most are seriously harmful to human health since they are toxic and sometimes carcinogenic (Henschler 1994; Volpe et al. 2007; Matteucci et al. 2015). One of the major problems with these chemicals polluting groundwater systems is the creation of dense non-aqueous phase liquids (DNAPLs). Until they hit a non-permeable layer, DNAPLs created by PCE and TCE migrate downward through permeable groundwater aquifers. The normal distribution of DNAPL is highly complex and irregular. Trapped DNAPL mass slowly dissolves into mobile groundwater, acting as a long-term source of groundwater contamination (Aulenta et al. 2006).

1.5.2.1 Age Dating PCE and TCE

Acetylene Process: Acetylene is chlorinated to produce 1,1,2,2-tetrachloroethane (1,1,2,2-TeCA) with ferric chloride and, occasionally, phosphorus and antimony chloride as catalysts for the production of PCE, which uses acetylene and chlorine as the main raw materials (Morrison and Hone 2010; Schobert 2014). Following its dehydrochlorination, 1,1,2,2-TeCA is converted to TCE, frequently in aqueous bases like calcium hydroxide or by thermally cracking it with barium chloride on activated carbon, silica, or aluminum gels. Pentachloroethane is created by chlorinating TCE, and PCE is subsequently created by dehydrochlorination. Notably,

1,1,2,2-TeCA also undergoes abiotic degradation to TCE, which subsequently undergoes hydrogenolysis to liberate 1,1-DCE, vinyl chloride (VC), cis/trans-1,2-DCE, and ethene (Lorah 2003).

Acetylene was the primary source of TCE production before 1950 [19]. Only 8% of the TCE produced in the USA by the early to mid-1970s contained acetylene (Ochsner et al. 1980). The manufacture of TCE switched from acetylene to ethylene by the late 1970s, ostensibly producing a purer product.

Ethylene Process: In the presence of oxygen and catalysts, ethylene undergoes further chlorination to yield PCE and TCE after first forming 1,2-dichloroethane. Since 1972, the most widely used method has been to add chlorine to ethylene to create 1,2-dichloroethane, which is then chlorinated to create 1,1,1,2-tetrachloroethane, and finally, hydrochloric acid is removed to make TCE. Another result of chlorinating ethylene to 1,2-DCA is 1,1,2-trichloroethane (Milchert et al. 1995).

1.5.3 Polychlorinated Biphenyls (PCBs)

Polychlorinated biphenyls are ones of the most stable organic compounds with low dielectric constant, high heat capacity, chemical inertness, nonflammability, high electrical resistivity, and low acute toxicity, which render them ideal for industrial applications such as such as dielectric fluids, heat transfer fluids, hydraulic fluids, flame retardants, solvent extenders, and organic diluents (Urbaniak 2007).

1.5.3.1 Sources of Polychlorinated Biphenyls

Atmospheric buildup is most likely the source of PCBs' extensive environmental dispersal. Research indicates that whereas low chlorinated PCBs are more common in filtered air, congeners with a high degree of chlorination are mostly detected in aerosols and rains. Due to their weak water solubility, mono- and dichlorobiphenyls, which are frequently found in the gas phase, are not eliminated by atmospheric rainfall. Raindrop aerosols can be used to remove higher chlorinated biphenyls from the atmosphere since they are adsorbed on particles. A relative buildup of the less chlorinated PCBs in the atmosphere is the outcome of these two actions (Klecka et al. 2000).

Three steps make up the process of PCB transport: first, PCBs are released into the atmosphere from the source region through combustion, incineration, or volatilization; second, they are transported through the atmosphere to farther-flung locations (Imamoglu et al. 2002; Wania 2003; Wania and Dugani 2003; Offenberg et al. 2005); and third, they are deposited (dry or wet). PCBs can occasionally re-emit from environmental compartments (soil, water, and plant).

1.6 Legislative Framework and Regulatory Aspects in Environmental Forensics

Investigations to determine the source of the crime and the perpetrator are becoming too crucial for fair justice, and environmental forensics includes a wide variety of crimes pertaining to all species found in the environment. For national and international authorities, the creation of legislation to handle instances involving environmental forensics is crucial.

1.6.1 National and International Environmental Regulations

To ensure proper investigation and workflow, national and international institutions have an important role in establishing regulations for environmental forensics.

1.6.1.1 National Environmental Regulations

1.6.1.1.1 The National Green Tribunal Act, 2010

Act to establish a National Green Tribunal to dispose of cases related to environmental protection, conservation of forests, and other conservation of natural resources in an efficient and timely manner. This involves enforcing any environmental legal rights, granting relief and compensation for property and person damages, and issues incidental or relating to these (Samad et al. 2015). The law established a specialized court of environmental issues known as the National Green Tribunal (NGT). Any issue concerning the enforcement of environmental legislation and environmental disputes falls within the domain of the NGT (Varghese and Alappat 2012).

1.6.1.1.2 The Environment (Protection) Act, 1986

It attempts to conserve and improve the environment as well as prevent risk to individuals, other organisms, businesses, and property. It prescribes penalties for offenses and empowers the federal government to take steps to protect the environment.

1.6.1.1.3 The Air (Prevention and Control of Pollution) Act, 1981

The country has a challenge with air pollution, which is turning to the most lethal thing for all living organisms. This is due to the fact that all states and cities are witnessing a rise in air pollution because of all the developments, such as building, the high rate of demand for cars, and other manufacturing plants. The purpose of this regulation was to assist and regulate air pollution in India. It establishes guidelines for the emigration of pollutants from automobiles and diligence and calls for the creation of pollution control boards at the state and federal levels (Onyeabor 2024b).

1.6.1.1.4 The Water (Prevention and Control of Pollution) Act, 1974

The purpose of this law was to assist and regulate water pollution in India. It establishes guidelines for the release of adulterants into water bodies and calls for the creation of state and federal pollution control boards. Since India is experiencing a severe problem with water pollution, which includes the discharge of chemical waste and other contaminants into river channels or other locations, the beneficial water becomes hazardous and has the potential to trigger a mass tragedy or damage many species in the ecosystem (Onyeabor 2024a).

1.6.1.1.5 The Forest (Conservation) Act, 1980

This law regulates the diversion of forest land for uses other than forests and promotes forest conservation. It establishes sanctions for infractions and mandates that prior authorization from the federal government be sought before diverting forest property (Onyeabor 2024a).

1.6.1.1.6 The Biological Diversity Act, 2002

This law ensures that India's natural variety is preserved and used sustainably. It establishes guidelines for benefit sharing and gives the federal government and state governments the authority to control access to natural resources (Onyeabor 2024a).

1.6.1.1.7 The Wildlife Protection Act, 1972

This regulation controls hunting, poaching, and commerce while safeguarding India's wild animals and vegetation in animals. It provides for the creation of national parks, sanctuaries, and other protected places as well as sanctions for infractions (Onyeabor 2024a).

1.6.2 Legal Challenges in Environmental Forensic Investigations

The process of handling environmental crimes is difficult and full of obstacles that the investigator or any victims of the case may encounter. These obstacles can arise from a variety of sources and can occasionally become more difficult to overcome. In several instances, scientific studies were conducted to determine the industries' culpability for pollution accidents. For illustration, in 1993, environmental experts from the Ministry of Environment and Forest investigated the Rajasthani village of Bichhri at the Supreme Court's request to ascertain how four industries that generated H-acid (1-amino-8-hydroxynaphthalene-3,6-disulfonic acid) accounted for to the contamination of the village's soil and groundwater. Based on the findings of the experts' research, the Supreme Court used the common law "Polluter Pays" concept to require the industries to cover the expenses of remediation (Varghese and Alappat 2012).

1.6.2.1 Lack of Powerful Legislation

Criminal responsibility is the main tenet of India's environmental redressal procedure. It is widely acknowledged that this has not been successful in resolving the numerous environmental problems the nation faces (Singh 2024). The prosecution must prove its case beyond a reasonable doubt in criminal proceedings, with very few exceptions. In contrast, a less demanding standard is used in civil cases, where the claimant need only to establish their position by a preponderance of the evidence in order to prevail. Therefore, civil liability laws play a critical role in preserving the environment. Furthermore, environmental forensics may be particularly useful in civil liability claims involving pollution and the environment (Abraham and Abraham 1991). Without any formalized laws pertaining to civil liability, the law of torts serves as an appropriate remedy.

However, there are very few tort cases in India for a variety of reasons. Redress is typically sought in cases involving widespread pollution by filing a petition under Article 32 of the COI with the Supreme Court of India (Kirpal 2000). However, this solution approach is ineffective due to the magnitude of the pollution problem. Therefore, the primary reason environmental forensics did not develop in India was the absence of strong regulation (Varghese and Alappat 2012).

1.6.2.2 Delay in Delivering Justice

A significant critique of the Indian judiciary, particularly in civil cases, has been its tardiness in administering justice. Ten civil liability cases that were adjudicated between 1985 and 2010 were randomly reviewed by the Supreme Court of India, which discovered that it took anything from 4 to 34 years for a final decision to be made once litigation began in any of the lower courts. It is undoubtedly concerning because one of the possible application areas of environmental forensics is civil lawsuits resulting from pollution (Varghese and Alappat 2012).

1.6.3 Role of Forensic Science in Environmental Litigation

Environmental litigation relies heavily on forensic science, which tracks contaminants back to their sources using physical, chemical, and statistical methods. To identify the sources of contamination, such as industrial leaks, groundwater pollution, or improper waste disposal, this involves chemical fingerprinting, geographical mapping, and historical data analysis.

Similar to crime scene analysis, forensic investigations adhere to stringent procedures for gathering, preserving, and recording environmental evidence. This maintains a clean chain of custody and good documentation to withstand legal scrutiny, ensuring the integrity and admissibility of samples in court. Forensic specialists also recreate the environmental events in order to determine the source of contamination, use different tools and techniques for the litigation needs.

1.7 Conclusion and Future Directions

To better study environmental evidence, forensic science—namely, age dating and identifying the sources of contamination—has evolved into a multidisciplinary profession that includes the integration of geology, analytical chemistry, biology, and sophisticated computer software. By employing techniques such as chemical fingerprinting, isotopic investigation, and profiling microbial communities, forensic science offers strong methods for tracing the sources of a range of organic contaminants such as chlorinated solvents and petroleum products.

These methods are important in discriminating different sources of pollution and determining their temporal behavior. In even complex and multifaceted settings, precision and effectiveness in determining causes of environmental pollution have improved through the use of statistical models and simulation-optimization methods. In spite of these advancements, several challenges with computationally intensive requirements remain. The necessity for databases applicable at a global scale and merging data from numerous sources to enhance source attribution and age dating. Future developments in this area are likely to concentrate on increasing the range of readily accessible analysis platforms available, the precision and resolution of age-dating methods, and encouraging cross-disciplinary research to deal with novel contaminants and intricate pollution contexts. To further improve forensic methods, assist environmental monitoring, court cases, and the development of feasible remediation processes, continued research and technical development are essential.

References

Abdel-Shafy HI, Mansour MS (2016) A review on polycyclic aromatic hydrocarbons: source, environmental impact, effect on human health and remediation. Egypt J Pet 25:107–123

Abraham CM, Abraham S (1991) The Bhopal case and the development of environmental law in India. Int Comp Law Quart 40:334–365

Aeppli C, Berg M, Cirpka OA, Holliger C, Schwarzenbach RP, Hofstetter TB (2009) Influence of mass-transfer limitations on carbon isotope fractionation during microbial dechlorination of trichloroethene. Environ Sci Technol 43:8813–8820

Albrecht P, Hall TT Jr (2003) Atmospheric corrosion resistance of structural steels. J Mater Civ Eng 15:2–24

Ashley K (2003) Field-portable spectroscopy. Appl Occup Environ Hyg 18:10–15

Aulenta F, Majone M, Tandoi V (2006) Enhanced anaerobic bioremediation of chlorinated solvents: environmental factors influencing microbial activity and their relevance under field conditions. J Chem Technol Biotechnol 81:1463–1474

Balali-Mood M, Naseri K, Tahergorabi Z, Khazdair MR, Sadeghi M (2021) Toxic mechanisms of five heavy metals: mercury, lead, chromium, cadmium, and arsenic. Front Pharmacol 12:643972

Bartick EG (2012) In situ crime scene analysis. In: Infrared and Raman spectroscopy in forensic science, pp 171–184

Bartley DL, Slaven JE, Rose MC, Andrew ME, Harper M (2007) Uncertainty determination for nondestructive chemical analytical methods using field data and application to XRF analysis for lead. J Occup Environ Hyg 4:931–942

Bear J, Cheng AH-D (2010) Modeling groundwater flow and contaminant transport. Springer

Bednar AJ, Russell AL, Georgian T, Splichal DE, Hayes CA, Tackett P, Parker LV, Kirgan RA, Wells M, Justes D (2011) Field-portable gas chromatograph mass spectrometer (GC-MS) unit for semi-volatile compound analysis in groundwater. ERDC TR-11-11

Béen F (2015) Assessing the added value of wastewater-based epidemiology to monitor illicit drug use. Université de Lausanne, Faculté de droit, des sciences criminelles et d …

Bosco GL (2013) Development and application of portable, hand-held X-ray fluorescence spectrometers. TrAC Trends Anal Chem 45:121–134

Buchberger WW (2011) Current approaches to trace analysis of pharmaceuticals and personal care products in the environment. J Chromatogr A 1218:603–618

Carron K, Cox R (2010) Qualitative analysis and the answer box: a perspective on portable Raman spectroscopy. ACS Publications

Contreras JA, Murray JA, Tolley SE, Oliphant JL, Tolley HD, Lammert SA, Lee ED, Later DW, Lee ML (2008) Hand-portable gas chromatograph-toroidal ion trap mass spectrometer (GC-TMS) for detection of hazardous compounds. J Am Soc Mass Spectrom 19:1425–1434

Da Silva ML, Alvarez PJ (2002) Effects of ethanol versus MTBE on benzene, toluene, ethylbenzene, and xylene natural attenuation in aquifer columns. J Environ Eng 128:862–867

David E, Niculescu V-C (2021) Volatile organic compounds (VOCs) as environmental pollutants: occurrence and mitigation using nanomaterials. Int J Environ Res Public Health 18:13147

Duffy LM, Downing E, Huey BM, McKone TE (2000) Strategies to protect the health of deployed US forces: detecting, characterizing, and documenting exposures. National Academies Press

Eckenrode B (2001) Environmental and forensic applications of field-portable GC-MS: an overview. J Am Soc Mass Spectrom 12:683–693

Feng Y, Yang Y, Huang B (2019) Corrosion analysis and remaining useful life prediction for storage tank bottom. Int J Adv Robot Syst 16:1729881419877051

Fleyfel LM, Matta J, Sayegh NF, El Najjar NH (2024) Olive mill wastewater treatment using coagulation/flocculation and filtration processes. Heliyon 10:e40348

Frantz S (2017) Applications of compound-specific isotope analysis (CSIA) for determining persistent organic pollutant (pop) origin, transport, and transformation in the aquatic environment. University of Virginia

Fuqi C, Syamsuddin E, Umar EP (2019) Interpretation of paleodepositional environment using biomarkers and carbon isotope (δ13C) in Talang Akar Formation, South Sumatra Basin, Indonesia. In: IOP conference series: earth and environmental science. IOP Publishing, p 012024

Gałuszka A, Migaszewski ZM, Namieśnik J (2015) Moving your laboratories to the field–advantages and limitations of the use of field portable instruments in environmental sample analysis. Environ Res 140:593–603

He M, Moldowan MJ, Peters KE (2018) Biomarkers: petroleum. In: Encyclopedia of geochemistry, Encyclopedia of earth sciences series, vol 10. Springer, Cham, pp 978–973

Henschler D (1994) Toxicity of chlorinated organic compounds: effects of the introduction of chlorine in organic molecules. Angew Chem Int Ed Engl 33:1920–1935

Hewstone R (1994) Health, safety and environmental aspects of used crankcase lubricating oils. Sci Total Environ 156:255–268

Hollas JM (2004) Modern spectroscopy. Wiley

Imamoglu I, Li K, Christensen ER (2002) Modeling polychlorinated biphenyl congener patterns and dechlorination in dated sediments from the Ashtabula River, Ohio, USA. Environ Toxicol Chem 21:2283–2291

Jjunju FP, Maher S, Li A, Badu-Tawiah AK, Taylor S, Graham CR (2014) Analysis of polycyclic aromatic hydrocarbons using desorption atmospheric pressure chemical ionization coupled to a portable mass spectrometer. J Am Soc Mass Spectrom 26:271–280

Kaljurand M (2014) Analytical separations in harsh environments. Trends Environ Anal Chem 1:e2–e7

Kalnicky DJ, Singhvi R (2001) Field portable XRF analysis of environmental samples. J Hazard Mater 83:93–122

Kirpal B (2000) Public interest litigation: potential and problems. Oxford University Press

Klecka G, Boethling R, Franklin J, Grady L, Graham D, Howard PH, Kannan K, Larson R, Mackay D, Muir D (2000) Evaluation of persistance and long-range transport of organic chemicals in the environment. SETAC Pensacola

Kuntze K, Eisenmann H, Richnow H-H, Fischer A (2020) Compound-specific stable isotope analysis (CSIA) for evaluating degradation of organic pollutants: an overview of field case studies. In: Anaerobic utilization of hydrocarbons, oils, and lipids, pp 323–360

Kuppusamy S, Maddela NR, Megharaj M, Venkateswarlu K (2020) Total petroleum hydrocarbons: environmental fate, toxicity, and remediation. Springer
Kuppusamy S, Thavamani P, Megharaj M, Lee YB, Naidu R (2016) Kinetics of PAH degradation by a new acid-metal-tolerant Trabulsiella isolated from the MGP site soil and identification of its potential to fix nitrogen and solubilize phosphorous. J Hazard Mater 307:99–107
Kuppusamy S, Thavamani P, Venkateswarlu K, Lee YB, Naidu R, Megharaj M (2017) Remediation approaches for polycyclic aromatic hydrocarbons (PAHs) contaminated soils: technological constraints, emerging trends and future directions. Chemosphere 168:944–968
Lam R, Lennard C, Kingsland G, Johnstone P, Symons A, Wythes L, Fewtrell J, O'Brien D, Spikmans V (2018) Person-portable equipment in environmental forensic investigations: application to fire scenes. Aust J Forensic Sci 50:672–681
Landolfo R, Cascini L, Portioli F (2010) Modeling of metal structure corrosion damage: a state of the art report. Sustainability 2:2163–2175
Lauer N, Vengosh A (2016) Age dating oil and gas wastewater spills using radium isotopes and their decay products in impacted soil and sediment. Environ Sci Technol Lett 3:205–209
Lay-Ekuakille A, Palamara I, Caratelli D, Morabito F (2013) Experimental infrared measurements for hydrocarbon pollutant determination in subterranean waters. Rev Sci Instrum 84
Li Y, Chen B, Yang S, Jiao Z, Zhang M, Yang Y, Gao Y (2025) Advances in environmental pollutant detection techniques: enhancing public health monitoring and risk assessment. Environ Int 109365
Littlejohns JV, Daugulis AJ (2008) Kinetics and interactions of BTEX compounds during degradation by a bacterial consortium. Process Biochem 43:1068–1076
Ljungberg A (2002) Process measurement. Int J Phys Distrib Logist Manag 32:254–287
Lorah MM (2003) Anaerobic degradation of 1, 1, 2, 2-tetrachloroethane and association with microbial communities in a freshwater tidal wetland, Aberdeen Proving Ground, Maryland: Laboratory experiments and comparisons to field data. US Department of the Interior, US Geological Survey
Mach PM, Verbeck GF (2018) Analytical methods and trends in environmental forensics. Dev Environ:285–301
Matteucci F, Ercole C, Del Gallo M (2015) A study of chlorinated solvent contamination of the aquifers of an industrial area in Central Italy: a possibility of bioremediation. Front Microbiol 6:924
Matthews RL, Longworth TL, Ong KY, Zhu L, Brown CD (2006) Testing of Ahura's FirstDefender handheld chemical identifier against toxic industrial chemicals. US Army Edgewood Chemical Biological Center, Report ECBC-TR-528
Milchert E, Pazdzioch W, Myszkowski J (1995) Dehydrochlorination of waste 1, 1, 2-trichloroethane. Ind Eng Chem Res 34:2138–2141
Moran MJ, Zogorski JS, Squillace PJ (2007) Chlorinated solvents in groundwater of the United States. Environ Sci Technol 41:74–81
Morrison RD (1999) Environmental forensics. CRC Press
Morrison RD (2000) Application of forensic techniques for age dating and source identification in environmental litigation. Environ Forensic 1:131–153
Morrison RD, Hone J (2010) Age dating the release of PCE, TCE and TCA using stabilizers and feedstock impurities as indicators. In: Environmental Forensics Proceedings of the 2009 INEF Annual Conference. Royal Society of Chemistry, Cambridge, UK, pp 289–304
Mudge SM (2008) Environmental forensics and the importance of source identification. In: Environmental forensics
Mulligan CC, Justes DR, Noll RJ, Sanders NL, Laughlin BC, Cooks RG (2006) Direct monitoring of toxic compounds in air using a portable mass spectrometer. Analyst 131:556–567
Murphy B, Morrison RD (2006) Environmental forensics: contaminant specific guide. Academic Press
Murphy BL, Morrison RD (2014) Introduction to environmental forensics. Academic Press
Narayana TL, Venkatesh C, Kiran A, Kumar A, Khan SB, Almusharraf A, Quasim MT (2024) Advances in real time smart monitoring of environmental parameters using IoT and sensors. Heliyon 10:e28195

Ochsner J, Blackwood T, Micheletti W (1980) Status assessment of toxic chemicals: trichloroethylene. Industrial Environmental Research Laboratory, Office of Research and …
Offenberg J, Simcik M, Baker J, Eisenreich SJ (2005) The impact of urban areas on the deposition of air toxics to adjacent surface waters: a mass budget of PCBs in Lake Michigan in 1994. Aquat Sci 67:79–85
Oladimeji T, Oyedemi M, Emetere M, Agboola O, Adeoye J, Odunlami O (2024) Review on the impact of heavy metals from industrial wastewater effluent and removal technologies. Heliyon 10:e40370
Omokpariola DO, Nduka JK, Kelle HI, Mgbemena NM, Iduseri EO (2022) Chemometrics, health risk assessment and probable sources of soluble total petroleum hydrocarbons in atmospheric rainwater, Rivers State, Nigeria. Sci Rep 12:11829
Onyeabor E (2024a) Environmental law. Springer
Onyeabor E (2024b) Strategies for atmospheric pollution abatement and control. In: Environmental law: international and regional African perspectives on law and management. Springer, pp 255–374
Peters K, Walters C, Moldowan J (2005) The biomarker guide. Cambridge University Press, Cambridge, UK
Peters KE, Moldowan JM (1993) The biomarker guide: interpreting molecular fossils in petroleum and ancient sediments. Prentice Hall
Petrisor IG (2014) Environmental forensics fundamentals: a practical guide. CRC Press
Rad HD, Babaei AA, Goudarzi G, Angali KA, Ramezani Z, Mohammadi MM (2014) Levels and sources of BTEX in ambient air of Ahvaz metropolitan city. Air Qual Atmos Health 7:515–524
Ramsey CA, Hewitt AD (2005) A methodology for assessing sample representativeness. Environ Forensic 6:71–75
Roberge PR (2000) Handbook of corrosion engineering. Mcgraw-hill
Samad MSA, Varghese GK, Alappat BJ (2015) Environmental forensics in India–four years after the National Green Tribunal Act, 2010. Procedia Environ Sci 30:91–96
Schobert H (2014) Production of acetylene and acetylene-based chemicals from coal. Chem Rev 114:1743–1760
Sedyaaw P, Pandey A, Kawade SS, Ade PJ, Nimbarte PV, Bhaladhare DR, Kamble PA (2024) A review on acid rain it's causes, effects and management measures. Bhartiya Krishi Anusandhan Patr 12(4):2320–2882
Sekhwama M, Mpofu K, Sivarasu S, Mthunzi-Kufa P (2024) Applications of microfluidics in biosensing. Discov Appl Sci 6:303
Singh A (2024) Environment justice and role of national green tribunal in India: an analysis. School of Law, UPES, Dehradun
Spikmans V (2015) Complexity of scientific evidence in environmental forensic investigations. J Criminol Res Policy Pract 1:223–232
Spikmans V (2019) The evolution of environmental forensics: from laboratory to field analysis. Wiley Interdiscip Rev Forensic Sci 1:e1334
Stout S, Wang Z (2016) Standard handbook oil spill environmental forensics: fingerprinting and source identification. Academic Press
Tong X, Mohapatra S, Zhang J, Tran NH, You L, He Y, Gin KY-H (2022) Source, fate, transport and modelling of selected emerging contaminants in the aquatic environment: current status and future perspectives. Water Res 217:118418
Urbaniak M (2007) Polychlorinated biphenyls: sources, distribution and transformation in the environment–a literature review. Acta Toxicol 15
Valcárcel M, Cárdenas S (2005) Vanguard-rearguard analytical strategies. TrAC Trends Anal Chem 24:67–74
Varghese G, Alappat BJ (2012) National Green Tribunal act: a harbinger for the development of environmental forensics in India? Environ Forensic 13:209–215
Volpe A, Del Moro G, Rossetti S, Tandoi V, Lopez A (2007) Remediation of PCE-contaminated groundwater from an industrial site in southern Italy: a laboratory-scale study. Process Biochem 42:1498–1505

Wang Z, Fingas M, Yang C, Christensen JH (1964) Crude oil and refined product fingerprinting: principles. In: Environmental forensics. Elsevier, pp 339–407
Wania F (2003) Assessing the potential of persistent organic chemicals for long-range transport and accumulation in polar regions. Environ Sci Technol 37:1344–1351
Wania F, Dugani CB (2003) Assessing the long-range transport potential of polybrominated diphenyl ethers: a comparison of four multimedia models. Environ Toxicol Chem 22:1252–1261
Zheng C, Bennett GD (2002) Applied contaminant transport modeling. Wiley

Principles of Forensic Ecology

2

Shivam Arya and Neeharika Srivastava

Abstract

Forensic ecology is an interdisciplinary and applied field that encompasses forensic geology, forensic botany, soil analysis, forensic entomology, wildlife forensics, forensic microbiology, etc. Principles, methodology, and application of forensic ecology provide a holistic approach to investigating and solving a wide range of criminal cases. By seeking ecological principles and techniques, experts contribute significantly to the criminal justice system and the ecological consequences of *Homo sapiens* activities. The complex relationship between ecology and forensic science can be established by incorporating principles of ecology like habitat of different flora and fauna, ecosystem dynamics, life cycle of flies for postmortem interval, growth of necrobiomic microorganisms at the cadaver-soil interface, diatomic examination, wildlife evidences, palynology, botany, archaeology, organism symbiosis geology, and isotopic analysis in criminal investigation. Principles of forensic ecology have proved to be effective in linking objects, people, and places. Characterization of these dynamic changes has been made possible by recent developed principles and concepts in ecology like temporal ecology, abrupt shifts in ecological systems, ecological memory, lag hypothesis for community dynamics, asymptotic environmentally determined trajectories, etc., which made a framework for streamlining research in forensic science, which promotes quantitative approaches to collect and applying data to forensic investigations. Forensic experts have always been familiar with sampling and analysis of evidences obtained from environmental sources to pro-

S. Arya (✉)
Department of Forensic Science, ICFAI University, Jaipur, India

Department of Forensic Science, Galgotias University, Greater Noida, India

N. Srivastava
Department of Forensic Science, Galgotias University, Greater Noida, India

M. S. Sankhla et al. (eds.), *Introducing the Synergy Between Forensic Science and Environmental Analysis*, https://doi.org/10.1007/978-981-95-0199-1_2

vide information about past incidents. The emergence and applications of these novel laws can be specific, precise, or interchangeable, sometimes depending on who is using them and the context. These recent ecological principles and concepts have a significant impact on forensic science and will be a useful tool in forensic ecological studies for decades to come.

Keywords

Forensic ecology · Ecological principles · Criminal investigation · Necrobiomic · Evidences · Isotope · Ecological dynamics · Habitat

2.1 Introduction

Ecology term originally taken from German word "Öekologie," which was first defined in 1866 by Ernst Haeckel, an enthusiastic and influential disciple of Charles Darwin. In 1973, Ricklefs defined ecology as "the study of the natural environment, particularly the interrelationships between organisms and their surroundings" in his textbook (Colin et al. 2008). Forensic ecology is an emerging discipline in the field of forensic science that utilizes the principles of ecology and environment to investigate or solve the criminal cases. It involves the principles of palynology, zoology, archaeology, anthropology, botany, pedology, geology, and entomology. This discipline often deals with the analyzing these evidences to determine the location, time, and circumstances surrounding a crime, which leads to the establishment of corpus delicti (M'arquez Grant and Robert 2012). This chapter discusses the importance of ecological principles and factors in understanding the nature, application and utilization of ecological evidence in forensic investigations.

Connection of victim and offender with a crime scene may be determined using the examination of fragments of ecological evidences. Different findings can be drawn, like:

- *Time of Death Determination:* Ecological evidences, like the activity of insects or plant growth, can help to estimate how long a body has been at a particular location.
- *Locating Crime Scenes:* Different ecological conditions (such as specific types of flora, fauna, or soil characteristics) can narrow down the potential crime scene locations.
- *Movement Tracking:* Flora-fauna or pedological evidences found on suspects or objects can be traced back or linked to specific locations, providing evidence of movement.
- *Locating Graves or Hidden Bodies:* By examining changes in the composition of vegetation or soil, forensic ecologists can be utilized to locate clandestine graves.
- *Development of* Corpus Delicti*:* Soil, botanical, insects, or other ecological evidences found on or near the suspect and victim can connect them to a specific location, supporting other forensic evidences.

Forensic ecology is thus an emerging and valuable discipline in crime scene investigation, which can be utilized to get critical information that might not be obtainable through more conventional forensic methods.

2.2 Implementation of Various Principles in Forensic Ecology

The use of forensic ecology in legal investigations is guided by a number of fundamental concepts. These guidelines assist forensic ecologists in methodically gathering, evaluating, and interpreting ecological data to bolster criminal prosecutions. Various core principles used in the field of forensic ecology are listed below:

2.2.1 Principle of Locard's Exchange

A well-known principle of forensic science, often paraphrased as "every contact leaves a trace," asserts that when two objects come into contact, there is always an exchange of material. Forensic ecology utilizes this concept for environmental evidences, such as soil, pollen, or plant material, etc., which can be transferred between a suspect, victim, or object and a specific location. Forensic ecologists analyze remnants left after contact to lay the foundation of links between individuals or articles and crime scenes.

2.2.2 Principle of Ecological Niche

The principles of an ecological niche play an important role in forensic science, especially in forensic entomology, as it helps forensic investigators to understand how living organisms interact with the environment and how these interactions can be helpful in solving crimes. The ecological niche is important in forensics for the following reasons:

Crime Scene Reconstructions: This model can help to reconstruct past climatic and environmental conditions at the time of death. The niche preferences of various organisms, including environmental tolerances, can assist investigators about the environmental conditions when a crime was committed. This might lead to narrowing down the timeframe for the crime.

Locating Crime Scenes or Evidence Movement: Certain species of arthropods or plants occupy very specific niches in particular ecosystems, like urban, rural, aquatic, forest, etc. If a forensic investigator finds an organism on a corpse or location that could not have come from the immediate environment, it suggests that the evidences were moved from the crime scene (M'arquez-Grant and Robert 2012). For example, finding insects or plants traces native to a coastal area on a body found inland could suggest relocation from the coastal area to inland.

Understanding Insect Succession in Forensic Entomology: Forensic entomologists may use the ecological niches of insects to examine the chronology of decomposition. Different stages of a corpse's decomposition attract different insects, such as blowflies. Every insect species has a distinct ecological niche that it fills. The order in which different species appear on a body can reveal important information on the postmortem interval (PMI), or the amount of time that has passed since death.

Identifying Poisoning or Contaminants: The ecological niche of some organisms could also help in identifying contaminants, toxins, or poisons in the environment.

Taphonomy and Decomposition: The niche of scavengers and decomposers can provide crucial information that affects how a body is broken down in different environments, which influences the taphonomic processes (post-death changes). Forensic taphonomists can also consider the interaction between a body and its environment, including animals, insects, and microbes in their niches, to estimate decomposition rates and the time since death (Colin et al. 2008).

2.2.3 Principle of Ecological Specificity

Bornstein's Specificity Principle explains that human development always involves specific outcomes in specific individuals occurring in specific places at specific times in specific ways. As defined at present, the specificity principle has five main terms: setting, person, time, process, and outcome. The specificity principle states that processes of development involve mutually influential relations between specific individuals and their specific contexts, which can be used in the determination of corpus delicti.

Different ecosystems have unique ecological characteristics, including specific plant and animal species, soil types, and other ecological markers. These unique specifications can be used to narrow down the geographical location of a crime scene as a primary, secondary, and tertiary crime scene. It also follows the law of individuality (first principle of forensic science) (Mildenhall and Smith 2005).

A study of ecological evidence (such as pollen or insect species) can be helpful for forensic ecologists in linking the evidences to a specific area or environment, which may help to establish the location of an event.

2.2.4 Principle of Succession

Ecological succession refers to the sequential shift in species association and community structure that occurs within a given ecosystem with time. It is especially useful in forensic entomology, as instances of insect colonization on a dead body can be employed to estimate the postmortem interval, or the time elapsed since death. Forensic ecologists can identify the time and date of an incident or the duration for which a dead body has lain in a particular spot using knowledge of the stages of natural succession (Colin et al. 2008).

2.2.5 Principle of Environmental Context

Context is highly required for the examination of ecological data that may be presented by the environment in which the evidences are located. Weather and climatic conditions, topography, and human impact are some of the parameters that influence the presence of ecological evidences (Tibbett and Carter 2008). In order to make a proper assessment of the evidences, forensic ecologists must consider the environmental context. For instance, the occurrence of certain plant and animal species will be more likely to be attributed to current weather conditions or human behavior than to the natural state.

2.2.6 Principle of Representative Sampling

In order to effectively analyze ecological evidence, it is important to maintain the best sampling practices, that is, collecting samples that will be representative of the diversity and distribution of ecological characteristics in a given area of sampling (Sampson 2012). Forensic ecologists must take care to select and collect samples that accurately represent the criminal(s) ecological evidence within the crime scene to ensure the results are valid and reliable. This frequently involves the systematic sampling strategies to capture the full range of ecological evidences. Important components of this principle include:

1. Randomness: In order to prevent bias and make sure the samples are an accurate representation of the overall features of the evidences, they must be chosen at random.
2. Sample Size: The size of the sample must be large enough in order to capture differences between evidences, i.e., medicine combinations or soil compositions.
3. Relevance: The samples must cover the essential aspects in which the perpetrator is involved in the crime and relate to the investigation.

For purposes of enabling the results obtained from the study to be generalized to the crime scene as a whole or even a collection of evidences, representative sampling is utilized.

2.2.7 Principle of Comparative Analysis

In forensic ecology, the process is usually that one compares to known reference samples or databases when attempting to come to a conclusion. These are comparisons of ecological evidence between a known sample and evidence found at the crime scene.

The comparison study method is used by forensic ecologists to ascertain whether the soil, pollen, or other ecological aspects of one location are compatible with another location's soil, pollen, or other ecological aspects in an effort to confirm or refute association with a crime scene. For instance: comparing fingerprints, tool marks, tire prints, bullet striations, or identical DNA profiles. Chemical analysis involves analyzing chemicals, like narcotics or explosives, for their makeup and source. Comparative analysis solidifies the results by eliminating certain people as suspects or trying to match the evidence to a suspect.

2.2.8 Principle of Interdisciplinary Collaboration

Various intersections exist between forensic ecology and scientific disciplines that include botany, entomology, soil science, and environmental science. Interdisciplinary collaboration between experts is essential to provide careful examination and valid interpretation of environmental data. Forensic experts collaborate with other professional experts to integrate multiple lines of evidence and establish their findings more forcefully and confidently. Experts, if they collaborate with each other, produce a fuller insight into the data to allow valid analysis, scene reconstruction, and legal conclusions. Such collaboration enhances the accuracy of the inquiry and fosters an integrated method of addressing complex problems through merging coupled information and the principles of scientifically important-led prioritization, to ultimately provide the assurance that justice can be achieved.

2.2.9 Principle of Temporal Variation

Ecological conditions are changing over time, and these changes affect the integrity and relevance of the evidences discovered that influences the result of forensic evaluation. It follows and proves the concept of the "law of analysis" and "law of progressive changes," etc. Temporal variations like seasonal diversity, weather patterns, and ecological succession must be considered at the time of evidence examination. Forensic ecologists must consider these changes to accurately interpret the age or history of ecological evidence, such as the determination of time since death based on entomological activity or plant growth. The effects of temporal variations on evidences and the crime scene are:

Biological evidences, like blood, bodily fluids, DNA, etc., may degrade or decompose due to environmental factors like heat, moisture, or microbial activity. Trace evidences, such as fingerprints, hair, or fibers, may become less detectable as time passes, particularly due to environmental exposure or physical disturbance. If digital evidences, such as data on devices, is not immediately safeguarded, it may be lost or manipulated, particularly because of system changes, human error, or automated deletion procedures.

A crime scene can undergo significant changes as time progresses. For example, weather conditions may alter the outdoor crime scenes, washing away or obscuring the evidences.

Eyewitnesses and testimonials can differ as their memory fades over time.

The principle of temporal variation extends to the chain of custody also, as it is very important to track when evidence is collected and handled and any changes it undergoes. Timely documentation of all the evidence helps in maintaining the integrity of the evidence for court proceedings.

There are time limits on legal proceedings, too, for example, the statute of limitations, which emphasizes the importance of collecting and keeping evidence to prosecute within time.

Certain forensic methods also have to be performed immediately if they are to be effective. For example, identification of some chemical residues can be done in a short while before they disappear or get masked out.

Forensic ecology is also one of the specialized branches of forensic science that applies ecological as well as environmental knowledge for legal inquiry purposes. Forensic ecology deals with the examination of organism-environment interactions to develop evidence in favor of crime-solving. Forensic ecologists apply a range of ecological techniques to connect a suspect, victim, or article to a location or event on the basis of the available ecological evidence.

2.3 Key Areas of Forensic Ecology

2.3.1 Forensic Entomology

Forensic entomology uses the systematic patterns of infestation of insects on a decomposing body to assist in criminal investigations. Flies and beetles would infest a body first, and their life stages can be highly determinative of the estimation of the time of death (Byrd and Castner 2009). Insect infestation of a body is a sequential process that occurs predictably. Blowflies (Calliphoridae) would infest the body first, laying eggs that, in hours, would develop into larvae. Forensic experts will be better placed to estimate the postmortem interval (PMI) more accurately by examining the presence and stages of specific species of insects (Haskell and Van Laerhoven 2003). Forensic entomology, and its rich contribution, has countless challenges. Weather, location, and availability of other decomposing bodies will affect the activity and development of insects and thus complicate analysis (Gordon and Grayson 2009). Even variations in methodologies and protocols will affect the accuracy of forensic entomological analyses, and the necessity for uniform and standardized protocols to increase accuracy is dictated (Haskell and VanLaerhoven 2003).

Applications

Forensic entomologists can estimate a timeline for the postmortem period by examining the types of insects present and their respective stages of development (Byrd and Castner 2009).

Insect evidence is a great investigative tool used to associate suspects with a crime scene (Sukontason et al. 2006).

In cases of environmental crimes (i.e., illegal disposal of hazardous waste), insect evidence can also be used in determining the type and timing of contamination (Smith et al. 2010).

2.3.2 Forensic Palynology

The principle that excites forensic palynology is that pollen and spores have different characteristics that vary by species or locality. This gives forensic scientists (investigators or practitioners) the ability to establish the geographical and ecological origins of evidence. In forensic palynology, it is essential to comprehend pollen signatures, that is, the recognizable combination of pollen types found in a specific area or during a defined time period. Pollen signatures allow forensic scientists to connect evidence from a crime scene to a known reference sample by comparing pollen profiles (Bryant et al. 2006). Pollen and spore samples are routinely obtained from many locations, including soil, clothing, and objects within a crime scene (Léon et al. 2011). Once pollen and spore materials are collected, they are placed under a microscope where forensic palynologists will identify and count the types of pollen grains present to create a composite profile of the evidence sample (Jankowski and Carter 2008). Forensic palynology can be very useful, yet it has several challenges. The abundant number of species of pollen and the microscopic nature of them complicate the analysis and interpretation of evidence (Léon et al. 2011) to establish pollen type, and pollen grains, over time or at death, disintegrate or are contaminated with extraneous materials (Jankowski and Carter 2008).

Applications

The geographic source of evidence can be determined very much by forensic palynology (Jankowski and Carter 2008). Forensic palynologists can determine many aspects of a crime scene through pollen and spore evidence, including the season, type of environment, and changes to the scene over time (Léon et al. 2011).

Additionally, pollen evidence can link different crime scenes; if similar pollen profiles are found at different locations, there is a connection between the crime scenes (Smith et al. 2010).

2.3.3 Forensic Botany

Assuming that the presence of plant products can provide salient evidence in criminal cases, forensic botany The first premise of forensic botany is plant identification and analysis. The identification of a species and its geographical origin is based upon an analysis of its distinctive morphological characteristics. Forensic botanists rely upon these characteristics to connect plant material at a crime scene with known reference materials (Sampson 2012). Forensic botanists will collect seeds, wood products (Mildenhall and Smith 2005), and leaves and other plant materials. In the identification of plant species, the main focus is typically on the study of reproductive structures, such as flowers and seeds, as well as morphological characteristics, such as the shape, texture, and size of the leaves. Forensic botanists can refer to taxonomic keys and reference collections to ensure accurate plant identification (Kamal et al. 2013). Forensic botany can be difficult. For instance, plant materials are the most sensitive or fragile forms of evidence and can degrade over time under a variety of environmental conditions, which can hinder the investigation (Sampson 2012). Similarly, the diversity of species, along with the different forms of a plant species, makes identification difficult (Mildenhall and Smith 2005).

Applications

Forensic botany assists in reconstructing crime scenes. Plant evidence can provide information about the scene and help to establish if a crime occurred inside or outside to identify the plant types available (Sampson 2012). Furthermore, plant material found on suspects' property, including clothing, can be connected back to a crime scene and provide a relevant association (Mildenhall and Smith 2005). Forensic botanists can also add more value to an investigation by using plant remains to provide information about the season and timing of a crime (Kamal et al. 2013).

2.3.4 Forensic Pedology

This field of knowledge incorporates ideas from pedology, geology, and forensic science, offering insightful viewpoints on numerous aspects of criminal investigations. The concepts, techniques, applications, and challenges of forensic pedology will be considered in this chapter. Every site has a soil profile that is characterized by the weight of organic matter, soil texture, and mineral composition. Forensic pedologists match evidence soil samples from a crime scene to a potential source based on this unique feature (Pye and Croft 2004). Soil profiles are compared to established reference profiles using statistical analysis to help with determining similarity to samples taken from the crime scene (Vass 2001).

There are many barriers to the use of forensic pedology. Factors associated with the environment result in soils being similar, but not identical, due to the rapid change that happens over short distances, making precise matches very difficult (Burt 2004). In addition, soil samples can easily become contaminated from outside sources, which complicates the accuracy of research and can subsequently hinder reliable conclusions (Tibbett and Carter 2008).

Applications

By examining the composition and texture of the soil, forensic pedologists can infer details about the surroundings in which a crime took place (Burt 2004).

Comparing the soil on suspects' clothing, cars, or personal items with soil from crime scenes is conceivable. Either the suspect was at the crime scene or not can be identified by soil carried from the scene by him (Pye and Croft 2004).

Forensic pedology may also help to determine the provenance of items found at crime scenes, hence generating additional investigation leads (Lal 2003).

2.3.5 Forensic Aquatic Ecosystems

Systems of forensic water forensic aquatic ecosystems assist police investigations by taking advantage of the particular traits of aquatic habitats and their life. Studying the health and composition of marine species to obtain data on environmental conditions and changes across time (Karr 1991), aquatic biomonitoring is a basic tenet. Unique chemical and biological indicators of aquatic ecosystems help to track the source of evidence and recreate crime scenes. pH, dissolved oxygen, and contaminants are among the parameters checked for in water samples. Along with biological samples, including maritime plants, insects, and fish (Hellawell 1986), physical and chemical properties of sediment samples are assessed. Forensic investigators utilize taxonomic keys and reference books to categorize fish, insects, and vegetation species found in the samples. This knowledge could shed light on the environment and the environment at the crime scene (Miller et al. 2002). Furthermore, knowledge of the source and timing of pollution comes from studies of toxins and pollutants found in maritime ecosystems. Studying their concentrations and distributions helps forensic scientists to identify any correlations to criminal activity (Carter et al. 2007).

How pollutants like heavy metals and pesticides affect the water environment. For forensic aquatic habitats, problems abound. Due to seasonal shifts, weather, and human activity, among other reasons, aquatic habitats are extremely dissimilar; conditions vary. This variance makes interpretation of evidence difficult and calls for careful investigation of environmental elements to guarantee accurate analysis (Karr 1991). Additionally, external sources can contaminate specimens from aquatic habitats; biological evidence may disintegrate over time, therefore complicating the forensic investigation (Hellawell 1986).

Applications

Certain aquatic plants or pollutant levels offer great indications on whether a body had been buried in the water or had been underwater for an extended period (Weber et al. 2006).

Likewise, if plant or animal species found on an object or victim match those discovered in a particular aquatic habitat, it might imply that the item or victim was positioned in that area (Judd et al. 2004).

By analyzing the presence and concentration of pollutants in aquatic systems, forensic specialists can also trace the origin of pollution and identify possible perpetrators (Sullivan et al. 2008).

2.4 Challenges in Forensic Ecology

Environmental Variations: Natural settings exhibit great variation, and many elements—such as climate, human action, or seasonal changes—can impact ecological data and confound or complicate interpretations.

Evidence Degradation: Over time, evidence such as plant remnants or bug remains might deteriorate, making it challenging to get accurate results (Vass 2001).

Interdisciplinary Expertise: Forensic ecology sometimes entails cooperation with professionals in several areas, including botany, entomology, and pedology; thus, it calls for a thorough knowledge of several ecological and biological systems.

2.5 Conclusion

It entails research of species interactions with their surroundings to find evidence that could help police solve crimes. Particularly in situations when conventional forensic techniques could be constrained, it gives a strong and well-organized structure for studying offenses. By understanding and employing these concepts, forensic ecologists can link a suspect, victim, or object to a specific place or event, depending on the availability of ecological evidence, using several ecological techniques. An investigating officer may grasp the connections between species and their surroundings. Hence, forensic ecologists may discover important information from these sources that assists to recreate the events of a crime and hence make it a useful and expanding area in forensic science.

References

Bryant VM, Jones RR, Hill RC (2006) Forensic palynology: principles and applications. CRC Press

Burt R (2004) Soil survey laboratory methods manual. Soil Survey Investigations Report No. 42. United States Department of Agriculture

Byrd JH, Castner JL (2009) Forensic entomology: the utility of arthropods in legal investigations. CRC Press

Carter J, Brown C, Smith M (2007) Environmental forensics: contaminant investigation and remediation. CRC Press
Colin R Townsend, Begon M, Harper JL (2008) Essentials of ecology, 3rd edn. Blackwell Publishing
Gordon KD, Grayson RE (2009) Soil analysis in forensic investigations: a review. J Forensic Sci 54(1):32–42
Haskell NH, Van Laerhoven SL (2003) Entomology and death: a procedural guide. CRC Press
Hellawell JM (1986) Biological indicators of fresh water pollution
Jankowski RJ, Carter GC (2008) The role of forensic palynology in criminal investigations. J Forensic Sci 53(2):290–298
Judd KA, Miller SR, Weber MS (2004) Aquatic forensic analysis: techniques and applications. Forensic Sci Int 139(2–3):203–212
Kamal A, Smith MR, Brown JA (2013) Forensic botany: principles and applications. Academic Press
Karr JR (1991) Biological integrity: a long-neglected aspect of water resource management. Environ Manag 15(4):551–556
Lal R (2003) Soil erosion and the global carbon cycle. Nutr Cycl Agroecosyst 66(1):7–20
Léon J, Girard P, Guérin L (2011) Forensic palynology: applications to crime scene investigation. Palynology 35(2):131–142
M'arquez Grant N, Robert J (2012) Forensic ecology handbook: from crime scene to court, 1st edn. Wiley
Mildenhall DC, Smith MJ (2005) Forensic botany and crime scene investigation. J Forensic Sci 50(2):343–350
Miller AC, Miller T, Smith JA (2002) Field guide to aquatic organisms: identification and analysis
Pye K, Croft D (2004) Soil analysis for forensic investigations
Sampson SM (2012) Forensic botany in criminal investigations. Bot Rev 78(2):110–123
Smith KGV, Anderson GS, Hall M (2010) Forensic entomology: the role of insects in crime scene investigations. J Appl Entomol 134(5):367–378
Sukontason K, Sukontason K, Piangjai S (2006) The use of blowflies in forensic investigations. J Forensic Sci 51(4):820–823
Sullivan R, Johnson A, Green M (2008) Forensic investigations of aquatic pollution. J Environ Manag 88(3):391–401
Tibbett M, Carter DO (2008) Soil analysis in forensic taphonomy: Chemical and biological effects of buried human remains. CRC Press
Vass AA (2001) Beyond the grave: understanding human decomposition. Microbiol Today 28:190–192
Weber M, Wright M, Carter J (2006) Aquatic ecosystem analysis in forensic investigations

Forensic Science in Ecosystems: Techniques for Ecosystem Analysis

3

Chitra Jangid, Mansha Pragyan Ray, Suryapratap Ray, Priyadharshini Kalyanasundaram, and Jyoti Dalal

Abstract

Over the years, environmental forensics has been important in investigating crimes to find vital evidence and solve difficult cases. Because of the growing problems of environmental degradation, reduced species, and pollution, forensic science now plays a vital part in environmental protection. As a result, finding new ideas to understand and restore the environment is becoming more important. Ecosystem forensics greatly improves our ability to study and protect the environment by adding forensic methods to ecological research. The chapter points out several forensic techniques for studying ecosystems, including eDNA analysis, stable isotope studies, toxicology measures, and others. They make it possible to identify species properly, track pollutants, and evaluate ecological activities, offering key information on how ecosystems change. Using these methods helps us face environmental concerns and improve how ecological

C. Jangid
Department of Forensic Science, Lovely Professional University, Phagwara, Punjab, India

M. P. Ray
School of Forensic Biotechnology, Department of Biotechnology, National Forensic Science University, Gandhinagar, Gujarat, India

S. Ray
Department of Biosciences, School of Biosciences and Technology, Vellore Institute of Technology, Vellore, Tamil Nadu, India

P. Kalyanasundaram
Department of Social Sciences, School of Social Sciences and Languages, Vellore Institute of Technology, Vellore, Tamil Nadu, India

J. Dalal (✉)
Department of Forensic Science, Lovely Professional University, Phagwara, Punjab, India

Amity School of Forensic Sciences, Amity University, Punjab, Mohali, India

M. S. Sankhla et al. (eds.), *Introducing the Synergy Between Forensic Science and Environmental Analysis*, https://doi.org/10.1007/978-981-95-0199-1_3

research is done, enabling ecological and sustainable relationships with our natural environment.

Keywords

Ecosystem analysis · Environmental DNA (eDNA) · Sustainability · Forensic

3.1 Introduction

Wildlife crime investigations, just like human crime scenes, happen at a specific point where evidence is abundant, the "death scene" (Horswell 2004). The best way to save any evidence is by managing a crime scene properly. Many procedures used in forensic science come from those created for human investigations (Fox and Cunningham 1973; Adrian 1996), simplifying the discovery, documentation, and collection of evidence left at scenes. With the invention of the microscope in the 1600s, forensic investigators have been able to use microscopy for many tasks. For almost a century, researchers have depended on optical and scanning microscopy to examine animal hair, skin features, bones, and plant anatomy (Housman 1920; Hardy and Wallace 2012). Experts have shown that applying forensic entomology to cases involving protected species such as harbor seals from the United States in the northeast and impalas within Africa is very effective (Lord and Burger 1984; El-Kady 1999). The use of contemporary molecular tools, especially with DNA as the main object, has led to major changes in wildlife forensics by eliminating the obstacles found in standard morphological and microscopic techniques. Forensic DNA analysis became important for prosecuting wildlife crimes after it started in the mid-1980s. In many cases, crime against wildlife has been resolved thanks to minisatellites (VNTRs), mitochondrial markers (mtDNA), and real-time PCR. Wasser et al. (2015) used genetics to pinpoint areas where elephant poaching takes place in Africa, which benefits law enforcement. Using isotope ratios, it is possible to discover trading routes, detect those who poach, and evaluate the length of time between a poaching event and when the trophy was seized (Cerling et al. 2016; Coutu et al. 2016). Although there have been major improvements, there are still difficulties in wildlife forensics because not all scientists agree on the best methods for analyzing wildlife DNA. Fortunately, groups like OSAC are making progress in developing stable procedures. Many problems with technology still exist, mainly in taxonomic ID, exact location, and differentiating captive-bred fish from those found in the wild. This review looks at the ways in which forensic science helps with addressing today's ecological problems. Through looking at how forensic science assists ecologists, the study's aims are to highlight how forensics can explain changes in ecosystems, monitor biodiversity, identify polluters, and set up conservation efforts. It will also outline improvements in forensic technology, provide examples of successful applications, and highlight where more work can be done. This chapter aims to highlight the main role of forensic science in helping the environment and in promoting living side by side with nature.

3.2 Ecosystem Forensics: Concept and Significance

In this area, forensic science applies both scientific methods and theories to help detect, explain, and correct ecological problems, environmental harm, and the relationships among parts of the ecosystem. For the last half-century, wildlife forensics has been developing to investigate crimes against plant and animal wildlife. Using information from biology, chemistry, and anthropology, this area of forensic science digs into and prosecutes wildlife crime cases within the limits of the law. (Wallace and Ross (2012) describe how border agencies and street-level police share intelligence.) Forensic ecosystem analysis is especially important for tackling the worldwide problem of wildlife crime, which is very profitable, harms many species, brings about the collapse of ecosystems, and affects the finances of numerous countries (INTERPOL 2018; UNODC 2016, 2020). Poaching, smuggling, and marketing wildlife are part of illegal wildlife trade, which often focuses on animals already vulnerable to extinction (Kahler and Gore 2017). Pangolins (*Manis* spp.) and sea cucumbers are still being exploited, and uncontrolled overharvesting played a role in the extinction of the passenger pigeon (*Ectopistes migratorius*) (Wyatt 2021; Maxwell et al. 2016). Despite some achievements, wildlife crime remains a serious problem since as much as 20% of terrestrial vertebrates sold in markets are endangering the survival of vulnerable wildlife. Besides being punished by law, wildlife crimes rarely cause serious outcomes as it is not easy to apply the regulations and speed through the process. Such simple solutions inspire criminals, which makes wildlife crime more likely to increase rapidly (Warchol 2018). Those organizations have an advantage, as it's not hard for new members to join, and, because the patterns evolve, they don't deal with tough laws and can still make good profits. The wildlife trade includes famously known animals such as elephants and rhinos, as well as rarer species. Many people take black-winged mynas, gray-backed mynas and gray-rumped mynas as pets, so these species are now included in the critically endangered category (Nijman et al. 2018). Traditional medicine is one reason why the number of pangolins has decreased across the globe (Challender et al. 2019). Most of the loss of biodiversity in Southeast Asia comes from major wildlife hunting for global trade and extensive damage to their habitats. It demonstrates why it's important to create stronger laws and efforts to keep endangered species from going extinct. The use of wildlife in medicine and cosmetics and for luxury items has increased their demand more than at any other time. While many traditional techniques use parts from tigers, rhinos, and bears, others are sacrificed to make bushmeat (Aisher 2016; Harrington et al. 2018). Yet, trading farmed wildlife products for those taken from the wild has not worked to decrease the need for wild species, because people still prefer those in the wild for their believed superior quality. The fact that wildlife trade, cultural practices, financial situations, and challenges faced by enforcers are linked together complicates conservation.

Because of higher pressures caused by wildlife trade and poaching, a number of NGOs have formed to address the problems of unsustainable wildlife harvesting. TRAFFIC and Free the Bears aim to fight overharvesting by urging tougher rules and actively helping save wildlife that is being taken out of the wild for trade.

Confiscated animals are rehabilitated, and then the centers try to place them in their natural habitats. Despite this, reintroducing species takes a lot of knowledge and careful planning because both in situ and ex situ environments need to be well understood (Cheyne 2009). In addition, the discussion about returning hidden wildlife to nature and how ethical it is is ongoing and questions if it really helps conserve their species (Palmer 2020). A large share of global rubber is produced in Southeast Asia, and the increasing demand for rubber around the world is expected to cause more rubber plantations to appear in the region (Warren-Thomas et al. 2015). Lately, oil palm farming has become more popular on rubber plantations, mainly to produce palm oil, an important item in global trade. The expansion of both rubber and oil palm plantations is linked to large-scale deforestation, and a major number of oil palm plantations in Southeast Asia began development in former forest areas (Vijay et al. 2016; Koh and Wilcove 2008). Concluding whether oil palm is responsible for deforestation or acts on ground already cleared for alternative use is hard, but the illegal use of reserve land by these businesses reveals oil palm's key role in destroying the environment (Fitzherbert et al. 2008).

This reduction in biodiversity occurs in secondary forests too, since changing rubber plantations into oil palm plantations often leads to more species becoming threatened (Koh and Wilcove 2008). Even though selective logging is not as damaging to biodiversity as removing a whole forest, it doesn't totally avoid the loss of species and can increase the forest's vulnerability to growing agriculture (Wilcove et al. 2013; Edwards et al. 2014). In Southeast Asia, mining, fires, along with the development of infrastructure, help cause deforestation (Alamgir et al. 2019). Habitat loss jeopardizes endemic species, and it is a major danger for orangutans. Both Bornean and Sumatran orangutans are now found only in breaking forest areas, while troubles for the Tapanuli orangutan are immediate due to hydropower as well as road development (Nater et al. 2017; Wich et al. 2019). Wildlife conflicts are more common now as orangutans have to leave their forests to find something to eat. Many people view these primates as problems and capture or kill them to sell on the illegal pet market (Meijaard et al. 2011). Because their habitat is still being destroyed, efforts to conserve them and their natural environment should become stronger and more widespread.

Wildlife law enforcement will greatly benefit from the introduction of smart devices, web applications, and the use of cloud computing. With these advances, data processing can be done more quickly, and this will help conservation efforts work better. GPS collars, DNA tracking, chips for analyzing distance, hidden cameras, alarmed fences, and drones are being designed for use in wildlife conservation. It should make use of new technologies and also depend on coordinating with law enforcement, local residents, and supportive government. Success in wildlife law enforcement depends largely on empowering local people, who are the main guards against crime against wildlife. Taking part in community life and receiving local assistance are necessary for enhancing intelligence as well as for making enforcement more efficient (Gray et al. 2016). Financially strong governments must also supply the training and tools needed for effective conservation work.

3.3 Techniques for Ecosystem Analysis

3.3.1 Molecular Techniques

Because of research and development, many molecular techniques have changed how we analyze ecosystems by allowing us to better see and study living things and their interactions. DNA barcoding, environmental DNA (eDNA) analysis, and metagenomics are powerful ways to assess and keep track of ecosystems. As a result, scientists can now identify species, study the activities of microbes, and explain how different places store biodiversity. We have described a few of the significant and popular techniques further (Table 3.1).

Table 3.1 Summary of various techniques along with their advantages and disadvantages

Technique	Application	Advantages	Disadvantages	Citation
DNA barcoding and genetic fingerprinting	Species identification and biodiversity assessment	Rapid, accurate species cataloguing; analysis of genetic variation, gene flow, and evolutionary relationships	Limited to regions of DNA; requires reference libraries	Kalendar and Kairov (2024)
Environmental DNA (eDNA) analysis	Biodiversity monitoring in various ecosystems	Non-invasive, cost-effective, sensitive to rare species, applicable across ecosystems	May miss low biomass species; affected by environmental factors	Millard-Martin et al. (2024); Jia et al. (2024); Kaulbarsch et al. (2024); Michael et al. (2024)
Metagenomics	Microbial community analysis	Provides insights into genetic diversity, metabolic potential, ecological roles	Data analysis complexity; expensive sequencing	Sangwan et al. (2024); Kalendar and Kairov (2024)
Stable isotope analysis (SIA)	Tracing pollutants and food web dynamics	Reveals pollutant movement, bioaccumulation patterns, trophic interactions	Requires specialized equipment; interpretation complexity	Kümmel et al. (2025); Pintado-Herrera et al. (2024)
Position-specific isotope analysis (PSIA)	Origins and degradation processes of pollutants	Detailed pollutant migration and transformation insights	Expensive and technically demanding	Julien (2023)
Atomic absorption spectroscopy (AAS) & inductively coupled plasma mass spectrometry (ICP-MS)	Heavy metal detection and quantification	High sensitivity and accuracy	Costly equipment; skilled operation required	Capar et al. (2007); Mahanta et al. (2022)

(continued)

Table 3.1 (continued)

Technique	Application	Advantages	Disadvantages	Citation
Fourier transform infrared spectroscopy (FTIR)	Identification of organic pollutants	Effective for complex mixtures and functional group detection	Less effective for very low concentrations	Mahanta et al. (2022)
Gas chromatography-mass spectrometry (GC-MS)	Organic pollutant profiling and trophic transfer studies	Detailed pollutant composition and biomagnification insights	High operational cost; extensive sample preparation	Meher and Zarouri (2025)
Remote sensing and GIS	Habitat and landscape analysis	Provides spatial data for environmental management	Data resolution limitations; requires expertise	Lyon et al. (2001); Wegmann et al. (2020)
Scanning electron microscopy (SEM)	Particulate matter analysis	High-resolution imagery of particles	Requires specialized training; costly	Huang et al. (2023); Priyadarshi et al. (2023)
Plants as bioindicators	Detecting ecosystem changes and pollutants	Practical, economical, sensitive to environmental changes	Limited to specific pollutants or environments	Tarish et al. (2024); Dalvand et al. (2016)
Animals as bioindicators	Monitoring ecosystem health and pollution	Diverse applications; sensitive to changes	Requires species-specific knowledge	Chowdhury et al. (2023)
Microbes as bioindicators	Monitoring soil and water contamination	Highly sensitive; useful in genotoxicity assays	Challenging to isolate and identify specific responses	Ogidi et al. (2024); Ma et al. (2022)

3.3.2 DNA Barcoding and Genetic Fingerprinting

This technique involves sequencing a short, standardized region of selected DNA to identify the species. It is most commonly used for species identification purposes and biodiversity assessment. The technique, no doubt, provides a rapid and accurate approach toward cataloguing life forms in an ecosystem (Kalendar and Kairov 2024).

Similarly, genetic fingerprinting was reported to be used for analysis of genetic variation within and between populations. It provides a better resolution on the population structure, gene flow and evolutionary relationships among species (Kalendar and Kairov 2024).

3.3.3 Environmental DNA or eDNA Analysis for Biodiversity Assessment

Environmental DNA (eDNA) analysis has become a major revolution in biodiversity assessment, allowing scientists to observe species without physically interacting with them. This method involves getting DNA that organisms put into their environments, which scientists can then explore to find out how many and which species are present.

These methods are so sensitive that they can find and identify numerous types of organisms, including rare or hardly noticeable species. In British Columbia, eDNA was more successful than beach seining in spotting various fish species (Millard-Martin et al. 2024). eDNA was successful in a middle-sized reservoir in China as well, detecting 42 fish taxa and 188 types of benthic organisms, confirming its usefulness in aquatic ecosystems (Jia et al. 2024). With eDNA, Italy and Albania gathered key biodiversity information more efficiently and economically, alongside conventional biological surveys (Kaulbarsch et al. 2024). The process is made even more efficient by using LAMP or Loop-Mediated Isothermal Amplification (LAMP), which makes it possible to detect species more rapidly (Jothinarayanan et al. 2024). Its methods can be used in diverse ecosystems, such as marine environments, freshwater places, and rock pools (Michael et al. 2024). Rivers have been examined to check vertebrate species diversity, and it has been found that habitats have greatly varying amounts of life (Picq et al. 2024).

We should note the limitations as discussed above to achieve the best outputs from data science. Though eDNA can sample species over large distances, it may not capture small populations that are present locally, just as was noticed in the British Columbia study (Millard-Martin et al. 2024). It seems that using eDNA is best for large-area studies, not detailed ones. DNA extracted from water may change because of environmental factors such as how water flows, its temperature, and how salty it is. Jia et al. found that the way fish communities were structured in a reservoir was mainly affected by water temperature and salinity levels (Jia et al. 2024). Even though eDNA has many strengths, it is most powerful when associated with other approaches. When these two methods are applied together, we learn more about how biodiversity works, as seen in both benthic macroinvertebrates and rock pool ecosystems (Bae et al. 2024; Michael et al. 2024).

3.3.4 Metagenomics for Microbial Community Analysis

Metagenomics involves sequencing the total genomes from the mixture of microbes that live in one area. Using this approach, scientists can learn important details about the genes, metabolic roles and roles in the environment of different microorganisms (Sangwan et al. 2024). In addition, it makes it possible to find new microorganisms and analyze their functions in nutrient and element flows in the environment. Oxford nanopore sequencing and other similar approaches allow for extensive study of DNA from the environment, helping to study both the community and interactions of intricate microorganisms (Kalendar and Kairov 2024).

3.3.5 Chemical Analysis

Experts in chemical analysis use isotopic analysis, spectroscopy, and chromatography to trace pollutants and understand how food webs evolve. They help researchers understand how pollutants are formed, how they change, and where they gather, along with the results these changes have on nature and people's health. The following sections will look at the special advantages and uses of each technique.

3.3.5.1 Isotopic Analysis for Tracing Pollutants and Food Web Dynamics

Stable isotope analysis or SIA: SIA can reveal how pollutants flow and alter different species in food webs. For example, α-HCH shows clear isotopic fractionation when studied, which reveals varied ways it converts as it moves up and down the food chain (Kümmel et al. 2025). Analyses have also used SIA to study bioaccumulation of heavy metals and organic pollutants in tuna, revealing that what these creatures eat and other factors play a role in where and how pollutants accumulate in their bodies (Pintado-Herrera et al. 2024).

Position-specific isotope analysis or PSIA: Highly accurate information about how pollutants like petroleum hydrocarbons and solvents enter the environment and what leads to their degradation can be obtained by examining position-specific isotope ratios (Julien 2023). Pollutant movement and changes in the environment are better understood using this technique.

3.3.6 Spectroscopy for Heavy Metal (HM) and Toxin Detection

Atomic absorption spectroscopy (AAS) and inductively coupled plasma- mass spectrometry (ICP-MS): Many environmental and food samples are analyzed for heavy metals using various spectroscopic tools. Consequently, the results from these methods are accurate and sensitive, allowing them to be trusted for monitoring lead, cadmium and mercury levels in food (Capar et al. 2007; Mahanta et al. 2022).

3.3.7 FTIR or Fourier Transform Infrared Spectroscopy

FTIR is employed to identify organic pollutants and their concentrations in natural sources. It is particularly useful for analyzing complex mixtures and detecting specific functional groups associated with pollutants (Mahanta et al. 2022).

3.3.8 Gas Chromatography-Mass Spectrometry (GC-MS) for Pollutant Profiling

Pollutant profiling: GC-MS is an excellent method for studying the spectrum of organic contaminants, including many persistent organic pollutants (POPs) and micropollutants in environmental substances. This technique allows for careful separation and identification of complex mixtures which gives us a lot of information about the quantity and nature of pollutants ((Meher and Zarouri 2025).

Trophic transfer studies: Gas chromatography-mass spectrometry (GC-MS) has been employed to investigate the trophic transfer of micropollutants and their metabolites within ecological food webs, elucidating patterns of biomagnification and trophic dilution for a myriad of compounds, including polychlorinated biphenyls (PCBs) and polycyclic aromatic hydrocarbons (PAHs) (Meher and Zarouri 2025).

3.4 Physical Techniques

Analyses of habitats, landscapes, and particulate matter in ecosystems rely heavily on remote sensing, GIS, scanning electron microscopy (SEM), and imaging. Habitat and landscape analysis make widespread use of remote sensing and GIS methods, offering spatial information that is used in decision-making for the environment. Fine particulate matter can be analyzed in detail with SEM, since it gives detailed pictures at a microscopic scale. For addressing environmental pollution, especially by microplastics, using advanced imaging is very important.

3.4.1 Remote Sensing and GIS for Habitat and Landscape Analysis

Data about the environment are managed well by using remote sensing and GIS which helps create thematic maps and analyze types of land and vegetation. Using this integration is necessary for managing the environment and making decisions, as revealed in studies of protected areas and land capability maps. Image classification, change detection, and spatial analysis are used to obtain ecologically useful information. These approaches let us learn about changes in land use, habitats, and the useful services offered by ecosystems, as exhibited in works ranging from wetland studies to analyzing more urban areas (Lyon et al. 2001; Wegmann et al. 2020; Wang et al. 2022). Regularly, users work with open-source QGIS and R programs to organize and study spatial data. They support the design of maps and models necessary for both working outdoors and explaining results (Wegmann et al. 2020).

3.4.2 Scanning Electron Microscope (SEM) for Particulate Matter Analysis

Using detailed imaging is one of the main ways to examine particulate matter. The high-resolution images produced by SEM make it easier to examine both the makeup and the surface details of particles. To learn about the physical composition of particles in samples, this approach is very important (Huang et al. 2023). SEM is often teamed up with different tests to find and count microplastics and other sorts of pollutants. When these technologies are used together, environmental assessments become more accurate and new pollution control strategies can be created (Priyadarshi et al. 2023).

3.4.3 Imaging Techniques for Microplastic Identification

Microplastics are also identified in images from microscopes, with further help from techniques such as ATR-FTIR and Raman spectroscopy. Using these methods, scientists can determine the shape, size, and materials in microplastics with more detail, cutting down the chance of mistaken identification (Huang et al. 2023; Priyadarshi et al. 2023). New methods involve portable micro-Raman and SEM-FTIR, which increase our ability to analyze microplastics in the field. These methods help to correctly measure and identify both aquatic and terrestrial ecosystems (Priyadarshi et al. 2023).

3.5 Plants as Bioindicators

Researchers use bioindicators and biomonitors to find out about changes happening in ecosystems. The results also show information about the health of the environment and the amount of pollution. These indicators are plants, animals, and microbes and all of them add something special to the process of tracking changes in ecosystems. Because they are simple to use, inexpensive, and effective, bioindicators are major contributors to ecological research and handling environmental problems. Because plants are in touch with soil, water and air, they are a good way to detect changes in the environment. Some indications include detecting substances such as heavy metals, agrochemicals, and radioactive substances (Tarish et al. 2024). Soil and air tests often make use of the plants *Allium cepa*, *Vicia faba*, and *Nicotiana tabacum*. Lichens and mosses excel at spotting both air pollution and traces of radioactive materials, for example, cesium-137 found in disasters at nuclear plants.

3.5.1 Animals as Bioindicators

Due to their broad range of species and strong reaction to changes in the environment, insects play a huge role in bioindication. Scientists rely on aquatic biomonitoring to assess the physical state of water and the strength of heavy metals in water bodies. Beetles, ants, and butterflies detect even minor changes in pollutants and can signal when the ecosystem's health changes (Chowdhury et al. 2023). Monitoring pollution and overall vitality of an aquatic ecosystem is based on analyzing fish and benthic macroinvertebrates. They help discover contaminants and check how well the environment is being affected (Serafín Muñoz et al. 2022).

3.5.2 Microbes and Bioindicators

Scientists are increasingly recognizing that bacteria, fungi, and algae respond to changes in their surroundings. They are able to detect contamination in both land and water settings which allows for detailed insight into which contaminants are affecting these places (Ogidi et al. 2024; Ma et al. 2022). Microbial indicators are used in both genotoxicity studies and toxicogenomics to check the risks to ecosystem health due to pollutants. Propagation of environmental pollution by petroleum hydrocarbons and many kinds of industrial chemicals can be detected with them (Ogidi et al. 2024).

Scientists use lichens to check for radioactive contamination, such as cesium-137, whenever a nuclear accident occurs, such as at Chernobyl. Their tendency to store radioactive substances makes them useful for indicating pollution with radionuclides (Dalvand et al. 2016). Specifically, *Spirogyra* sp. and similar algae are applied to observe heavy metal contamination in waters. The presence of nickel, copper, and zinc tells us about pollution and its origins (Wacławek et al. 2022).

3.6 Forensic Databases and Technologies

Using forensic approaches and technology makes a big difference in wildlife trade reduction and environmental protection, especially in the analysis of ecosystems. Bringing together CITES and new technologies such as AI and machine learning has made it easier to observe, investigate, and address both wildlife crime and changes in the environment. With these tools, information and insights necessary for conservation are available to guide actions by law enforcement.

3.6.1 Role of Databases in Wildlife Trade

3.6.1.1 CITES Trade Database

Governments and wildlife groups rely on the CITES Trade Database to watch and manage international wildlife trade. The dashboard gathers information from CITES

Parties about wildlife and wildlife product trades, with over 40 years of details on each shipment. The database supports the evaluation of legal wildlife trade, providing key information on how and when trends change, which is essential for planning conservation projects (Liew et al. 2021).

3.6.1.2 Data Accessibility and Analysis

With the help of the "citesdb R" package, analysts can download and look up large CITES Trade Database records more quickly. As a result, researchers and policymakers can better interpret trade patterns and locate potential issues (Liew et al. 2021).

3.6.1.3 Open-Access Data Directories

These new data directories let anyone explore places where wildlife is frequently trafficked and the patterns involved. Thanks to these directories, it's possible to carry out data-based analyses, including ones for predicting links and intercepting networks that play a key role in preventing illegal wildlife trade (Gore et al. 2022).

3.6.1.4 AI and Machine Learning in Ecosystem Forensics

Using machine learning and computer vision, platforms like Conservation AI find and classify both animals and things related to poaching. They use CNNs and transformers to examine both visual and thermal camera data, allowing for monitoring of species and habitats on the spot or later (Fergus et al. 2024).

SCI resorts to image and sound analysis to assist in both species testing and monitoring the condition of these species' habitat. Modelling how different factors influence animals makes it possible to forecast damage to animal communities and plan actions in advance. AI is used in wildlife management to provide current tracking and distress signals for poaching and harm to habitats. With the help of environmental sensors, these systems provide a good understanding of the ecology and help with making conservation decisions based on real data (Chimwemwe 2024).

With LSTM neural network models and others, machine learning helps forecast decreases in illegal wildlife trade thanks to particular programs. Models explain how using strategies can support both the conservation of ecological systems and sustainable progress (Madhasu and Pande 2024).

3.7 Case Studies and Applications

3.7.1 The Augusta Harbor

Along the eastern coast of Sicily, the Augusta Bay is a semi-closed marine basin covering about 30 kilometers. Approximately 4000 hectares of water make up this basin (Colombo et al. 2005; Jiang et al. 2019). During the 1960s, the section named Augusta Harbor in the north was changed through the construction of breakwaters, which linked the area to the open sea by means of two inlets called Scirocco (300 meters wide and 13 meters deep) and Levante (400 meters wide and 40 meters deep)

(Jiang et al. 2019). The harbor covers 24 square kilometers, with depths typically at 15 meters and a maximum of 40 meters found at the Levante inlet (Carle et al. 2015). Because of turbulence in the Adriatic Sea, the majority of water shifts into and out of the lagoon through tidal flows, with ingress via the Levante inlet and egress via the Scirocco inlet (Sprovieri et al. 2011). Contribution from nearby stream discharges is generally low during the wet months (Sprovieri et al. 2011). Because the harbor is restricted, it holds onto pollutants, which can be a cause of eutrophication (Sprovieri et al. 2011; Romano et al. 2021). When industry started in the 1950s, oil refineries were built, and in the 1960s, an important mercury cell chlor-alkali plant (MCCAP) was added (Sprovieri et al. 2011). Many activities, such as dredging, dumping garbage, and industry discharges, have caused major changes to the environment (Ingersoll et al. 2001). During the 1970s, the area became well-known internationally because its pollution from industrial, agricultural, and urban waste posed serious threats to the environment and people's health (Jiang et al. 2019; Sprovieri et al. 2011; Benedetti et al. 2022). Mercury coming from the MCCAP is of biggest concern, as there are thought to be about 500 tons released into the harbor environment (Jiang et al. 2019; Sprovieri et al. 2011; Di Leonardo et al. 2009). Tackling pollution started in the late 1970s but the overall remediation efforts are still incomplete (Manta et al. 2016; Croudace et al. 2015). In 2003, CNR-ISMAR Bologna began research to check how much contamination was in the area. Radionuclides, mercury, and hexachlorobenzene in the sediment cores were examined using a series of analyses to understand the history of pollution along the river (Bellucci et al. 2012; Gasperini and Stanghellini 2009). Mercury levels in the sediment cores changed from 0.07 to 575 mg kg^{-1}, and the highest levels were seen during the time when industries made mercury products and at the surface of the sediment core (Di Leonardo et al. 2009; Bergamin and Romano 2016; Manta et al. 2016; Croudace et al. 2015; Bonsignore et al. 2015). According to the traces of HCB, the times of mercury deposition show that resuspending sediments when dredging and boating bring contamination to the surface more than direct mercury from industry (Bellucci et al. 2010; Colombo et al. 2005). Forensic studies confirmed that recommendations reducing sediment resuspension and pollutant redistribution during dredging and sea activities are effective (Colombo et al. 2005; Jiang et al. 2019; Carle et al. 2015). Other research revealed that transport of mercury to close marine areas is possible through both sediment formation and the release of effluents (Sprovieri et al. 2011; Romano et al. 2021).

3.7.2 Elephant Poaching in Africa

Wildlife crime is happening all around the world; based on Interpol data, it brings in more than $20 billion annually. The largest share of the market is held by China, with the United States and Japan occupying the next two largest places (McMurray 2008). Even though wildlife trafficking is highly profitable, law enforcement agencies frequently view it as not being very important and thus take few legal actions. The available evidence indicates that organized crime is an important cause of this illegal trade (Clark 2008), as is shown by a stronger increase in the confiscation of ivory, coral, and various wildlife products (Clark 2008). The height of wildlife trafficking is seen through a reawakening of the ivory trade from African elephants. During this period from August 2005 to August 2006, over 25,000 kg of ivory was seized by law enforcement—about four times as much as was captured in the previous 3 years (Wasser et al. 2007). The cost of ivory has increased greatly, going from $200/kg in 2004 to $850/kg in 2006, showing that many consider it now to have investment potential (Mainichi Daily News 2007). It appears that with roughly a 10% seizure rate, the 25,000 kg detected could represent 250,000 kg of ivory and more than 38,000 dead elephants, which could exceed 8% of all African elephants in a given year (Blanc et al. 2007). Trade in illegal ivory is on the rise due to rising demand in China, the United States, and Japan, as well as discussions about sustainable ivory trade at CITES meetings (Loder 2008). A fundamental aspect of solving this issue is to pinpoint which part of the world the poached animals come from. By analyzing DNA in ivory, it is possible to see where these products come from which can guide law enforcement to focus on regions of greatest risk for seizures. Analyses of DNA taken from ivory captured in two significant seizures in Singapore and Hong Kong indicated that the ivory came from some parts of Africa, highlighting the ways criminal syndicates poached their supply (Wasser et al. 2007). The study noticed that out of the ivory found in Singapore, part was from Zambia, whereas all ivory in Hong Kong was exported from southern Gabon. Thanks to this approach, law officers can find where the crime takes place and destroy the networks supporting illegal trade.

3.7.3 Monitoring Biodiversity Using Environmental DNA (eDNA)

Monitoring biodiversity by means of eDNA adds important insights to the analysis of aquatic ecosystems. Finding Pseudocrangonyx Akatsuka and Komai 1922, in a subterranean habitat in Japan after more than 90 years shows that eDNA approaches are effective for studying rare species. The examination of 16S rRNA gene sequences in eDNA samples taken in Kyoto revealed that the local groundwater contains two taxa distinct from one another and previously seen in the region (eDNA Kyoto 1 from lineage A and eDNA Kyoto 2 from lineage B). There are many different species of stygobitic amphipods near Lake Biwa, and their similar habitats show the complex biological features of the family. *P. kyotonis* is known to differ from the holotype based on its first antennal articulation and the proportions of uropod 3, information recorded by Uéno (1927). The phenotypic features of *P. kyotonis* agree with what is known about *P. komaii* and *P. uenoi* (Tomikawa et al. 2019), yet its placement in the phylogeny requires further genetic and morphological study. These investigations are necessary to find the pattern of evolution and to see where species live around the world.

3.8 Challenges in Ecosystem Forensic Science

A number of problems stand in the way of environmentally-focused forensics being able to properly identify and handle contamination. The existing problems in imaging include technical limits with the methods used, no agreed-upon national protocols, and difficulties with resources in less-developed nations, along with issues of ethics and law. Such difficulties prevent environmental forensics from reliably and usefully exploring environmental crimes and incidents of contamination.

3.9 Limitations in Current Techniques

Forensic technicians frequently have problems achieving both sensitivity and specificity in determining and measuring pollutants. An example is that predictive accuracy in environmental forensics can change with seeding method, noise and sample size, due to issues with receptor models (Rose et al. 2024). Compound-Specific Isotope Analysis (CSIA) and chemical fingerprinting work for identifying pollution sources but need good and accurate data to do so (Morrison 2013). Because the chemical mix of things like petroleum is so complex, it is hard to identify their nature and who caused the contamination, which adds difficulty (Stout and Wang 2007).

3.10 Lack of Standardized Protocols

Because there are no common methods in environmental testing, outcomes and conclusions may differ from case to case. These different approaches are made even more distinct by using unusual methods, which often makes it hard to produce credible environmental measures for environmental law cases (Wait 2002). Since there are not many standardized ways to do investigations in aquatic cases, the discipline becomes more complicated. Because there is no standard system, the wrong diagnosis or unclear reasons for environmental incidents could happen.

3.11 Resource Constraints in Developing Countries

A shortage of resources usually keeps developing countries from carrying out detailed investigations into environmental incidents. Testing in forensic investigations in Nigeria is usually held back by difficulties in using the latest technology and knowledge from experts (O'Sullivan and Megson 2015). The major reason these regions have issues using sophisticated forensics and establishing important laws is that they lack finances and the proper infrastructure (Morrison and O'Sullivan 2012).

3.12 Ethical and Legal Challenges

Proper safeguarding of environmental forensic data matters a great deal, since fraud with data is still a major challenge. Another problem is that legal rules about admitting forensic evidence have changed, including the Daubert ruling (Wait 2002). Debates among experts over environmental issues can delay finding who is responsible for different kinds of contamination (Stout and Wang 2007).

Even so, many efforts are being made to resolve these issues, such as using systems that can be installed fast and using chemometrics are recent approaches aimed at making environmental forensics more correct and productive (Kolobaric et al. 2023). Experts are recommending that the forensic community use a team approach to improve methods and training, especially in sectors like examining cases in aquatic environments. Such work points to what can be achieved, even with the existing obstacles in the field.

3.13 Future Perspectives

Modern technology is changing how we manage and fight wildlife crime and conservation. With the aid of complex networks, advanced means of communication, and advanced skills in finding rare animals, these groups design supply chains that drop off the radar. They commonly operate within international markets for wildlife products, so they can stay ahead of government institutions with restricted funds (Kretser et al. 2015). Because the economy favors criminals, they can quickly test and use the newest technology, making it hard for agencies to respond. At the same time, tools from advanced technology are used by conservationists to reduce these risks and help protect biodiversity.

New technologies play a key role in improving forensic ecosystem analysis. Having learned from precision agriculture and livestock farming, such technologies now support the continuous monitoring of animal health. Just as an example, tags and GPS collars on animals collect important information about their health and movements in different locations. Even with their requirement for physical connections and the money needed, they have proven capable of finding and helping to stop wildlife crime (Webb 2019; UNEP 2014).

Biologgers and GPS collars are becoming more important tools in fighting poaching. The devices use the accelerometers and magnetometers to both follow movement and indicate possible stress or death. A link between these sensors and bio-telemetry systems makes it easy to get near real-time data that helps respond to any dangers quickly. Using such resources, predictive modeling and geofencing may make it simpler to find high-risk areas in future forensic ecosystem studies (Wall et al. 2014).

The use of night vision, thermal imaging, and UAVs is becoming more valuable as great nonanimal options in the military. UAVs stand out by being able to offer superb observational abilities, supporting what is done on the ground with images from the air (Wich 2015). Autonomous recording units (ARUs) are becoming popular because they can identify acoustic events like gunshots or sounds from animals over a wide area. Basic technologies teamed with machine learning and AI allow for more efficient detection of crime and more precise investigations (Blumstein et al. 2011).

The key to forensic ecosystem analysis is to link live data analysis with conservation programs. The use of GIS and resource selection models in predictive mapping identifies parts of a forest that are most likely to be targeted by poachers. It is shown through SMART and similar advanced tools how data-based understanding of wildlife crime prevention is effective (Krester et al. 2017). When models of species distribution are combined with maps of poacher risk, conservationists know which areas to deal with first (Dzialak et al. 2011).

Today, prevention is not the priority for most tools, but it ought to be in the future. It is important for forensic ecosystem analysis to start using technology that can help predict crimes. For instance, by reviewing both environmental and animal behavior patterns, AI can detect and stop illegal activities before they happen. More importantly, cloud computing and online applications will handle data and enable quick distribution of resources (Wall et al. 2014).

Trying to conserve wildlife is made challenging by the interactions between wildlife trade, cultural beliefs, economic factors, and law enforcement. As a result, we need to strengthen policies, engage local communities in action and learning, strengthen law enforcement, and ensure the laws safeguarding wildlife are followed (Biggs et al. 2017; Lee et al. 2005). Integrating agriculture and eco-tourism as alternative livelihoods has reduced some people's dependence on poaching, but additional studies are required to know if these strategies are really reducing the rate of poaching.

3.14 Conclusion

Thanks to molecular techniques and eDNA analysis, forensic ecology can now more accurately detect live and dead organisms and clarify how they take part in decomposition. Because of these new tools, subtle changes in nature can be identified in any environmental type, helping to strengthen forensic investigations. However, several long-standing difficulties persist, covering how ecological responses vary in different places and at different times, as well as the effects of temperature, humidity, and the structure of soil. For this reason, the ability of forensic ecology increases when it is united with forensic entomology, taphonomy, and bioinformatics. With new technology and improved cooperation in research, forensic ecological tools will be used even more by law enforcement and courts everywhere.

References

Adrian WJ (1996) Wildlife forensic field manual. Association of Midwest Fish and Game Law Enforcement Officers, Colorado1

Aisher A (2016) Scarcity, alterity, and value: Decline of the pangolin, the world's most trafficked mammal. Conserv Soc 14(4):317–329

Akatsuka K, Komai TE (1922) *Pseudocrangonyx*, a new genus of subterranean amphipods from Japan. Annot Zool Jpn 10:119–126

Alamgir M, Campbell MJ, Sloan S, Suhardiman A, Supriatna J, Laurance WF (2019) High-risk infrastructure projects pose imminent threats to forests in Indonesian Borneo. Sci Rep 9(1):140. https://doi.org/10.1038/s41598-018-36594-8

Bae M-J, Chun Y-C, Han S-C, Kim S-Y, Kim E-J (2024) Comparisons of eDNA and traditional sampling methods for enhanced assessment of benthic macroinvertebrate diversity in lotic environments with spatially varying land use. Res Sq. https://doi.org/10.21203/rs.3.rs-4428098/v1

Bellucci LG, Giuliani S, Mugnai C, Frignani M, Paolucci D, Albertazzi S, Fernandez ACR (2010) Anthropogenic metal delivery in sediments of Porto Marghera and Venice Lagoon (Italy). Soil Sediment Contam Int J 19:42–57

Bellucci LG, Giuliani S, Romano S, Albertazzi S, Mugnai C, Frignani M (2012) An integrated approach to the assessment of pollutant delivery chronologies to impacted areas: Hg in the Augusta Bay (Italy). Environ Sci Technol 46:2040–2046

Benedetti M, Romano E, Ausili A, Fattorini D, Gorbi S, Maggi C, Salmeri A, Manta DS, Sesta G, Sprovieri M et al (2022) 10-year time course of Hg and organic compounds in Augusta Bay: bioavailability and biological effects in marine organisms. Front Public Health 10:968296. https://doi.org/10.3389/fpubh.2022.968296

Bergamin L, Romano E (2016) Suitable sediment fraction for paleoenvironmental reconstruction and assessment of contaminated coastal areas based on benthic foraminifera: a case study from Augusta Harbour (Eastern Sicily, Italy). Ecol Indic 71:66–78

Biggs J, von Fumetti S, Kelly-Quinn M (2017) The importance of small waterbodies for biodiversity and ecosystem services: implications for policy makers. Hydrobiologia 793:3–39. https://doi.org/10.1007/s10750-016-3007-0

Blanc JJ, Barnes RFW, Craig GC, Dublin HT, Thouless CR, Douglas-Hamilton I, Hart JA (2007) African elephant status report 2007. IUCN, Gland

Blumstein DT, Mennill DJ, Clemins P, Girod L, Yao K, Patricelli G, Deppe JL, Krakauer AH, Clark C, Cortopassi KA, Hanser SF, McCowan B, Ali AM, Kirschel ANG (2011) Acoustic monitoring in terrestrial environments using microphone arrays: applications, technological considerations, and prospectus. J Appl Ecol 48:758–767

Bonsignore M, Tamburrino S, Oliveri E, Marchetti A, Durante C, Berni A, Quinci E, Sprovieri M (2015) Tracing mercury pathways in Augusta Bay (Southern Italy) by total concentration and isotope determination. Environ Pollut 205:178–185

Capar SG, Mindak WR, Cheng J (2007) Analysis of food for toxic elements. Anal Bioanal Chem 389:159–169. https://doi.org/10.1007/s00216-007-1433-6

Carle J, Oliver G, Carter S, Usui H (2015) Ichthys LNG project: successful dredging campaign in Darwin Harbour, Northern Territory. In: Australasian Coasts & Ports Conference 2015: 22nd Australasian Coastal and Ocean Engineering Conference and the 15th Australasian Port and Harbour Conference., Auckland, New Zealand, 15–18 September 2015. Engineers Australia and IPENZ, pp 147–153

Cerling TE, Barnette JE, Chesson LA, Douglas-Hamilton I, Gobush KS, Uno KT, Wasser KT, Xu X (2016) Radiocarbon dating of seized ivory confirms rapid decline in African elephant populations and provides insight into illegal trade. PNAS 113(47):13330–13335

Challender D, Willcox DHA, Panjang E, Lim N, Nash H, Heinrich S, Chong J, Manis javanica. (2019) The IUCN red list of threatened species. IUCN. https://doi.org/10.2305/IUCN.UK.2019-3.RLTS.T12763A123584856.en

Cheyne SM (2009) The role of reintroduction in gibbon conservation: opportunities and challenges. In: Whittaker D, Lappan S (eds) The gibbons. Developments in primatology: progress and prospects. Springer, New York, NY. https://doi.org/10.1007/978-0-387-88604-6_23

Chimwemwe C (2024) Wildlife watch. i-manager's J Inf Technol 13(2):24–30. https://doi.org/10.26634/jit.13.2.20819

Chowdhury S, Dubey VK, Choudhury S, Das A, Jeengar D, Sujatha B, Kumar A, Kumar N, Semwal A, Kumar V (2023) Insects as bioindicator: a hidden gem for environmental monitoring. Front Environ Sci 11. https://doi.org/10.3389/fenvs.2023.1146052

Clark W (2008) Testimony of William Clark to the U.S. House of Representatives Committee on Natural Resources (5 March). Hearing on Poaching American Security: impacts of Illegal Wildlife Trade. U.S. House of Representatives, Washington, DC

Colombo F, Frignani M, Bellucci LG (2005) Origine Storica della Contaminazione dei Sedimenti della Rada di Augusta. Public Attorney Office of Siracusa, Siracusa. (In Italian)

Coutu AN, Lee-Thorp J, Collins MJ, Lane PJ (2016) Mapping the elephants of the 19th century east African ivory trade with a multi-isotope approach. PLoS One 11(10):e0163606

Croudace IW, Romano E, Ausili A, Bergamin L, Rothwell RG (2015) X-ray core scanners as an environmental forensics tool: a case study of polluted harbour sediment (Augusta Bay, Sicily). In: Croudace IW, Rothwell RG (eds) Micro-XRF studies of sediment cores. Springer, Dordrecht, pp 393–421

Dalvand A, Jahangiri A, Iranmanesh J (2016) Introduce lichen *Lepraria incana* as biomonitor of Cesium-137 from Ramsar, northern Iran. J Environ Radioact 160:36–41. https://doi.org/10.1016/j.jenvrad.2016.04.018

Di Leonardo R, Bellanca A, Neri R, Tranchida G, Mazzola S (2009) Distribution of REEs in box-core sediments offshore an industrial area in SE Sicily, Ionian Sea: evidence of anomalous sedimentary inputs. Chemosphere 77:778–784

Dzialak MR, Olson CV, Harju SM et al (2011) Identifying and prioritizing greater sage-grouse nesting and brood-rearing habitat for conservation in human-modified landscapes. PLoS ONE 6(10):e26273. https://doi.org/10.1371/journal.pone.0026273

Edwards DP, Gilroy JJ, Woodcock P, Edwards FA, Larsen TH, Andrews DJR et al (2014) Land-sharing versus land-sparing logging: Reconciling timber extraction with biodiversity conservation. Glob Chang Biol 20(1):183–191. https://doi.org/10.1111/gcb.12353

El-Kady EM (1999) Problems facing the application of forensic entomology. Pak J Biol Sci 2:280–289

Fergus P, Chalmers C, Longmore S, Wich S (2024) Harnessing artificial intelligence for wildlife conservation. Conservation 4(4):685–702. https://doi.org/10.3390/conservation4040041

Fitzherbert EB, Struebig MJ, Morel A, Danielsen F, Brühl CA, Donald PF, Phalan B (2008) How will oil palm expansion affect biodiversity? Trends Ecol Evol 23(10):538–545. https://doi.org/10.1016/j.tree.2008.06.012

Fox RH, Cunningham CL (1973) Crime scene search and physical evidence handbook. U.S. Department of Justice, National Institute of Justice, Washington, DC

Gasperini L, Stanghellini G (2009) SeisPrho: an interactive computer program for processing and interpretation of high-resolution seismic reflection profiles. Comput Geosci 35:1497–1507

Gore ML, Schwartz LR, Amponsah-Mensah K et al (2022) Voluntary consensus-based geospatial data standards for the global illegal trade in wild fauna and flora. Sci Data 9:267. https://doi.org/10.1038/s41597-022-01371-w

Gray TN, Hort S, Lefter E, Grosu R, Kong K, Keo O, Gauntlett S (2016) A decade of zero elephant poaching in the Cardamom Rainforest Landscape, Cambodia. Gajah 45:35–38

Hardy CR, Wallace JR (2012) Algae in forensic investigations. In: Hall DW, Byrd JH (eds) Forensic botany: a practical guide. Wiley, London, pp 145–173

Harrington LA, D'Cruze N, Macdonald D (2018) Rise to fame: events, media activity, and public interest in pangolins and pangolin trade, 2005–2016. Nat Conserv 30:107

Horswell J (2004) The practice of crime scene investigation. International Forensic Science and Investigation Service, CRC Press, Boca Raton, FL

Housman LA (1920) Structural characteristics of the hair of mammals. Am Nat 5:496–523

Huang Z, Hu B, Wang H (2023) Analytical methods for microplastics in the environment: a review. Environ Chem Lett 21(1):383–401. https://doi.org/10.1007/s10311-022-01525-7

Ingersoll CG, MacDonald DD, Wang N, Crane JL, Field LJ, Haverland PS, Kemble NE, Lindskoog RA, Severn C, Smorong DE (2001) Prediction of sediment toxicity using consensus-based freshwater sediment quality guidelines. Arch Environ Contam Toxicol 41:8–21

INTERPOL (2018) Global Wildlife Enforcement. Strengthening Law Enforcement Cooperation Against Wildlife Crime. INTERPOL General Secretariat. https://www.interpol.int/en/Crimes/Environ%20mental-crime/Wildlife-crime

Jia H, Xia M, Zhang T, Zhang H (2024) Rapid monitoring of aquatic organism biodiversity based on environmental DNA in a medium-sized reservoir. Fishes 9(10):396. https://doi.org/10.3390/fishes9100396

Jiang J, Han H, Karunarathna A (2019) Best practice for sediment plume dispersion model application. In: Australasian Coasts and Ports 2019 Conference: Future Directions from 40° S and Beyond., Hobart, 10–13 September 2019. Engineers Australia and IPENZ, Auckland, pp 665–671

Jothinarayanan N, Pham CH, Karlsen F, Roseng LE (2024) eDNA-based survey of fish species in water bodies using loop-mediated isothermal amplification (LAMP) for application of developing automatic sampler. Methods Protoc 7(6):85. https://doi.org/10.3390/mps7060085

Julien M (2023) Position-specific isotope analysis as a fingerprint for pollutants. In: Yoshida N, Gilbert A, Foriel J (eds) Handbook of isotopologue biogeochemistry. Springer, Singapore, pp xx–xx. https://doi.org/10.1007/978-981-10-7048-8_43-1

Kahler JS, Gore GL (2017) Conservation crime science. In: Gore ML (ed) Conservation criminology. Wiley Blackwell, West Sussex, pp 27–44

Kalendar R, Kairov U (2024) Genome-wide tool for sensitive *de novo* identification and visualisation of interspersed and tandem repeats. Bioinform Biol Insights 18. https://doi.org/10.1177/11779322241306391

Kaulbarsch S, Diana F, Estrada GCD, Pavanel E, Ross-Gillespie V, Sali J, Marsicano A, Saladini V, Gerone D, Gioia G, Sanetti S, Florio M, Cassi M, Marcantonio G, La Gala S, Shumka S (2024) Environmental genomics (eDNA) in comparison with onventional methods for biodiversity field surveys: case studies from Italy and Albania. In: SPE International Health, Safety, Environment and Sustainability Conference and Exhibition. https://doi.org/10.2118/220258-MS

Koh LP, Wilcove DS (2008) Is oil palm agriculture really destroying tropical biodiversity? Conserv Lett 1(2):60–64. https://doi.org/10.1111/j.1755-263X.2008.00011.x

Kolobaric A, Cozzolino D, Chapman J (2023) Environmental forensics, pp. 254–293. https://doi.org/10.1039/bk9781839166099-00254

Krester H, Stokes E, Wich S, Foran D, Montefiore A (2017) Technological innovations supporting wildlife crime detection, deterrence and enforcement. In: Gore ML (ed) Conservation Criminology. Wiley Blackwell, West Sussex, UK, pp 157–177

Kretser H, Wong R, Roberton S, Pershyn C, Huang J, Sun F, Kang A, Zahler P (2015) Mobile decision-tree tool technology as a means to detect wildlife crimes and build enforcement networks. Biol Conserv 189:33–38

Kümmel S, Ottosen CF, Olsson ME, Broholm MM, Bjerg PL, Richnow HH (2025) Development of a $\delta^{13}C$ and $\delta^{34}S$ isotope analysis method for sulfadimidine and its potential to trace contaminant transformation in groundwater systems. Anal Chem 97(7):4014

Lee TM, Soh MCK, Sodhi N, Koh LP, Lim SLH (2005) Effects of habitat disturbance on mixed species bird flocks in a tropical sub-montane rainforest. Biol Conserv 122(2):193–204. https://doi.org/10.1016/j.biocon.2004.07.005

Liew JH, Kho ZY, Lim RBH, Dingle C, Bonebrake TC, Sung YH, Dudgeon D (2021) International socioeconomic inequality drives trade patterns in the global wildlife market. Sci Adv 7(19):eabf7679. https://doi.org/10.1126/sciadv.abf7679

Loder N (2008) Trade bans and conservation. The Economist Mar 8–14:85–87

Lord WD, Burger JF (1984) Arthropods associated with harbor seal (Phoca vitulina) carcasses stranded on islands along the New England coast. Int J Entomol 26:282–285

Lyon JG, Lopez RD, Lyon LK, Lopez DK (2001) Wetland landscape characterization: GIS, remote sensing and image analysis. CRC Press. https://doi.org/10.1201/9781420022681

Ma F, Wang C, Zhang Y, Chen J, Xie R, Sun Z (2022) Development of microbial indicators in ecological systems. Int J Environ Res Public Health 19(21):13888. https://doi.org/10.3390/ijerph192113888

Madhasu N, Pande SD (2024) Revolutionizing wildlife protection: a novel approach combining deep learning and night-time surveillance. Multimed Tools Appl 84:23465. https://doi.org/10.1007/s11042-024-19876-4

Mahanta S, Sinha S, Shree J (2022) Determination and quantification of chemical pollutants by spectroscopic techniques: a possible use in chemical analysis of food crops under stress. In: Response of field crops to abiotic stress. CRC Press, pp 297–311. https://doi.org/10.1201/9781003258063-24

Mainichi Daily News (English edition) (2007) Customs and police show off 3 tons of ivory confiscated from smugglers. Mainichi Daily News, 1 March. Available from http://mdn.mainichi-msn.co.jp/national/news/p20070301p2a00m0na017000c.html

Manta DS, Bonsignore M, Oliveri E, Barra M, Tranchida G, Giaramita L, Mazzola S, Sprovieri M (2016) Fluxes and the mass balance of mercury in Augusta Bay (Sicily, southern Italy). Estuar Coast Shelf Sci 181:134–143

Maxwell SL, Fuller RA, Brooks TM, Watson JE (2016) Biodiversity:tThe ravages of guns, nets, and bulldozers. Nat News 536(7615):143–145

McMurray C (2008) Testimony of Claudia McMurray to the U.S. House of Representatives Committee on Natural Resources (5 March). Hearing on: Poaching American security: impacts of illegal wildlife trade, before the Committee on Natural Resources. U.S. House of Representatives, Washington, D.C.

Meher AK, Zarouri A (2025) Environmental applications of mass spectrometry for emerging contaminants. Molecules 30(2):364. https://doi.org/10.3390/molecules30020364

Meijaard E, Buchori D, Hadiprakarsa Y, Utami-Atmoko SS, Nurcahyo A, Tjiu A et al (2011) Quantifying killing of orangutans and human-orangutan conflict in Kalimantan, Indonesia. PLoS One 6(11):e27491. https://doi.org/10.1371/journal.pone.0027491

Michael DR, Murray A, Murray P, Murray B, Hagen S, McTernan R, Furst DJ, Nimmo DG (2024) Use of eDNA and conventional sampling methods to survey rock pool (gnamma) biodiversity on granite inselbergs. Austral Ecol 49(8). https://doi.org/10.1111/aec.13562

Millard-Martin B, Sheridan K, Morien E, Lemay MA, Hessing-Lewis M, Clemente-Carvalho RBG, Sunday JM (2024) Effect of environmental DNA sampling resolution in detecting near-shore fish biodiversity compared to capture surveys. PeerJ 12:e17967. https://doi.org/10.7717/peerj.17967

Morrison RD (2013) Overview, analysis, and interpretation of environmental forensic evidence, pp. 45–50. https://doi.org/10.1016/B978-0-12-382165-2.00118-5

Morrison RD, O'Sullivan G (2012) Environmental forensics: proceedings of the 2011 INEF Conference

Nater A, Mattle-Greminger MP, Nurcahyo A, Nowak MG, de Manuel M, Desai T et al (2017) Morphometric, behavioral, and genomic evidence for a new orangutan species. Curr Biol 27(22):3487–3498.e3410. https://doi.org/10.1016/j.cub.2017.09.047

Nijman V, Langgeng A, Birot H, Imron MA, Nekaris KAI (2018) Wildlife trade, captive breeding and the imminent extinction of a songbird. Global Ecol Conserv 15:e00425. https://doi.org/10.1016/j.gecco.2018.e00425

O'Sullivan G, Megson D (2015) Environmental forensics: proceedings of the 2014 INEF Conference. https://doi.org/10.1039/9781782625070

Ogidi OI, Angaye TCN, Carbom HE (2024) Microorganisms in the monitoring of terrestrial and aquatic ecosystems: applications, challenges, and future prospects. In: Biomonitoring of pollutants in the global south. Springer, Singapore, pp 639–665. https://doi.org/10.1007/978-981-97-1658-6_18

Palmer A (2020) Ethical debates in orangutan conservation. 1st edn. Routledge, London. https://doi.org/10.4324/9780429060533

Picq S, Curtis AN, George A, Larson ER, Thompson ME, Willink PW, McMahan CD, de Souza LS (2024) Environmental DNA metabarcoding for whole community inventories of vertebrates in rivers of the Midwestern United States. J Freshw Ecol 39(1). https://doi.org/10.1080/02705060.2024.2382454

Pintado-Herrera MG, López-López JA, Lara-Martín PA, Medina A, Cadenas I, Giansiracusa S, Corada-Fernández C, Varela JL (2024) Assessment of metal and organic pollutants in combination with stable isotope analysis in tunas from the Gulf of Cadiz (east Atlantic). Mar Environ Res 196:106432. https://doi.org/10.1016/j.marenvres.2024.106432

Priyadarshi G, Prajapati S, Trivedi M (2023) Tools and techniques to analyse microplastic pollution in aquatic and terrestrial ecosystems. In: Microplastic pollution: causes, effects and control. Bentham Science Publishers, pp 1–17. https://doi.org/10.2174/9789815165104123010004

Romano E, Bergamin L, Croudace I, Pierfranceschi G, Sesta G, Ausili A (2021) Measuring anthropogenic impacts on an industrialised coastal marine area using chemical and textural signatures in sediments: a case study of Augusta Harbour (Sicily, Italy). Sci Total Environ 755:142683. https://doi.org/10.1016/j.scitotenv.2020.142683

Rose ND, Negley TL, Bock MJ (2024) Predictive accuracy of polytopic vector analysis in environmental forensics: sensitivity to seeding methods, random noise, and sample size. Environ Forensic 25:534. https://doi.org/10.1080/15275922.2023.2297438

Sangwan S, Kumar M, Lamba R, Singh S, Kumari A, Wati L (2024) Bioindicators. In: Environmental nexus approach. CRC Press, pp 311–337. https://doi.org/10.1201/9781003408352-21

Serafín Muñoz AH, Guerrero Aguilar A, Rodríguez Castrejón UE, Schüth C, Noriega Luna B (2022) Bioindicators and biomonitoring: review of methodologies applied in water bodies and use during the Covid-19 pandemic. Acta Univ 32:10.15174/au.2022.3388

Sprovieri M, Oliveri E, Di Leonardo R, Romano E, Ausili A, Gabellini M, Barra M, Tranchida G, Bellanca A, Neri R et al (2011) The key role played by the Augusta basin (southern Italy) in the mercury contamination of the Mediterranean Sea. J Environ Monit 13:1753–1760. https://doi.org/10.1039/c1em10284j

Stout S, Wang Z (eds) (2007) Oil spill environmental forensics case studies. Academic Press. https://doi.org/10.1016/B978-0-12-369523-9.X5000-9

Tarish M, Ali RT, Shan M, Amjad Z, Rui Q, Akher SA, Al Mutery A (2024) Plant tissues as biomonitoring tools for environmental contaminants. Int J Plant Biol 15(2):375–396. https://doi.org/10.3390/ijpb15020030

Tomikawa K, Abe Y, Nakano T (2019) A new stygobitic species of the genus *Pseudocrangonyx* (Crustacea: Amphipoda: Pseudocrangonyctidae) from central Honshu, Japan. Species Divers 24:259–266

Uéno M (1927) Notes on some subterranean isopods and amphipods of Japan. Mem Coll Sci Kyoto Imp Univ Ser B 3:355–368

UNEP (United Nations Environment Programme) (2014) Emerging technologies: smarter ways to fight wildlife crime. https://na.unep.net/geas/archive/pdfs/GEAS_Jun2014_EmergingTechnologies_illegalwildlife.pdf

UNODC (2016) World wildlife crime report: trafficking in protected species. United Nations Office on Drugs and Crime, Austria, Vienna. https://globalinitiative.net/wp-content/uploads/2017/12/UNODC-World_Wildlife_Crime_Report_2016_final.pdf

UNODC (2020) World wildlife crime report: Trafficking in protected species. United Nations Office on Drugs and Crime, Austria, Vienna. https://www.unodc.org/documents/data-and-analysis/wildlife/2020/World_Wildlife_Report_2020_9July.pdf

Vijay V, Pimm SL, Jenkins CN, Smith SJ (2016) The impacts of oil palm on recent deforestation and biodiversity loss. PLoS One 11(7):e0159668. https://doi.org/10.1371/journal.pone.0159668

Wacławek M, Świsłowski P, Rajfur M (2022) The biological monitoring as a source of information on environmental pollution with heavy metals. Chem Didact Ecol Metrol 27(1–2):53–78. https://doi.org/10.2478/cdem-2022-0006

Wait AD (2002) Challenges in producing defensible environmental chemistry measurements for litigation. Environ Claims J 14(4):415–454. https://doi.org/10.1080/10406020291041910
Wall J, Wittemyer G, Klinkenberg B, Douglas-Hamilton I (2014) Novel opportunities for wildlife conservation and research with real-time monitoring. Ecol Appl 24:593–601
Wallace JR, Ross JC (2012) The application of forensic science to wildlife evidence. In: Huffman JE, Wallace JR (eds) Wildlife forensics: methods and applications. Wiley-Blackwell, pp 35–50
Wang S, Padmanaban R, Shamsudeen M, Campos FS, Cabral P (2022) Landscape impacts on ecosystem service values using the image fusion approach. Land 11(8):1186. https://doi.org/10.3390/land11081186
Warchol G (2018) Wildlife crime and criminal organizations: can the theory of enterprise help explain the ivory and rhino horn trade? In: Moreto WD (ed) Wildlife crime: from theory to practice. Temple University Press, pp 81–105
Warren-Thomas E, Dolman PM, Edwards DP (2015) Increasing demand for natural rubber necessitates a robust sustainability initiative to mitigate impacts on tropical biodiversity. Conserv Lett 8(4):230–241. https://doi.org/10.1111/conl.12170
Wasser SK, Mailand C, Booth R et al (2007) Using DNA to track the origin of the largest ivory seizure since the 1989 trade ban. Proc Natl Acad Sci USA 104(10):4228–4233. https://doi.org/10.1073/pnas.0609714104
Wasser SK, Brown L, Mailand C, Mondol S, Clark W, Laurie C, Weir BS (2015) Genetic assignment of large seizures of elephant ivory reveals Africa's major poaching hotspots. Science 349(6243):84–87
Wegmann M, Schwalb-Willmann J, Dech S (2020) An introduction to spatial data analysis. Pelagic Publishing. https://doi.org/10.53061/HCED6492
Webb SL (2019) Animal tech that could help you make decisions on the ranch. Noble News Views 37(8):16–11
Wich SA (2015) Drones and conservation. In: Kakaes K (ed) Drones and aerial observation: new technologies for property rights, human rights, and global development: a primer. New America, pp 64–70
Wich SA, Fredriksson G, Usher G, Kühl HS, Nowak MG (2019) The Tapanuli orangutan: status, threats, and steps for improved conservation. Conserv Sci Pract 1
Wilcove DS, Giam X, Edwards DP, Fisher B, Koh LP (2013) Navjot's nightmare revisited: logging, agriculture, and biodiversity in Southeast Asia. Trends Ecol Evol 28(9):531–540
Wyatt T (2021) Wildlife trafficking: a deconstruction of the crime, victims, and offenders, 2nd edn. Palgrave, Cham. https://doi.org/10.1007/978-3-030-83753-2

Tracing Environmental Changes Over Time

4

Saba Rashid, Anu Singla, Baljeet Yadav, Shakti Prakash, Mahipal Singh Sankhla, and Shobha Ahi

Abstract

Climate change is a long-term alteration in Earth's climate system, primarily caused by greenhouse gas accumulation from human activities, like fossil fuel combustion, forest clearing, and industrial operations. Rising temperatures, heatwave events, and melting ice caps are key signs of climate change. This leads to increased sea-level rise, threatening coastal areas and disrupting human populations, ecosystems, agriculture, and wildlife populations. Climate change also causes severe weather phenomena, such as hurricanes, extended dry seasons, and heavier precipitation, causing damage to infrastructure and making clean water harder to obtain. Oceans are also affected, absorbing excess carbon dioxide and heat, increasing temperatures and decreasing pH levels, and damaging coral reefs and shellfish, which are essential for marine ecosystems and economic activities. Ecosystems are under stress, with species relocating to lower-temperature locations and rising mountains to survive. This has led to the extinction of some biodiversity elements and serious threats to food security and natural resources. Global action is needed to combat global warming.

S. Rashid (✉) · B. Yadav · M. S. Sankhla
School of Basic and Applied Sciences, K. R. Mangalam University, Gurugram, Haryana, India

A. Singla
APJ Abdul Kalam Institute of Forensic Science and Criminology, Bundelkhand University, Jhansi, Uttar Pradesh, India

S. Prakash
School of Behavioural Sciences and Forensic Investigation, Rashtriya Raksha University, Gandhinagar, Gujarat, India

S. Ahi
Kings College London, London, UK

M. S. Sankhla et al. (eds.), *Introducing the Synergy Between Forensic Science and Environmental Analysis*, https://doi.org/10.1007/978-981-95-0199-1_4

Keywords

Climate change · Global warming · Habitat fragmentation · Greenhouse gases

4.1 Introduction

Climate change refers to long-term alterations in temperature, precipitation, wind patterns, and other elements of the Earth's climate system. Climate change results mainly from greenhouse gas accumulation, including carbon dioxide alongside methane, which emerges from human practices, such as fossil fuel combustion and forest clearing, and industrial operations (IPCC 2021). The traditional atmosphere demonstrates rising temperatures as a primary sign of climate change. The Earth has experienced an extensive temperature rise, which generates more frequent and serious heatwave events (NASA 2024). The melting of 2024 glaciers and Antarctic ice caps particularly affects Greenland. Meltwater from polar ice affects global sea-level increase, thus creating more threats to coastal areas from flooding risks. The temperature increase disrupts both the human population and affects ecosystems while disturbing agricultural activities alongside wildlife populations. A significant result stems from ice melting, which directly impacts sea level elevations.

The planet experiences heightened devastating weather phenomena, including severe hurricanes, together with extended dry seasons and heavier precipitation, because of climate change. Such incidents generate extreme damage to building infrastructure alongside population relocation and make clean water harder to obtain (UNEP 2023). The existing ocean networks face substantial negative impacts. Oceans take in surplus carbon dioxide and heat, which causes them to increase temperature while decreasing their pH levels. The increase in ocean temperatures damages two important marine organisms: coral reefs and shellfish, as they form the base of marine ecosystems and economic activities (NASA 2024). Additionally, ecosystems are under stress. Various species choose to relocate toward lower-temperature locations along with elevating themselves up mountains to maintain survival. The quick adaptation capabilities of species vary across the board; thus, some biodiversity elements experience extinction because of slow adaptation (IPCC 2021). Food security and natural resources that humans need to survive face serious threats from these disturbances. Climate change has started its damaging effects globally, which continue to expand throughout every region on Earth.

The planet needs immediate worldwide action to fight global warming since it modifies environmental conditions, including temperatures and weather systems and ocean elevations and system structures. Figure 4.1 compares GHG emissions (in $MtCO_2e$) from 1990 to 2021 for China, the USA, the EU, India, Russia, and Brazil. Due to fast industrial development, China's emissions have gone up sharply, making it the world's top emitter by 2021. Because of cleaner energy methods and better efficiency, the USA has seen a slow decrease in carbon emissions since the mid-2000s. Because of strict climate policies, the EU has consistently lowered its emissions. Emissions in India are rising because of the country's economic and

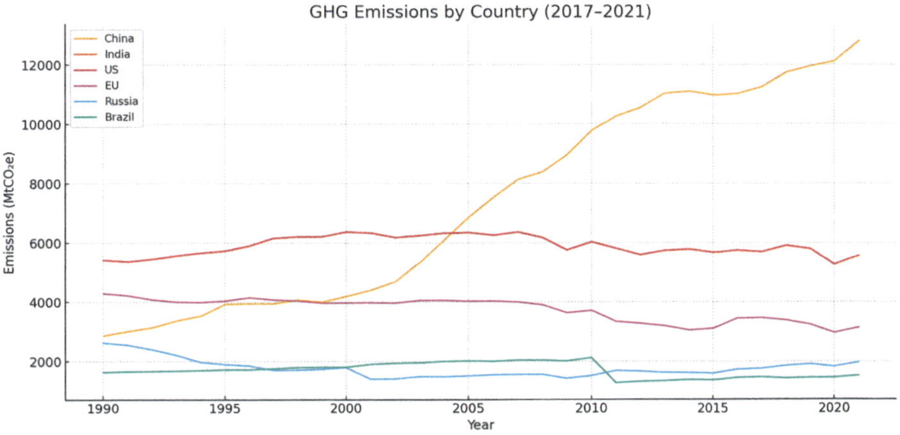

Fig. 4.1 Line chart titled "GHG Emissions by Country (2017-2021)" showing greenhouse gas emissions in MtCO2e from 1990 to 2021. The chart includes data for China, India, US, EU, Russia, and Brazil. China's emissions rise sharply, while India's and the US's remain relatively stable. EU, Russia, and Brazil show slight fluctuations. The y-axis represents emissions in MtCO2e, and the x-axis represents the year. (Climate Watch)

energy needs, but they are still smaller than China's and the USA's. After a sharp fall in the 1990s, Russia's emissions did not go up again, and Brazil's emissions today are still relatively low and steady, mainly because of activities related to land use. According to the figure, there are bigger differences in emissions from developed and developing countries, pointing to a worldwide difficulty with equal climate policies.

4.2 Land Use and Land Cover Changes Over Time

The physical characteristics that define Earth's surface, combined with human-based use of these characteristics, form the concept of land use and land cover (LULC). Natural features, which include forests and water bodies, and soil comprise land cover distinctions from human activities, which consist of urban development and agriculture and mining practices that constitute land use. LULC changes have brought substantial transformations to landscapes because of population growth combined with economic development together with climate change (Lambin et al. 2001). The clearing of forests remains the most fundamental alteration of land cover, which primarily affects tropical rainforests located in the Amazon and Southeast Asia. Open forest areas have been destroyed for agricultural purposes, along with cattle ranching and logging, to release carbon emissions and diminish biodiversity. During the period spanning from 1990 to 2020, developing countries witnessed the destruction of more than 420 million hectares of forest land according to FAO (2020). The process of urban growth represents one of the main factors contributing to land use transformation. The growth of urban areas leads to the development of built structures, including roads alongside housing and commercial outlets, where natural territories used to exist. The conversion of natural land into built structures causes habitat breakup and diminishes green zones and heightens surface water runoff, which intensifies urban water inundation and produces heat island effects (Seto et al. 2012). Likewise to other factors, agricultural assets have significantly modified areas of land use. The food demand has triggered the conversion of previous forestland and grassland territory into farmland and grazing areas. The enhanced agricultural practices for elevated food production lead to destructive impacts, including soil damage alongside water contamination and enhanced emission of greenhouse gases (Foley et al. 2005).

Scientists now possess greater precision for tracking time-based land use and land cover alterations through the combination of remote sensing alongside geographic information systems (GIS). Managers benefit from satellite imagery because this technology delivers precise data about vegetation evolution and urban growth and desert areas (Weng 2012). The relationship between human development and environmental sustainability operates through changes in land management and land state observations. The analysis of these modifications serves as essential background for proper natural resource management and ecosystem preservation, together with climate change reduction efforts.

4.3 Changes in River Flow Over Time

River flow plays a fundamental role in the global hydrological cycle because it enables ecosystem conservation, together with agricultural assistance and domestic and industrial water services. During the past several decades, river flow patterns underwent substantial modifications worldwide because of land use transformation and hydropower development and water extraction combined with climate change (Gudmundsson et al. 2021). Recent worldwide measurements indicate that river flow alterations occur across different geographical areas. Using over 7000 gauging stations, Gudmundsson et al. (2021) conducted a comprehensive analysis, which showed Northwestern Europe, together with parts of Central Africa and Eastern North America, have registered rising annual streamflow measurements. The water flow in southern European regions and western North American areas, along with certain African and Australian southern territories, experienced substantial decreases. The modifications occur because of variations in precipitation and evapotranspiration rates. Rivers in geographic areas where precipitation increased and temperature remains stable have experienced growing water flow rates. An increase in temperature and the variability of rainfall in arid and semiarid regions results in river flow reduction according to Huang et al. (2017). Human-built dams and reservoirs have produced extensive modifications to river operational patterns. The water flow patterns and sediment movement and aquatic ecosystem health have suffered due to the nearly 60% fragmentation rate that dams cause to the world's major rivers (Grill et al. 2019). Agricultural use of water stands as a significant additional factor.

Excessive agricultural water extraction from the Indus Nile and Colorado River areas has resulted in reduced river water discharges particularly during dry months.

The effects of climate change result in heightened fluctuations of river water levels. Research indicates that river flooding, together with dry periods, now occurs more frequently and strongly. Scientists have observed that rising temperatures in snow-heavy areas push peak water discharges to advance their timing in spring months before summer starts. Glacier-fed rivers, including those found in the Himalayas and the Andes, face changes in their water flow patterns. The first effect of glacier melt results in higher river flow, yet future water supply reductions will appear when glaciers disappear according to Pritchard (2019). HydroSHEDS and Global Streamflow Indices and Metadata Archive (GSIM) support large-scale hydrological modelling according to Do et al. (2018). Scientific instruments enable researchers, alongside public officials, to anticipate upcoming water reserves and run proper water management techniques for climate change.

4.4 Changes in Groundwater Over Time

The systematic patterns of water need specialization for water resource operations and ecosystem defence, along with climate adjustments. Human requirements rely heavily on groundwater because it stands as the largest available source of freshwater that supports agriculture and delivers drinking water and operates industrial activities. Future water resource sustainability faces severe risks from these water utilization trends, which produce broad environmental effects on ecosystems and food production systems (Famiglietti 2014). Several geographical areas exhibit major groundwater depletion from excessive water extraction rates that affect dry and partially dry regions. India, along with China, leads the list of countries, followed by the United States and Pakistan, which then join forces with Iran as world leaders in groundwater extraction. Northern India illustrates an alarming groundwater decline rate through NASA's GRACE satellite observations that mainly connected to increased irrigation activities (Rodell et al. 2009). The Central Valley of California experiences a rapid decline in aquifer levels during droughts because of increasing groundwater overdraft. The North China Plain presents an identical pattern of groundwater depression cone formation because irrigation activities exceed natural recharge rates, according to Feng et al. (2013).

Groundwater levels are influenced negatively by climate change because of modifications in recharge rates. The natural process of aquifer recharge is reduced due to modified precipitation patterns, alongside modified snowmelt timing and increases in evapotranspiration throughout many regions worldwide. The increased temperatures within arid regions create more evaporation while simultaneously decreasing infiltration capabilities (Taylor et al. 2013). Urban development combined with changes in landscape coverage produces lower groundwater infiltration rates through the creation of human-made non-porous surfaces, including buildings and roads. The restricted rain infiltration triggers surface runoff, which decreases the amount of water that replenishes aquifers. The expanding urbanization drives increasing groundwater use demands among the growing population totals (Wada et al. 2010). Remote satellite surveillance utilizes GRACE and GRACE-FO missions to study Earth's gravity field shifts as they relate to groundwater mass changes across the planet. Through their observations, the satellites have generated worldwide evidence that shows that groundwater depletion occurs over extended time periods (Rodell et al. 2018). Groundwater levels around the world decrease because of excessive extraction combined with decreased natural recharge. The situation creates multiple dangers that endanger drinking water supplies and agricultural systems as well as the natural ecosystem. The worldwide increase in temperature observed over several decades has affected freshwater reservoirs, together with saltwater bodies, across the Earth. The changing ocean temperatures are causing negative effects to aquatic organisms and weather systems, with consequences that lead to increased ice melting and rising sea levels.

4.5 Changes in Lake and Ocean Temperature Over Time

Since the 1970s, the global oceans have taken in more than 90% of greenhouse gas emission-induced heat increases, thus making ocean warming the most obvious indicator of climate change (IPCC 2021). According to data collected by the National Oceanic and Atmospheric Administration (NOAA), the upper 700 metres of ocean water experienced a temperature increase of 0.18 °C per decade from 1955 until present (NOAA 2023). The SST of the three major oceans, including the Atlantic, Pacific, and Indian ocean heating, has been recorded along with noticeable, swift temperature rises seen specifically in North Atlantic ocean areas. Ocean warming leads to stronger tropical storms as well as coral bleaching and marine heatwaves that damage marine species and coastal populations (Hoegh-Guldberg et al. 2014). Surface water warming stands as only one aspect of ocean temperature rises. Research indicates that heating processes are occurring at all levels of the deep ocean. Although ocean depth heats up gradually, the ocean's large thermal capacity contributes to sustained climate regulation alongside rising sea levels through thermal expansion (Purkey and Johnson 2010). The worldwide heating of lakes takes place at speeds that exceed ocean temperature increases. Lakes Superior and Tanganyika, located in northern regions and high elevations, demonstrate exceptional temperature rise patterns. The temperature increases cause disruptions to thermal layers and oxygen depletion, which harms fish populations according to O'Reilly et al. (2015). Warming waters affect ecosystems profoundly. The growing ocean heating triggers coral bleaching and causes species to relocate their habitats while decreasing fishery output. Lake environments experience three significant effects from temperature change, which include altered algal blooms accompanied by enhanced methane releases and biodiversity damage (Woolway et al. 2020). Heating of water bodies produces various climate-warming cycles that further boost global heating. Global warming creates a decrease in lake and ocean ice that diminishes reflective properties (albedo) and results in solar radiation absorption over reflection (IPCC 2021).

The heightened water temperatures in lakes and oceans result directly from climate change, creating environmental and economic problems, together with disruptions to climate regulation. Systematic change observation is essential to create adaptation plans and stop more environmental damage from occurring. Environmental health as well as human well-being heavily depend on water quality being at its best state. The state of water determines all its applicable uses for everyday purposes and industrial operations as well as environmental system preservation. Multiple human-made processes and changing climate, together with growing populations, cause water quality decline in rivers, lakes, and groundwater and coastal zones (UNEP 2016). Natural water bodies receive the highest damage from human activities. Global rivers exhibit poor water quality due to quick population growth and developments in both agriculture and industries, which generate heavy compounds and plastic impurities alongside nutrient and microbial contamination. The Ganges River in India and the Yangtze River in China encounter nutrient overenrichment resulting from untreated sewage and agricultural runoff, according to Vörösmarty et al. (2010). The excessive use of fertilisers in the Mississippi River Basin leads to hypoxic zones in the Gulf of Mexico, according to research by Rabalais et al. (2002).

The worldwide deterioration of lake water quality occurs because of climate change, together with land use modifications and pollution sources. Research indicates Lake Erie and Lake Geneva, alongside other European and North American lakes, display growing harmful algal bloom (HAB) events because of temperature increases, along with excessive nutrient deposits (Michalak et al. 2013). The combined effects of land degradation and urban water flow into Lake Victoria in Africa have led to increased turbidity and biological water pollution, according to Sitoki et al. (2010). The modifications to aquatic ecosystems cause both biodiversity decline and problems with water availability for drinking, together with decreased fishery productivity. The perception of groundwater as safer and less vulnerable has not stopped its pollution rates from growing. Over-extraction and poor waste management lead to contamination with nitrates, arsenic, fluoride, and pathogens. Research shows that shallow groundwater arsenic pollution affects more than 35 million people throughout Bangladesh (Smedley and Kinniburgh 2002). The application of fertilizers results in nitrate levels surpassing WHO standards throughout numerous aquifers throughout European and US agricultural regions (Rivett et al. 2008). Plastic debris, together with nutrients, oil spills, and heavy metals, represent the main land-based pollutants, which cause harm to coastal and marine ecosystems. The coral reef ecosystems in the Caribbean, together with Southeast Asia, face heavy threats from land-discharge pollutants that also bring in ocean-borne sediment (Burke et al. 2011). Eutrophication caused by massive river pollution from the Nile, Amazon, and Yangtze results in dead marine zones where most aquatic life cannot survive due to diminished oxygen levels (Diaz and Rosenberg 2008).

4.6 Habitat Fragmentation and Observed Changes Over Time

Global water quality degradation creates problems that impact water bodies of every type across the world. This environmental crisis needs worldwide cooperation with systems to control pollution and practices to manage land and water sustainably as well as complete monitoring capabilities. The separation of extensive continuous habitats into smaller segmented areas through human development practices constitutes habitat fragmentation, as shown in Fig. 4.2. The breaking down of habitats causes lowered biodiversity levels, together with disturbed ecosystem functions while making species more susceptible to death (Fahrig 2003). The partitioning of natural habitats has two major consequences which disconnect species populations from each other while preventing genetic mixing between groups and intensifying confrontations between humans and wildlife. This ecological disruption changes local climate conditions and disrupts natural ecological processes, which positions it among the chief contributors to worldwide biodiversity decline (Haddad et al. 2015).

Research conducted in 2015 demonstrates that approximately 70% of existing forest areas exist within a 1 kilometer distance from forest edges, leaving them vulnerable to various edge effects, including wind exposure and sunshine and invasive species and human activities (Haddad et al. 2015). Between 2000 and 2012, substantial growth became evident in the tropical forest fragmentation rate, which concentrated mainly inside the Amazon and Southeast Asia and Central Africa. Ecoalignments from tropical forests lead to a yearly core forest area decline totaling 3.2 million hectares according to DeFries et al. (2005) and Hansen et al. (2013). The

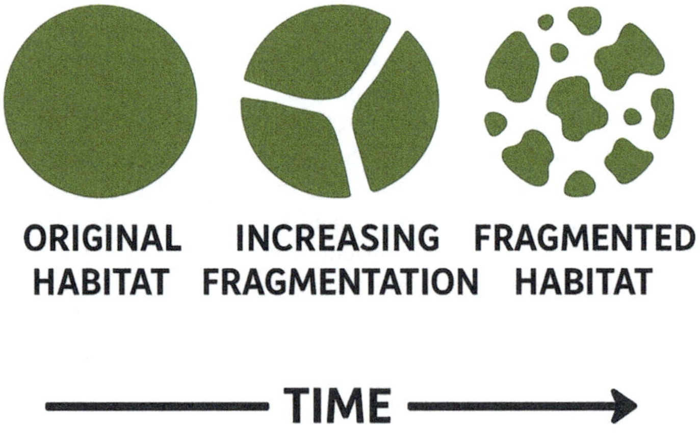

Fig. 4.2 Diagram illustrating habitat fragmentation over time. The first circle labeled "Original Habitat" is solid green, representing an intact habitat. The second circle, labeled "Increasing Fragmentation," is divided into three segments, indicating partial fragmentation. The third circle, labeled "Fragmented Habitat," shows multiple small, irregular green shapes, depicting a highly fragmented habitat. An arrow labeled "Time" points from left to right, indicating progression

United States has experienced a decline in its forested territory to a point where greater than 60% of forested regions now exist near human-designed areas that are under 100 meters wide because of suburban growth encroachment and lumber industry operations (Riitters et al. 2016). Europe stands as one of the most divided continents because humans have been manipulating the land since centuries ago. Only 1% of European forests are considered "undisturbed" and free from fragmentation. The extinction of pollinators and seed dispersers causes plants to reproduce less and limits forest natural development. Edge forests generate carbon emissions because the degradation process speeds up and forests become more prone to clearances (Laurance et al. 2011). In fragmented areas, certain species display adaptation while specialist species, which comprise big carnivores and migratory birds, together with amphibians, generally demonstrate steep population declines (Crooks et al. 2017). The global spread of habitat fragmentation grows more severe because of increasing human population growth and alterations to land management practices.

The graph in Fig. 4.3 shows the amount of greenhouse gas (GHG) emissions from each sector each year from 1990 to 2021. The larger the chart, the greater the total amount of emissions released that year. The graph groups energy (orange), agriculture (red), industrial activities (yellow), land use change and forestry (green), and waste (purple). For several years, GHG emissions from energy have exceeded all other areas and increased their proportion in the rise of emissions. At the same time, the parts played by both agriculture and waste do not change much, while the contributions from industrial activities and changing land use fluctuate a little. Year after year, the doughnut charts show a bigger outer ring, proving that total emissions have increased greatly in the last three decades, thanks to the energy sector.

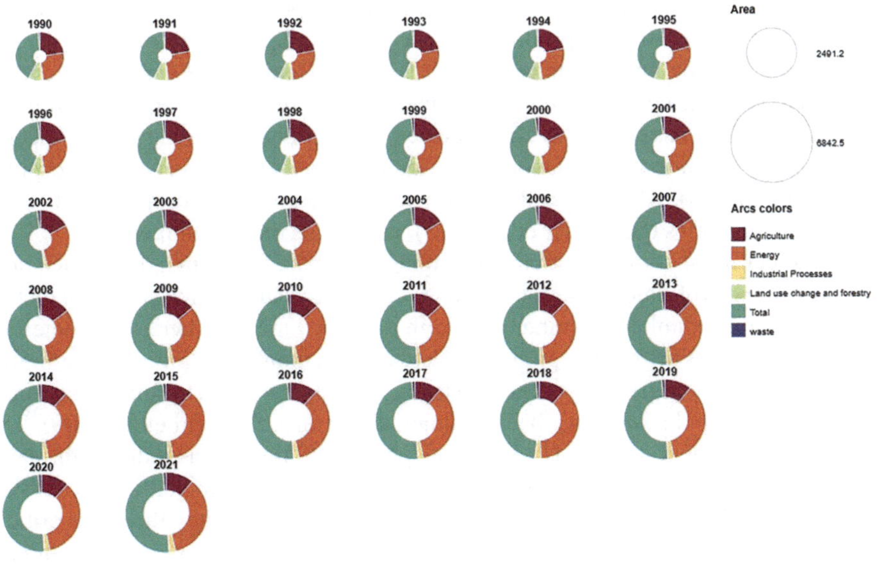

Fig. 4.3 India climate change data from the year 1990–2021

4.7 Changes in Biodiversity, Species Distribution, and Abundance Over Time

Multiple factors like climate change, habitat destruction, and pollution are responsible for the fast decline of biodiversity which includes species, ecosystems, and genetic diversity (Sala et al. 2000). Numerous species now face extinction alongside major population reductions because research shows their populations have dropped across both terrestrial regions and marine and freshwater areas. According to the Millennium Ecosystem Assessment, the ongoing deterioration of ecosystem services, which enable life on Earth, reached 60% resulting in major decreases of global biodiversity. Human activity creates an ongoing threat to millions of species, and numerous species already face drastic population decreases according to IPBES (2019).

4.8 Changes in Species Distribution Over Time

Species movements throughout habitats have noticeably altered since the arrival of climate change, together with human-caused ecological changes and urbanization. Global temperature rises have caused severe effects on how species distribute themselves across geographic areas. Species migrate toward colder northern regions as well as elevation increases to adapt to warming environmental temperatures. Scientists determined through research that marine species in the Northern Hemisphere move their ranges 20 kilometers in each decade direction northward (Pinsky et al. 2013). Several types of birds throughout North America, alongside Europe, are shifting their locations north to keep suitable living environments (Both et al. 2009). Global warming is influencing the temporal aspects of species' life cycles when they perform reproductive activities and migrate their populations and see their flowers bloom. Plants across the Northern Hemisphere experience earlier spring blossoming because their environments show rising temperatures (Walther et al. 2002). The pattern of invasive species moving and replacing native species matches the conditions that favor warm temperatures. The Asian tiger mosquito (Aedes albopictus), which spreads into the temperate areas of European and North American territories (Schaffner et al. 2013), represents this phenomenon.

The sheer number of individuals belonging to a species within a designated area has undergone profound alterations because of human-caused environmental disturbances. Species numbers continue to rise in disturbed areas of land, but numerous populations across different regions are decreasing in numbers. Since the late 1980s, the global fish catch has maintained a steady level as researchers found that many tuna and cod populations have been dramatically reduced (Worm et al. 2009). The abundance of large mammals (including elephants and tigers) has reduced throughout the 1980s because of hunting combined with habitat fragmentation, according to Ripple et al. (2014). The amphibian population demonstrates the most critical loss in abundance because more than 40% of amphibian species now face threats stemming from habitat destruction, together with disease spread and climate change effects (Stuart et al. 2004). The substantial modifications in biodiversity and species distributions, along with population trends, stem from human activities that especially result from climate change and habitat reduction and pollution. These alterations in global species stability are speeding up because numerous species face complete extinction, alongside major population reductions. Ecosystem services supporting life on Earth suffer major effects when biodiversity continues to decrease. The essentiality of both effective conservation practices and multinational cooperation exists to prevent additional decreases and foster species revival.

4.9 Changes in Air Quality Over Time

The quantity of airborne substances affecting human health and ecosystems, together with the climate, defines air quality. Stricter laws and modern technological developments have created better air quality conditions in various parts of the world, especially in industrialised nations. For example: the United States witnessed air pollution decreases through the implementation of the Clean Air Act, which Congress approved in 1970. Following the implementation of the European Union Ambient Air Quality Directive, Europeans witnessed better environmental air conditions. Many urban areas maintain air pollution quantities higher than WHO standards, especially in terms of PM2.5 counts (European Environment Agency [EEA] 2020). The challenges of poor air quality continue to affect developing nations because of how fast their cities are expanding and the environmental pollution generated by industry and vehicle traffic (Chan et al. 2018). Rainwater becomes acidified through atmospheric pollutants released during fossil fuel combustion and primarily consisting of sulfur dioxide (SO2) and nitrogen oxides (NOx). Water vapor joins with pollution to produce sulfuric acid and nitric acid before the acidic precipitation occurs. The Acid Rain Program launched in the United States during 1990 collaborated with the EU's National Emissions Ceilings Directive to decrease sulfur and nitrogen oxide pollutant releases throughout Europe, and the US acid rain deposition measurements reveal half the levels detected during the 1980s following sulfur and nitrogen oxide emission control programs, according to US EPA and National Research Council (2004) reports. The amount of acid rain in Asia has been rising because of increasing industrial activities and coal combustion in China and India (Galloway et al. 2015). The degradation of forests and lakes and soil health in these areas persists as a result of acid rain deposition.

The atmospheric concentration of dust and aerosols presents changing levels according to geographical areas and annual seasons. The factors contributing to dust emissions are desert areas, alongside agricultural operations and building construction. Aerosol particles have natural origins, such as sea spray and volcanic eruptions, and human-made origins, like industrial pollution and vehicle exhaust. Dust storms that occur frequently within the Saharan desert territory release significant amounts of airborne particles. Dust storms that originate from deserts have the ability to move thousands of miles in all directions while disturbing worldwide air quality standards. According to Prospero et al. (2010), Sahara dust travels up to the

Caribbean and Amazon regions. Several environmental factors, including land degradation combined with deforestation and climate change aspects, have resulted in increased occurrences, alongside higher intensity of dust storms during the last few decades. The NASA MODIS satellites reveal, through satellite data, a rising number of dust occurrences, especially in North African arid areas (Moulin et al. 1998). The levels of aerosol components have experienced major shifts across different time periods because of natural together with anthropogenic origins. Human industrial emissions and transportation activities have elevated black carbon and sulfate aerosols in the atmosphere, which cause air pollution and modify cloud formation together with atmospheric radiation as described by Bond et al. (2013).

The amount of aerosols was lower in places with strong pollution management policies, such as Europe and North America, from 1980 to 2010, yet rose throughout Asian territories experiencing rapid industrial development (Huang et al. 2016). Sunlight reflects from aerosols, which creates surface cooling, or the particles absorb solar radiation, leading to climate warming effects. The amounts of cooling produced by aerosols have worked to reduce the global warming effects from greenhouse gases during the past few decades, according to Kiehl (2007). Figure 4.3 demonstrates how things change from a "healthy Earth" to "global warming." A smiling, vibrantly colored planet under a green sky full of clean water is presented at the start, indicating an improved ecosystem. During the following frames, the impact of deforestation is seen as trees have been chopped down for the first step in disturbing natural habitats. After that, the animation adds industrial pollution, where factories pour smoke and fill water bodies with waste, meaning contaminants are carried into the environment. In the last picture, the planet is sad and overheated, and the ice is melting to warn about global warming and humanity's overall effects. It highlights that when we make environmental mistakes, they often create great harm to nature. Figure 4.4 demonstrates how things change from a "healthy Earth" to "global warming." A smiling, vibrantly coloured planet under a green sky full of clean water is presented at the start, indicating an improved ecosystem. During the following frames, the impact of deforestation is seen as trees have been chopped down for the first step in disturbing natural habitats. After that, the animation adds industrial pollution, where factories pour smoke and fill water bodies with waste, meaning contaminants are carried into the environment. In the last picture, the planet is sad and overheated, and the ice is melting to warn about global warming and humanity's overall effects. It highlights that when we make environmental mistakes, they often create great harm to nature.

4.10 Changes in Soil Degradation Over Time

Strengthened environmental regulations led to improved air quality and decreased acid rain quantities as well as aerosol concentrations yet provoked changes in dust levels throughout the years. The rising issues attributed to industrialization and urbanization, and climate change remain significant problems for developing regions. The process of soil quality deterioration alongside productivity reduction

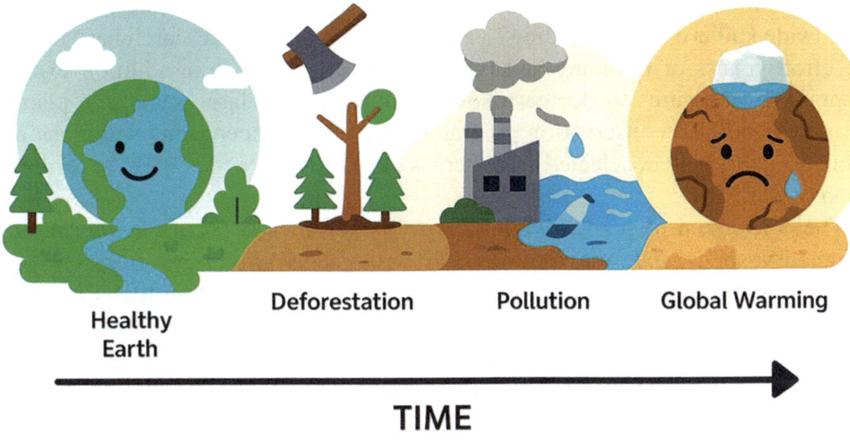

Fig. 4.4 Illustration depicting environmental changes over time. On the left, a smiling Earth with trees and a river represents a healthy planet. Moving right, deforestation is shown with an axe and a cut tree. Further right, pollution is illustrated with factories emitting smoke and water pollution. On the far right, a sad Earth with cracks and melting ice symbolizes global warming. Labels include "Healthy Earth," "Deforestation," "Pollution," and "Global Warming," with a timeline arrow labeled "Time" at the bottom

occurs from both human impact and natural causes. The key contributors to soil degradation include erosion together with nutrients disappearing from the soil and salinization and soil compaction. Land degradation produces major impacts on farming output as well as the diversity of species found in different terrains. The worldwide extent of land degradation reaches 33% of the world's total territory, while erosion acts as the primary cause of degradation. Water erosion caused by rainfall, along with wind erosion found in drylands, serve as the two main causes of degradation (Lal 2015). Erosion functions as one of the primary reasons that trigger soil degradation. The yearly worldwide soil loss attributed by Pimentel et al. (1995) study totals approximately 75 billion tonnes and causes a major decrease in agricultural productivity. Erosion proves to be the greatest concern for countries in Asia, Africa, and Latin America because intensive agricultural areas become especially prone to it. Soil degradation worsens because of climate change when excessive weather events, such as floods along with droughts, enhance soil loss rates according to Thomas et al. (2014). Data shows that extreme weather events have grown 30% more frequent in various locations during the last three decades, as IPCC (2021) reports.

Soil erosion emerges as a major concern because it targets areas undergoing agricultural growth and practicing land use approaches that prove unsustainable. Soil erosion exhibition varies substantially between different geographical areas on Earth. Lal shows that South Asian regions experience a soil erosion loss of 20–40 tonnes per hectare per year, but sub-Saharan Africa faces rates exceeding 50 tons per hectare per year. The process of soil erosion finds its main causes in agricultural practices which include tree clearing as well as field cultivation and excessive livestock grazing. The facts demonstrate that agriculture is responsible for 80% of worldwide soil erosion, according to Pimentel et al. (1995). Research has validated the effectiveness of terracing and reforestation programmes in addition to cover cropping as measures to decrease soil erosion. Chinese large-scale tree planting initiatives and land conservation programs succeeded in decreasing soil erosion by about 50% throughout selected areas across the nation since the 1980s, according to Zhou et al. 2019.

The conversion of productive land territory into desert occurs mainly through severe dry spells and tree removal and wrong agricultural practices. The UN Convention to Combat Desertification (UNCCD) reports that desertification affects 24% of Earth's land surface, primarily located in hot, dry, and semiarid zones. The expansion of desertification has become more widespread throughout Africa and Asia since the middle of the twentieth century. During the last few decades, the Sahara Desert has migrated 10 kilometers per year toward the south (Reynolds and Stafford Smith 2002). Desertification in the Sahel subregion has caused substantial damage to productive lands, according to experts who estimate that 50% of this area suffers from degradation (Herrmann and Hutchinson 2005). During the 1990s, China established significant investments to combat desertification, thereby decreasing its expansion rate by 30% (Li et al. 2012). Climate change produces worsened desertification effects by reducing arid and semiarid region rainfall and increasing temperatures. The elevated CO_2 concentrations in the atmosphere negatively affect plant development, leading to intensified desertification risks (Lal 2015). Soil degradation, together with erosion and desertification, has increased significantly as problems that affect arid and semiarid regions. Some areas affected by vegetation index recovery may benefit from active restoration projects, yet general dryland and agricultural regions continue to experience worsening environmental deterioration.

4.11 Tools and Methods for Tracing Environmental Changes Over Time

4.11.1 Remote Sensing and Satellite Imagery

Scientists have created multiple tools and methods to track the levels of air quality, along with water quality and soil conditions and biodiversity and climatic variations. By using aerial or satellite technologies, scientists obtain imagery to detect modifications occurring in the environment throughout durations of time. Key remote sensing technologies include:

Landsat satellites collect extended information about land use transitions and land cover alterations, along with forest clearance and city sprawl, and agricultural land development started in 1972. Torrid imagery stands as a fundamental tool for scientists who investigate how people and natural disasters affect landscapes since 1972, according to Wulder (2012).

The MODIS platform launched on NASA's Terra and Aqua satellites collects information about vegetation indices through NDVI measurements as well as ocean color data alongside cloud observations and aerosol measurements, which allow researchers to monitor vegetation health and agricultural output, alongside atmospheric quality assessments (Justice 1998). Under the Copernicus programme of the European Space Agency, *Sentinel Satellites* serve as part of their observation network. The satellites provide data necessary for examining land surfaces as well as water conditions and city development needs. The radar image data from Sentinel-1 allows tracking of land subsidence, and Sentinel-2 produces optical imagery to evaluate vegetation health and forest cover (Drusch et al. 2012). By employing these tools, researchers acquire the ability to observe long-term patterns and evaluate land use modifications while performing essential spatial-temporal evaluations to measure the depth of environmental denigration incidents (e.g., deforestation and desertification) (Pettorelli 2014).

A Geographic Information System (GIS) exists as a fundamental technology for conducting spatial data analysis and environmental change mapping and modelling future environmental modifications. Applications of GIS focus on time-based land cover evaluation to detect forest clearance and monitor quantitative increases of urbanized areas and farming zones and environmental habitat partitioning. The development of predictive ecosystem spatial models becomes possible through GIS because the technology allows scientists to build predictive frameworks using environmental data and human-caused effects like deforestation and irrigation. The developed models assist researchers to forecast how species distributions alongside ecosystem services will evolve in the future (Goodall 2017). Researchers utilize maps to identify vulnerable territories and develop virtual projections through various climate change modelling systems (Fitzharris et al. 2007).

Ground-based monitoring combined with field surveys creates essential localized data, which needs the basics provided by satellite imagery and geographic information science. Environmental researchers employ these methods for measuring different environmental indicators, which include water and air quality, together with soil state and biodiversity levels. The stations incorporate sensors to offer moment-by-moment measurements of air quality and temperature and humidity and multiple climatic elements. The Air Quality Index (AQI) is calculated from ground-based sensors, enabling the detection of pollutants, including particulate matter (PM2.5), together with ozone and nitrogen dioxide (NO2), according to the EPA (2020). Ground-based water quality assessments consist of measuring essential parameters, including pH levels and salinity, together with nutrient counts and heavy metal detections, and microorganism presence in water bodies.

The tracking of pollution trends, together with assessment of groundwater quality impacts resulting from agricultural runoff and wastewater, along with industrial pollution, relies on data acquired from river and lake monitoring stations (Allan 2004). The field assessment of soil health and degradation, along with erosion levels, takes place through soil monitoring surveys. Climate *modeling instruments analyze the long-term climate alteration patterns and forecast global warming effects on multiple ecosystems. Future climate predictions generated from these models include data about temperature levels, together with precipitation data, in addition to greenhouse gas measurements. Scientists utilize General Circulation

Models (GCMs) for climate system simulation through atmospheric as well as oceanic and terrestrial physical process representation. The models developed by Solomon (2007) enable scientists to make forecasts about temperature increases and modifications in precipitation levels as well as severe weather occurrences. Regional Climate Models (RCMs) generate specific climate change modeling results at regional and local scales for improved forecasts of regional climate change effects (Giorgi et al. 2001). The integration of climate models with economic models within IAMs enables researchers to assess the environmental as well as economic effects that various quantities of greenhouse gas emissions would create on a global scale. Together, these models evaluate different methods to reduce climate change effects (Nordhaus 2007).

Professionally developed monitoring tools help researchers detect biodiversity evolution by analysing species locations with population patterns and ecosystem well-being. Remote wildlife population monitoring occurs through camera traps, which automatically photograph animals crossing their range. Camera traps serve as critical equipment for species tracking, especially when studying elusive or endangered species (O'Connell 2011). Scientists employ audio recording devices for bioacoustic monitoring of animal sounds, specifically in forests and wetlands. Wildlife population assessment, habitat loss evaluation, and conservation program tracking have become more efficient through drone technology, which combines camera sensors and sensors (Anderson and Gaston 2013).

4.12 Conclusion

How environmental changes have unfolded over the years highlights a detailed and worrying story of the way humans interact with nature. In this chapter, I have followed the development from a world with a balanced nature to one dealing with serious environmental problems now. At one time, Earth had a wide variety of life forms, was not affected by large climate changes, and managed itself without human influence. Yet, the start of massive deforestation was a key moment in history. Largely due to the increased use of land for farming, logging, and building, the deforestation of land removed vital sources of absorption for carbon, broke up water cycles, increased soil erosion, and caused a great reduction in diversity among species. With industrial progress, it became much easier to see the results of the activities that people caused. A lot of greenhouse gases emitted by burning fossil fuels, industrial activities, and using energy without care have resulted in air quality problems and pollution in both land and water systems. Also, the use of solid waste, plastic overspill, and unhandled industrial pollution overwhelmed nature's resources for recovery and natural regeneration. Due to these influences, we now have an environmental crisis that shows itself as warmer global temperatures, shrinking ice caps, seas getting higher, ocean acidification, more extreme weather, and ecological transformation. The main danger caused by these changes, global warming, is the result of a prolonged neglect of the environment and unplanned development. The consequences have become noticeable in less diversity of nature, smaller farm harvests, more people moving due to climate change, and stronger natural disasters.

This chapter emphasises that environmental changes are not abrupt but result from a step-by-step process, shared by many small events, tied to past decisions about how resources have been used and policies made. The story told in this book involves decay, but it also creates a chance for us to see the errors of history and work toward a better, fairer tomorrow. Fixing the current situation involves taking strong steps to save the environment, work with the rest of the world on climate, promote renewables, care for ecosystems, and replace wasteful practices with natural economic systems. Basically, environmental change is an invitation to take action. It encourages us to balance development with nature, find new ways to protect the environment, and really honor the natural world around us. We can only restore the well-being of our environment and ensure others will still have a home in the future if we study our past and start to implement sustainable changes.

References

Allan JD (2004) Landscapes and riverscapes: the influence of land use on stream ecosystems. Annu Rev Ecol Evol Syst 35:257–284. https://doi.org/10.1146/annurev.ecolsys.35.120202.110122

Anderson K, Gaston KJ (2013) Lightweight unmanned aerial vehicles will revolutionize spatial ecology. Front Ecol Environ 11(3):138–146. https://doi.org/10.1890/120150

Bond TC, Doherty SJ, Fahey D (2013) Bounding the role of black carbon in the climate system: a scientific assessment. J Geophys Res Atmos 118(11):5380–5552. https://doi.org/10.1002/jgrd.50171

Both C, Artemyev AV, Briedis M (2009) Climate change and population declines in a long-distance migratory bird. Nature 441(7089):81–83. https://doi.org/10.1038/nature04539

Burke L, Reytar K, Spalding M, Perry A (2011) Reefs at risk revisited. World Resources Institute. https://www.wri.org/research/reefs-risk-revisited

Chan CK, Yao X, Liu Y (2018) Air pollution in mega cities in China. Atmos Environ 37(1):33–45. https://doi.org/10.1016/S1352-2310(03)00300-6

Crooks KR, Burdett CL, Theobald DM (2017) Quantification of habitat fragmentation reveals extinction risk in terrestrial mammals. Proc Natl Acad Sci 114(29):7635–7640. https://doi.org/10.1073/pnas.1705769114

DeFries RS, Foley JA, Asner GP (2005) Land-use choices: balancing human needs and ecosystem function. Front Ecol Environ 2(5):249–257. https://doi.org/10.1890/1540-9295(2004)002[0249:LCBHNA]2.0.CO;2

Diaz RJ, Rosenberg R (2008) Spreading dead zones and consequences for marine ecosystems. Science 321(5891):926–929. https://doi.org/10.1126/science.1156401

Do HX, Gudmundsson L, Leonard M, Westra S, Seneviratne SI (2018) The global streamflow indices and metadata archive (GSIM)—part 1: the production of a daily streamflow archive and metadata. Earth Syst Sci Data 10(2):765–785. https://doi.org/10.5194/essd-10-765-2018

Drusch M, Del Bello U, Carlier S (2012) Sentinel-2: ESA's optical high-resolution mission for GMES operational services. Remote Sens Environ 120:25–36. https://doi.org/10.1016/j.rse.2011.11.026

European Environment Agency (EEA) (2020) Air quality in Europe – 2020 report. https://www.eea.europa.eu/publications/air-quality-in-europe-2020

Fahrig L (2003) Effects of habitat fragmentation on biodiversity. Annu Rev Ecol Evol Syst 34:487–515. https://doi.org/10.1146/annurev.ecolsys.34.011802.132419

Famiglietti JS (2014) The global groundwater crisis. Nat Clim Chang 4(11):945–948. https://doi.org/10.1038/nclimate2425

FAO (2020) Global Forest Resources Assessment 2020: Main report. Food and Agriculture Organization of the United Nations. https://www.fao.org/forest-resources-assessment/2020

Feng W, Zhong M, Lemoine JM, Biancale R, Hsu HT, Xia J (2013) Evaluation of groundwater depletion in North China using the Gravity Recovery and Climate Experiment (GRACE) data and ground-based measurements. Water Resour Res 49(4):2110–2118. https://doi.org/10.1002/wrcr.20192

Fitzharris BB, Houghton RW, Price JS (2007) Climate change and its impacts on the environment. Glob Chang Biol 13(7):1506–1528. https://doi.org/10.1111/j.1365-2486.2007.01383.x

Foley JA, DeFries R, Asner GP (2005) Global consequences of land use. Science 309(5734): 570–574. https://doi.org/10.1126/science.1111772

Galloway JN, Dentener FJ, Capone DG (2015) Nitrogen cycle: past, present, and future. Biogeochemistry 70(1):153–226. https://doi.org/10.1007/s10533-004-0370-0

Giorgi F, Hewitson B, Solomon S (2001) Regional climate information—evaluation and projection. Intergovernmental Panel on Climate Change (IPCC) Report

Goodall J (2017) The future of ecosystem management: strategies, challenges, and successes. Springer

Grill G, Lehner B, Thieme M, Geenen B, Tickner D, Antonelli F, Zarfl C (2019) Mapping the world's free-flowing rivers. Nature 569(7755):215–221. https://doi.org/10.1038/s41586-019-1111-9

Gudmundsson L, Do HX, Leonard M, Westra S (2021) Observed trends in mean and extreme river flow globally. Nature 592(7855):391–395. https://doi.org/10.1038/s41586-021-03406-6

Haddad NM, Brudvig LA, Clobert J (2015) Habitat fragmentation and its lasting impact on Earth's ecosystems. Sci Adv 1(2):e1500052. https://doi.org/10.1126/sciadv.1500052

Hansen MC, Potapov PV, Moore R (2013) High-resolution global maps of 21st-century forest cover change. Science 342(6160):850–853. https://doi.org/10.1126/science.1244693

Herrmann SM, Hutchinson CF (2005) The changing contexts of desertification: the case of the Sahel. Glob Environ Chang 15(1):16–26. https://doi.org/10.1016/j.gloenvcha.2004.10.008

Hoegh-Guldberg O, Cai R, Poloczanska ES (2014) The Ocean. In: Climate change 2014: impacts, adaptation, and vulnerability. Intergovernmental Panel on Climate Change (IPCC). https://www.ipcc.ch/report/ar5/wg2/

Huang J, Li J, Li L (2016) Aerosol climate effects on global radiative forcing. Nat Geosci 9(8):529–535. https://doi.org/10.1038/ngeo2756

Huang J, Yu H, Guan X, Wang G, Guo R (2017) Accelerated dry land expansion under climate change. Nat Clim Chang 7(2):166–171. https://doi.org/10.1038/nclimate3187

IPBES (2019) The IPBES global assessment report on biodiversity and ecosystem services. Intergovernmental Science-Policy Platform on Biodiversity and Ecosystem Services. https://ipbes.net/global-assessment

IPCC (2021) Climate change 2021: the physical science basis. Intergovernmental Panel on Climate Change. https://www.ipcc.ch/report/ar6/wg1

Justice CO (1998) The MODIS land data set. Remote Sens Environ 83(1–2):1–17. https://doi.org/10.1016/S0034-4257(02)00017-3

Kiehl JT (2007) Twentieth century climate model response and climate sensitivity. Geophys Res Lett 34(22). https://doi.org/10.1029/2007GL031324

Lal R (2015) Restoring soil quality to mitigate soil degradation. Sustainability 7(5):589–609. https://doi.org/10.3390/su7050589

Lambin EF, Geist HJ, Lepers E (2001) Dynamics of land-use and land-cover change in tropical regions. Annu Rev Environ Resour 28:205–241. https://doi.org/10.1146/annurev.energy.28.050302.105459

Laurance WF, Camargo JLC, Luizão RCC (2011) The fate of Amazonian forest fragments: a 32-year investigation. Biol Conserv 144(1):56–67. https://doi.org/10.1016/j.biocon.2010.09.021

Li J, Fu B, Liu Y (2012) Desertification control in China: past, present and future. Environ Sci Pol 15(1):1–9. https://doi.org/10.1016/j.envsci.2011.11.005

Michalak AM, Anderson EJ, Beletsky D (2013) Record-setting algal bloom in Lake Erie caused by agricultural and meteorological trends consistent with expected future conditions. Proc Natl Acad Sci 110(16):6448–6452. https://doi.org/10.1073/pnas.1216006110

Moulin C, Chiapello I, N'dong N (1998) Impact of human activity on the long-term variability of mineral dust in the Sahel region. Geophys Res Lett 25(15):2789–2792. https://doi.org/10.1029/98GL01070

NASA (2024) Climate change: vital signs of the planet. https://climate.nasa.gov

NOAA (2023) Climate change: ocean heat content. National Oceanic and Atmospheric Administration. https://www.climate.gov/news-features/understanding-climate/climate-change-ocean-heat-content

Nordhaus WD (2007) The challenge of global warming: economic models and global policies. Yale University Press

O'Reilly CM, Sharma S, Gray DK (2015) Rapid and highly variable warming of lake surface waters around the globe. Geophys Res Lett 42(24):10773–10781. https://doi.org/10.1002/2015GL066235

O'Connell AF (2011) Camera trapping for wildlife research. Biol Conserv 144(4):1267–1276. https://doi.org/10.1016/j.biocon.2011.01.030

Pettorelli N (2014) Satellite remote sensing for applied ecologists: opportunities and challenges. J Appl Ecol 51(4):4–8. https://doi.org/10.1111/1365-2664.12394

Pimentel D, Lach L, Zuniga R, Morrison D (1995) Environmental and economic costs of nonindigenous species in the United States. Bioscience 45(8):687–693. https://doi.org/10.2307/1312558

Pinsky ML, Worm B, Fogarty MJ (2013) Marine taxa track local climate velocities. Science 341(6151):1239–1242. https://doi.org/10.1126/science.1239352

Pritchard HD (2019) Asia's shrinking glaciers reveal climate extremes. Nature 569(7758):393–394. https://doi.org/10.1038/d41586-019-01492-0

Prospero JM, Ginoux P, Torres O (2010) Environmental characterization of global sources of atmospheric soil dust identified with the Nimbus 7 Total Ozone Mapping Spectrometer (TOMS) absorbing aerosol product. Rev Geophys 48(2). https://doi.org/10.1029/2009RG000282

Purkey SG, Johnson GC (2010) Warming of global abyssal and deep Southern Ocean waters between the 1990s and 2000s: contributions to global heat and sea level rise budgets. J Clim 23(23):6336–6351. https://doi.org/10.1175/2010JCLI3682.1

Rabalais NN, Turner RE, Scavia D (2002) Beyond science into policy: Gulf of Mexico hypoxia and the Mississippi River. Bioscience 52(2):129–142. https://doi.org/10.1641/0006-3568(2002)052[0129:BSIPGO]2.0.CO;2

Reynolds JF, Stafford Smith DM (2002) Global desertification: do humans cause deserts? Environ Sci Pol 5(1):23–33. https://doi.org/10.1016/S1462-9011(02)00004-7

Riitters KH, Wickham JD, Costanza JK, Vogt P (2016) A global evaluation of forest interior area dynamics using tree cover data from 2000 to 2012. Landsc Ecol 31(1):137–148. https://doi.org/10.1007/s10980-015-0270-9

Ripple WJ, Wolf C, Newsome TM (2014) Status and ecological effects of the world's largest carnivores. Science 343(6167):151–162. https://doi.org/10.1126/science.1241484

Rivett MO, Buss SR, Morgan P, Smith JWN, Bemment CD (2008) Nitrate attenuation in groundwater: a review of biogeochemical controlling processes. Water Res 42(16):4215–4232. https://doi.org/10.1016/j.watres.2008.07.020

Rodell M, Famiglietti JS, Wiese DN, Reager JT, Beaudoing HK, Landerer FW, Lo MH (2018) Emerging trends in global freshwater availability. Nature 557(7707):651–659. https://doi.org/10.1038/s41586-018-0123-1

Rodell M, Velicogna I, Famiglietti JS (2009) Satellite-based estimates of groundwater depletion in India. Nature 460(7258):999–1002. https://doi.org/10.1038/nature08238

Sala OE, Perez-Harguindeguy N, Sykes MT (2000) Global biodiversity scenarios for the year 2100. Science 287(5459):1770–1772. https://doi.org/10.1126/science.287.5459.1770

Schaffner F, Mathis A, van Bortel W (2013) The Asian tiger mosquito (Aedes albopictus) in Europe. European Mosquito Bulletin 31:1–14

Seto KC, Güneralp B, Hutyra LR (2012) Global forecasts of urban expansion to 2030 and direct impacts on biodiversity and carbon pools. Proc Natl Acad Sci 109(40):16083–16088. https://doi.org/10.1073/pnas.1211658109

Sitoki L, Kurmayer R, Rott E (2010) Spatial variation of phytoplankton composition, biovolume and resulting microcystin concentrations in the Nyanza Gulf (Lake Victoria, Kenya). Hydrobiologia 664:377–390. https://doi.org/10.1007/s10750-010-0601-6

Smedley PL, Kinniburgh DG (2002) A review of the source, behaviour and distribution of arsenic in natural waters. Appl Geochem 17(5):517–568. https://doi.org/10.1016/S0883-2927(02)00018-5

Solomon S (2007) Climate change 2007: the physical science basis. Cambridge University Press

Stuart SN, Chanson JS, Cox NA (2004) Status and trends of amphibian declines and extinctions worldwide. Science 306(5702):1783–1786. https://doi.org/10.1126/science.1103538

Taylor RG, Scanlon B, Döll P (2013) Groundwater and climate change. Nat Clim Chang 3(4):322–329. https://doi.org/10.1038/nclimate1744

Thomas DSG, Twyman C, Downing TE (2014) Climate change and desertification: can human adaptation help? Glob Environ Chang 28:119–130. https://doi.org/10.1016/j.gloenvcha.2014.06.003

U.S. Environmental Protection Agency (EPA) (2020) The status of air quality in the United States. https://www.epa.gov/air-quality

UNEP (2016) A snapshot of the world's water quality: towards a global assessment. United Nations Environment Programme. https://uneplive.unep.org/media/docs/assessments/unep_wwqa_report_web.pdf

United Nations Environment Programme (UNEP) (2023) Climate change. https://www.unep.org/explore-topics/climate-action

Vörösmarty CJ, McIntyre PB, Gessner MO (2010) Global threats to human water security and river biodiversity. Nature 467(7315):555–561. https://doi.org/10.1038/nature09440

Wada Y, Van Beek LPH, Van Kempen CM, Reckman JWTM, Vasak S, Bierkens MFP (2010) Global depletion of groundwater resources. Geophys Res Lett 37(20). https://doi.org/10.1029/2010GL044571

Walther GR, Post E, Convey P (2002) Ecological responses to recent climate change. Nature 416(6879):389–395. https://doi.org/10.1038/416389a

Weng Q (2012) Remote sensing of impervious surfaces in the urban areas: requirements, methods, and trends. Remote Sens Environ 117:34–49. https://doi.org/10.1016/j.rse.2011.02.030

Woolway RI, Kraemer BM, Lenters JD (2020) Global lake responses to climate change. Nat Rev Earth Environ 1(8):388–403. https://doi.org/10.1038/s43017-020-0067-5

Worm B, Barbier EB, Beaumont N (2009) Impacts of biodiversity loss on ocean ecosystem services. Science 314(5800):787–790. https://doi.org/10.1126/science.1170216

Wulder MA (2012) Landsat continuity: issues and opportunities. Remote Sens Environ 122:4–22. https://doi.org/10.1016/j.rse.2011.08.023

Zhou X, Zhuang D, Liu Y (2019) Impacts of land-use change and management on soil erosion in China. Environ Manag 63(5):766–774. https://doi.org/10.1007/s00267-019-01128-0

Source Identification Issues in Environmental Forensics

5

Benex K. Babu, Pundarikaksha Das, Meghna Bordoloi, and Harshita Shekhawat

Abstract

An environment is the external conditions, resources, stimuli etc., with which an organism can interact. The combination of analytical and environmental chemistry for the purpose of legal and justice establishment can be termed as environmental forensics. The field of environmental forensics has a gradual growth over the past research years and has expanded widely over the fields of oil spill identification and toxic metal contamination, even the advancement of techniques related to the identification of petroleum sources and contamination sites using radio isotopes. The dynamic techniques and newly found applications, such as As speciation, Pb isotopes, and PAHs molecular ratios, were useful to categorize the cluster groups on the basis of distinct pollution sources, lifting up the field into new opportunities and applications of forensics to sink with the environmental sciences. The chapter deals with the source identification issues related to environmental forensics, which in turn discusses the problems related to the field of environmental side via forensic support to enlist the sources that cause and the methods for the identification and analysis. Moreover identification, preservation, and analysis are the major stepping stones of any analytical research.

Keywords

Environmental forensics · Source identification · Pollutants · Litigation

B. K. Babu (✉) · P. Das · M. Bordoloi · H. Shekhawat
Department of Forensic Science, Royal School of Life Science, The Assam Royal Global University, Betkuchi, Guwahati, Assam, India

M. S. Sankhla et al. (eds.), *Introducing the Synergy Between Forensic Science and Environmental Analysis*, https://doi.org/10.1007/978-981-95-0199-1_5

5.1 Introduction

Environmental forensics deals in identifying pollution sources because of their significance in determining a range of factors such as the origin, pathways, and effects of pollution in different ecosystems. This field from cross-border is able to provide a basis for such issues as environmental restoration, pollution, and natural resource exploitation. The process of pollution source identification has become more sophisticated and effective than ever before. Detecting pollution sources has come a long way, from being concerned only in oil spillage to a wide variety of pollutants and different scenarios. The industry still remains sophisticated by way of continuously improving analytical techniques, accurate knowledge regarding chemical release points and, its historical context, transport paths and liability estimation. Environmental forensics is instrumental in the identification of the pollution sources and even the clean-up procedures by providing strong evidence, which will be needed legally and by the regulatory agencies in every country (Murphy and Morrison 2014).

Environmental forensics does not only equate to pollutants, it also entails the impact environmental issues leave, such as natural calamities like earthquakes, hurricanes, tornadoes, floods, tsunamis, and even volcanic eruptions. These activities create a lot of debris, which aggravates water pollution. (PAHO 1998). In addition, the destruction inflicted by radioactive contamination caused by the presence or deposit of radiological materials is of dire concern to human beings, weaker species, and vegetation. This scenario is further exacerbated by the fact that there is a lack of stability in radioactive particles as well as their damaging radiation (Rejah et al. 2018; Aarkrog 1962; Sharma and Bhattacharya 2017).

Surface water is one of the most common sinks for pollutants from various other human activities, including agriculture, which discharges metals, pesticides, pathogens, fertilizers, and salts into the water resources (USEPA 2005; Parris 2011). Urban activities further add these hazardous substances by means of construction refuse, raw sewage, and solid or liquid waste (Varol and Sen 2012). Industrial effluents mainly from factories, mines, and mills discharging wastes lead, mercury, cadmium, arsenic, and other chemicals into surface water bodies (Shahabudin and Musa 2018; Bhardwaj et al. 2017; YCoelho et al. 2015). Contaminants utilize soil and land use patterns to permeate groundwater systems and continue the pollution cycle (Sagasta et al. 2017). Furthermore, facilitating the movement of people and goods is a major contributor to both air and water pollution because rainfall washes away air pollutants, picks up road dust, and deposits them into water bodies (Akhtar et al. 2021).

Recognizing pollution sources is crucial for tracking the movement of contaminants like per- and polyfluoroalkyl substances (PFAS), which are harmful and persistent in nature. In their research, Charbonnet et al. (2021) outline the influence of environmental forensics in pollution origin detection and how it decreases the cost and time associated with remediation efforts. PAHs, heavy metals, and other toxic chemicals serve as potent pollutants that get stored in the human body and fully transform into severe health problems such as cancer, heart attack, and respiratory disorder (Landrigan et al. 2018; Atmadja and Bagtzoglou 2001). The risks are further intensified with the combination of tobacco-derived smoke along with plastic-derived contaminants such as bisphenol A and phthalates pollution (Ghorani-Azam et al. 2016; Manisalidis et al. 2020).

Emerging pollutants like hormones, pharmaceuticals, and biological contaminants pose challenges due to their recent introduction and lack of monitoring. These substances are accessible to the human body and can be consumed, inhaled, or absorbed through the skin, exhibiting the depth of pollution problems today (Grandjean and Landrigan 2006). Previous methods of identifying pollution sources depended on the rate of chemical emission and environmental fate of the pollutant, which was modeled using different metrics such as NOEC (No-Observed-Effect Concentration) and LC50 (lethal concentration for 50% of organisms) (Maki and Bishop 1985). The methodologies evolved to more elaborate frameworks, for instance, the maximum tolerable concentration (MTC) that also considered chemical characteristic of the substances including Log K_ow in an endeavor to determine potential impacts by contaminant within an aquatic system (Kooijman 1987; van Straalen and Denneman 1989; Aldenberg and Slob 1991; Wagner and Løkke 1991).

Modern developments in environmental forensics have advanced pollution source identification techniques further through an interdisciplinary approach by incorporating analytical chemistry, geochemistry, and statistical analysis (Baragaño et al. 2022). With increasing complexity from mixed sources of pollution and novel contaminants like microplastics and pharmaceuticals, cutting-edge tools such as machine learning and multivariate analysis are now employed for the identification of pollution patterns. GIS and remote sensing technologies improve the spatial and temporal resolution, hence enabling the delimitation of pollutant distribution and identification of hotspots (Hatipoğlu 2022). Integration of advanced methods has extended the scope of application beyond environmental research. Geochemical tracers and stable isotope studies have proven to be instrumental in linking pollution to industrial or agricultural practices in legal contexts, thereby giving strong evidence in court trials (Daw et al. 2022). Such findings point out the crucial role of identifying sources of pollution in environmental management, legal accountability, and sustainable utilization of natural resources. Environmental forensics continues to play an essential role in combating environmental degradation and safeguarding ecosystems by addressing complex pollution scenarios with precision and efficiency. The chapter addresses the challenges posed by complex pollution scenarios, mixed sources, and emerging contaminants while outlining future trends, including innovations in real-time monitoring, sensor technology, and advanced analytical methods. By the end of this chapter, readers will gain a holistic understanding of the methodologies, applications, and advancements in pollution source identification within the realm of environmental forensics.

5.2 Classification of Pollution Sources

Pollution sources can be classified according to their origin, types, and impacts on the environment. Broadly, they are categorized as natural sources and anthropogenic sources depending on the involvement of humans in the production of contaminants. It has a significant negative impact on the environment when released, and determination of this source is important for its proper litigation strategies. The classification of pollution according to the sources from which they are released is given in Fig. 5.1.

5.2.1 Natural Sources

Natural sources of pollution are the sources that arise naturally without any human interference. Rock weathering, soil erosion, earthquakes, landslides, floods, and hurricanes are some examples of natural sources of soil pollution. These events can lead to the elimination of the top fertile layer of the soil, causing various additions of chemicals, heavy metals, salts and radioactive substances.

5.2.1.1 Volcanic Emissions

Volcanic eruptions release various pollutants into the atmosphere, which can significantly impact air quality, ecosystems, and climate. Massive amounts of volatile materials are released into the atmosphere by active volcanoes as aerosols and gases

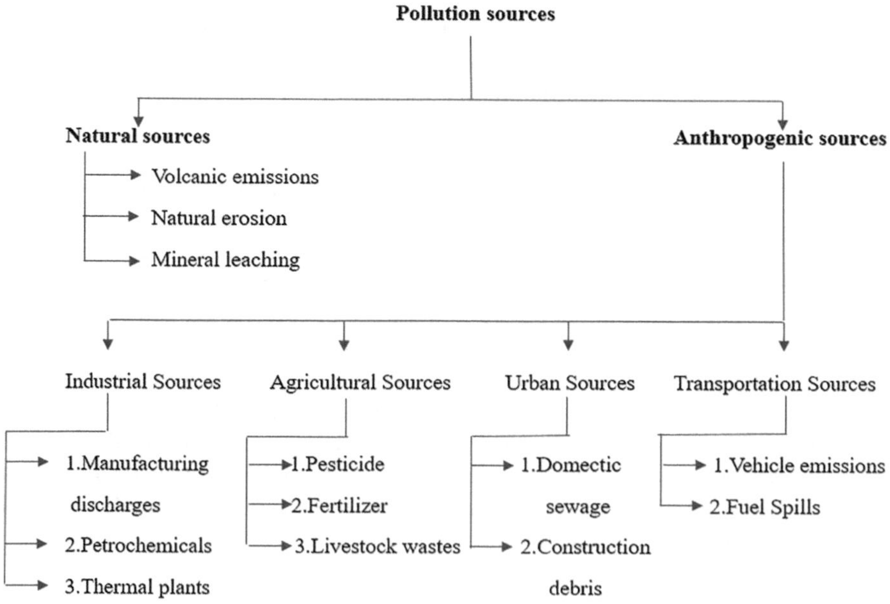

Fig. 5.1 Classification of pollution sources

(Oppenheimer et al. 2003; Mather et al. 2003). For many chemical species, volcanic emissions are an important natural source. Volcanoes release HCl, SO_2, and HF but also release large amounts of metalloids and poisonous volatile trace metals into the environment and oceans (Rubin 1997).

Aerosols from volcanoes, SO_2, combine with water vapor, and minute sulfuric acid droplets are produced, which results in a reflective layer in the atmosphere. Gases like sulfur dioxide (SO_2), hydrogen sulfide (H_2S), carbon dioxide (CO_2), and water vapors (H_2O) are released from volcanic emissions (Textor et al. 2004). Activities like acid rain and atmospheric cooling effects occur because sulfur dioxide reacts with water vapor to form sulfuric acid. Conversely, carbon dioxide leads to global warming and the greenhouse effect. Particulate matter remains suspended in the atmosphere for extended periods, reducing visibility and causing respiratory problems. Currently, active volcanism has a large impact on the planet's surface environment and our planet's atmosphere (Cole-Dai 2010).

5.2.1.2 Natural Erosion

The process by which soil, rock, and other surface materials are removed and transferred by natural forces, such as screaming wind, water, and ice, is known as natural erosion. Despite being a natural geologic process, erosion can pollute the environment by introducing pollutants, nutrients, and sediments into soil, water, and air ecosystems (Nearing et al. 2017).

One major significant environmental issue facing the globe today is land degradation. It is well-defined as any alteration to the land's initial state that will impact soil production, such as salinity, vegetation loss, and topsoil erosion (Hossini et al. 2022). Fertile land loss is not the only consequence of soil erosion. Fishes and other species have decreased because of the increased pollution and sedimentation in the rivers and streams (Kukuła and Bylak 2020).

The main causes of natural erosion are as follows:

Rainwater: It generally parts the soil up and washes it away.

Wind: Especially in arid areas, it lifts and moves soil particles.

Deforestation: The cut down of trees that holds the soil and prevent in erosion of the soil.

Heavy tillage: Uses large equipment to compact soil.

Overgrazing: Overgrazing results in poor ground cover, which makes it easier for wind and rain to erode and compact the land. This damages soil microorganisms and causes significant land erosion by limiting plant growth and water penetration (Balasubramanian 2017).

5.2.1.3 Mineral Leaching

It is the process by which water or other liquids dissolve minerals from rocks, soil, or other substrates and move them to a different area is known as mineral leaching. It can occur naturally or artificially. This process can be purposefully induced for industrial uses like metal extraction, or it can also happen naturally in ecosystems. Leaching happens in natural ecosystems when rainwater or groundwater percolates through the layers of soil and rock. As it travels, water dissolves soluble minerals and nutrients, allowing them to be transported to other sites. This mechanism contributes significantly to nitrogen cycling and soil formation (Kulkarni 2015).

5.2.2 Anthropogenic Sources

Anthropogenic source of pollution originates from human activities and are a major contributor to environmental degradation. These sources include domestic wastewater and industrial effluents, urban and agricultural runoff, atmospheric deposition, burning of fossil fuel, and antifouling paints from ships (primarily copper and tin) can increase above metal concentrations in the marine environment to higher than the background levels (Csuros and Csuros 2016).

5.2.2.1 Industrial Sources

One of the most significant contributors to anthropogenic sources is industrial sources. It releases a wide variety of pollutants into the air, soil, and water, which affects the ecosystem, the environment, and human health.

The pollution which was caused by human activity has a long history. Over 2000 years ago, early metal smelting was already causing serious environmental damage (Hong et al. 1996). Combustion of fossil fuels emitted toxins into the atmosphere, while untreated sewage carrying organic contaminants poisoned waterways (Halliday 1999; Davis et al. 2002).

Industrial pollution sources include:

5.2.2.2 Manufacturing Discharges

These are the pollutants released by the factories, namely nitrogen oxides, sulfur dioxide, carbon monoxide, and particulate matter are ejected into the environment (Odubo and Kosoe 2024). Water-containing effluents have heavy metals (e.g., mercury, lead, cadmium), dyes, and organic pollutants, which are released into the water bodies (Tripathy et al. 2025). Hazardous chemicals and metals from industrial waste frequently seep into the soil, decreasing its fertility and poisoning groundwater.

5.2.2.3 Petrochemical Industry

Oil spills that occur during extraction, shipping, or refinement damage coastal waters and oceans, negatively impacting marine ecosystems. Nearby water sources are contaminated by refineries which release oily wastewater (Lawan et al. 2023). Groundwater and soil can become contaminated when crude oil and petroleum products leak during transportation and storage which results in soil degradation (Elshafei and Mansour 2024).

5.2.2.4 Thermal Power Plants

One result of burning coal is fly ash, which pollutes air and soil. When hot water (thermal pollution) is released into rivers and lakes due to cooling procedures, the water temperature rises, and aquatic life is disturbed (Pędziwiatr et al. 2025). Heavy metals like arsenic and mercury can be found in contaminated effluent from power plants. Mercury is another harmful element released by coal-fired power plants. The combustion process emits mercury into the atmosphere, which can then settle into water bodies and terrestrial ecosystems, providing serious health concerns to humans and wildlife (Gingerich et al. 2018).

5.2.3 Agricultural Sources

Agricultural waste plays a very key role in the degradation of the water, air, and soil. Anthropogenic or man-made causes lead to water pollution (Maji et al. 2020). It includes human activities like the discharge of inorganic chemicals like pesticides, fertilizers, livestock wastes, and agricultural runoffs in the aquatic system, which impact the environment and human well-being in various ways (Pericherla et al. 2020a, 2020b). Untreated livestock effluent also contains chemicals that could harm both the environment and human health. The leaching of heavy metals and other contaminants from agricultural activities into water bodies exacerbates environmental degradation and endangers public health (Wu et al. 2024).

5.2.3.1 Pesticides

Pesticides (insecticides, weedicides, fungicides, herbicides, rodenticides, soil fumigants, etc.) are the chemicals that are used to control pests, weeds, and diseases in crops (Deolikar et al. 2021). They are basically used to improve crop yields, but in opposite, their improper use can result in pollution. Acute poisoning, hormonal abnormalities, malignancies, and neurological disorders can all result from human exposure to pesticide residues in food, drink, or the air (Kumar et al. 2012).

5.2.3.2 Fertilizers

Fertilizers (NPK fertilizers, urea, manures, etc.) are used to enhance soil fertility and crop production. However, the excessive application of the following can cause significant pollution. Excessive fertilizer application can cause considerable environmental degradation. Nitrates from fertilizers can seep into groundwater, and amounts greater than 10 mg/L are linked to methemoglobinemia, causing "blue baby syndrome" in newborns and causing water pollution (Majumdar 2003). Ammonia (NH_3) emissions from fertilizers contribute to particulate matter formation in the atmosphere (Pozzer et al. 2017).

5.2.3.3 Livestock Wastes

These represent the waste generated by livestock farming. Manure and urine are two examples of the waste products produced by livestock production that significantly contribute to pollution. During digestion, livestock release methane (CH_4), a strong greenhouse gas that fuels global warming (Gerber et al. 2013). Organic matter, nutrients, and pathogens are carried into the aquatic system by runoff from cattle operations, causing contamination and eutrophication (Pericherla et al. 2020a, 2020b).

5.2.4 Urban Sources

Human activity concentrates in urban areas, which frequently results in high levels of pollution from a range of sources. Transportation, industrial processes, human activity, and a dense population. These sources have a major effect on the urban environment and population health by contributing to noise and light pollution, as well as contamination of the air, water, and soil (Chen et al. 2023).

5.2.4.1 Domestic Sewage

Wastewater from household and communal activities, including cooking, washing, and bathing is referred to as domestic sewage. Contains high pathogen load, organic matter, detergents, nutrients like nitrogen and phosphorus, materials like heavy metals, and chemical substances like pesticides, which can result in eutrophication in water bodies (Save the Water 2022). Water pollution by domestic sewage takes place due to poor functioning and management of wastewater treatment plans. Domestic sewage and factory waste are the main causes of bacterial pollution. Sewage is often a mineral- and organic-rich diluted aqueous solution. Sewage, household trash, and food processing facilities account for around 75% of all water pollution (Harrison et al. 2006).

5.2.4.2 Industrial Waste

Approximately 2.8 billion gallons of industrial effluent, including those from offshore oil and gas operations and electric utilities, are thrown into the ocean every day (Bashir and Joshi 2019). The organic matter content, biological oxygen demand, and chemical oxidation demand of the relevant water body are raised by the significant quantities of heavy metals like lead, arsenic, mercury, pathogens, and nutrients that are released by wastewater treatment facilities, drainage pipes, and industries (Shah 2021).

5.2.5 Transportation Sources

The emissions and waste products produced by all forms of transportation, such as cars, trains, ships, and aeroplanes, are referred to as transportation sources of pollution. These sources are major contributors to pollution of water, air, soil, and noise, especially in cities with extensive transit systems.

5.2.5.1 Vehicle Emissions

The three primary types of vehicles are passenger vehicles and jeeps with four-stroke engines, two- and three-wheelers with small two-stroke gasoline engines, and large vehicles like buses, trucks, and commercial vehicles that run on four-stroke diesel engines. These sources of pollution have a range of negative effects on people, including eye discomfort, respiratory issues, circulatory issues, and asthma. Particulate matter, nitrogen oxides (NO_x), SO_x, PAN, ozone, and others are examples of outdoor air pollutants (Huo et al. 2009).

5.2.5.2 Fuel Spills

These are the spills that occur during fueling, transportation of fuels, or some accidents involving fuel tanks. It usually refers to accidental discharge or leakage of fuel, mainly petroleum-based products, for instance, diesel, gasoline, or oil, into the environment. These all have major health impacts, as well as environmental. The frequently encountered fuel spills include tank leaks, vehicular accidents, rupture of pipelines, overflows during fueling, and improper dumping of fuel (Michel and Fingas 2016).

5.3 Pollutant-Specific Source Identification

The emission of various pollutants from different sources due to oil, gas and industrial activities are common nowadays (Munshed et al. 2023). Identification of these pollution sources to implement effective mitigation strategies and pinpointing the origin of contaminant release are crucial in environmental litigation (Morrison 2000). Environmental forensic investigations are proposed mainly for source detection, identifying correlations between contaminants and their distribution patterns in a specific area with the objective of human health and remedial processes (Bari and Kindzierski 2018). The following are the specific pollutants present in environment that cause pollution at both the local and global scale.

5.3.1 Organic Pollutants

Organic pollutants are chemicals that are toxic, bioaccumulative, and obstinate in nature and can be obtained from different environmental matrices, like air, water, and soil (Megson et al. 2016). The Stockholm Convention on persistent organic pollutants (2001) aimed at the elimination and restriction of dozens of organic pollutants, including polychlorinated dibenzo-p-dioxins and dibenzofurans (PCDD and PCDF), hexachlorobenzene, and polychlorinated biphenyls (PCBs). The release of organic matter due to thermal processes along with the synthesis of chlorine as a byproduct poses risks to natural ecosystems and human life.

5.3.2 PAHs (Polycyclic Aromatic Hydrocarbons)

Polycyclic aromatic hydrocarbons (PAHs) are groups of organic chemical contaminants that are released due to the incomplete combustion of coal, oil, gas, and biomass. It contains multiple aromatic rings and tends to be solid, colorless to white or pale yellow-green in color. They are of significantly environmental concern due to their toxic, carcinogenic, and mutagenic risks (Lundstedt et al. 2007). They are listed individually as priority chemicals by US Environmental Protection Agency (2008): acenaphthene, acenaphthylene, and anthracene. Benzo(g,h,i)perylene, fluorene, phenanthrene, and pyrene. Source identification and distinguishing using high molecular weight (HMW), four- through six-ring PAHs ratio can be a productive preliminary approach to differentiate urban background sources of pyrogenic tar and related environments. In this method, the concept is based on the similarity in physicochemical characteristics (molecular weight, octanol-water partition coefficient (K_{OW}), or aqueous solubility) and biological factors (photooxidation, dissolution, evaporation, microbial degradation, emulsification) that retain its concentration in source sample even after undergoing severe weathering conditions (Costa et al. 2004). The weathering processes combine together to reduce the number of hydrocarbons in oil spills and alter the chemical composition, which makes it complicated to trace the residual spilled oil. Analysis of gas chromatographic fingerprints using chromatogram paved a useful method in detecting urban sites. The modified USEPA method 8270 C has proven productive approach for characterization and fingerprinting petroleum residues Costa and Sauer 2005). In addition to PAH, polycyclic aromatic sulfur heterocycle and alkylated homologues (C_1 to C_4) are also vulnerable and hazardous to environment. For example, the alkylated homolog of anthracene can be toxic in nature depending upon the amount of exposure level in human being (Fig. 5.2). The investigators have used the diagnostic ratios of PAHs, both individual and homologue specific for the past 40 years for distribution pattern recognition and the site litigation (Antle et al. 2013). Wang et al. (1999) stated in their review that the alkylated PAH homologues are the cornerstone for the chemical fingerprinting and recognition of the distribution of petroleum hydrocarbons in oil spill identification.

The major step in the investigation is to differentiate the pyrogenic and petrogenic PAHs using the temperature at which it is formed. The degree of alkylation in PAH is inversely proportional to formation temperature. In a single homologue,

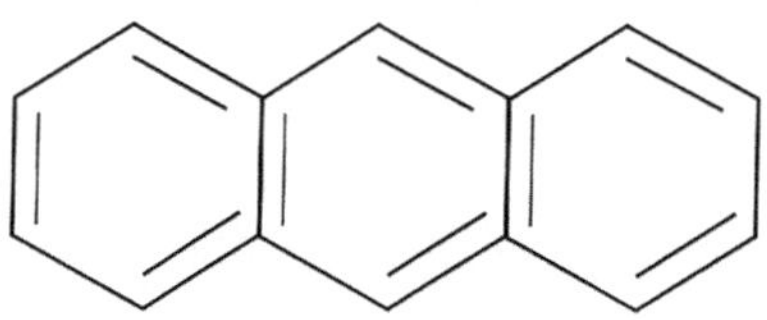

Anthracene (C_4H_{10})

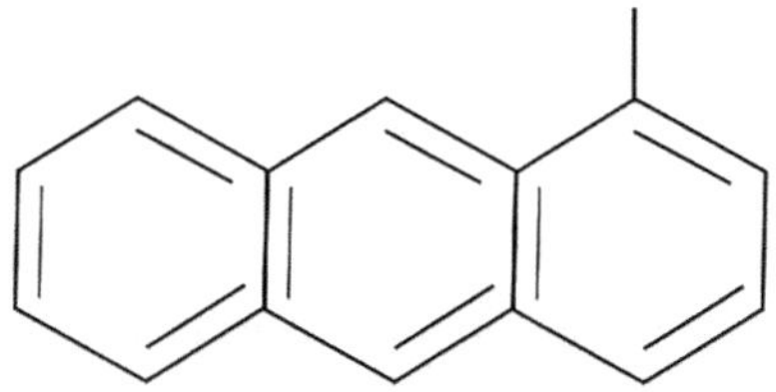

1-methyl anthracene ($C_{15}H_{12}$)

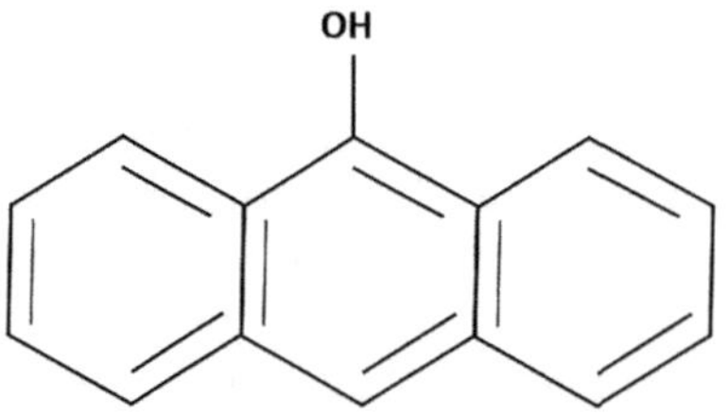

9-hydroxy anthracene ($C_{14}H_{10}O$)

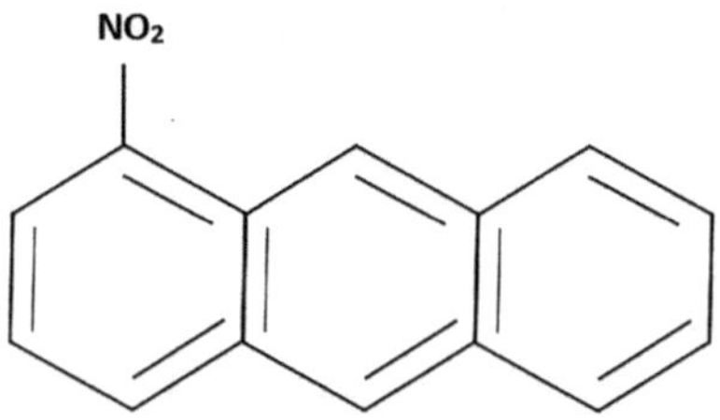

2-nitro anthracene ($C_{14}H_9NO_2$)

Fig. 5.2 Molecular structure of anthracene with different homologues (Costa and Sauer 2005)

there will be the formation of multiple alkylated PAHs with the difference in temperature. The profile comparison based on temperature can be used for finding the origin. Most investigations start its initial phase with existing data and later form a diagnostic source ratio (concentration of one PAH to the others). Varieties of forensic techniques like geochemical biomarkers, PAH compositional profiles, or multivariate analysis (e.g., principal component analysis, hierarchical cluster analysis, etc.) were adopted with consideration of weathering process in case of mixed sources (Shields et al. 2015).

5.3.3 PCBs (Polychlorinated Biphenyls)

Polychlorinated biphenyls (PCBs) are synthetic organic chemicals, consisting of two benzene rings attached with varying chlorine atoms, in different positions of carbon atoms. There are 209 different chemical PCB congeners with variable chemical and physical attributes due to the structural variations. They are resistant to chemical and biological degradation due to chemical stability, low volatility, and non-polarity. They are able to accumulate in the fatty tissues of fish and can lead to biomagnification during food chain process. Thermoplastic, adhesiveness, and low vaporization contributed to the use in paint industry, and their electric insulating properties, resistivity, and high dielectric constant attributed to the use in electrical industry (Johnson et al. 1964). The main aim of a forensic investigation is to determine and locate the historic source of the contaminant present in the polluted area and to take remediation decisions based on PCB concentration. Accurate identification is critical in forensic assessment in order to point out the responsible parties behind the act. Initial site visit and sample preparation are conducted for calculating PAH concentration after site assessment. Sampling must be done according to the objective of sample design. The analysis is conducted using EPA GC/MS methods, and a report is formulated for the total concentration in order to plot it on a map for distribution patterns and congeners ratio. Interpretation of data requires sophisticated numeric algorithms, and the forensic analyst must provide paradigm and documented rationale for the clear and unbiased decision made (Fig. 5.3). After the extensive site studies, remedial measures, site cleanup, and disposal process should be followed. From the evidence collected, the experts should formulate their opinions about who was responsible for the contamination. These are often interpreted with the purchase, usage, disposal, and discharge records of the disputed parties (Erickson 2020).

5.3.4 Dioxins and Furans

Dioxins and furans are known for their triple ring structure with two benzene rings attached by either one or two oxygen atoms. Dioxins are commonly known as polychlorinated dibenzo-p-dioxins (PCDDs) and furans as polychlorinated dibenzofurans (PCDFs). PCDD/Fs are a complex mixture with more stability and low

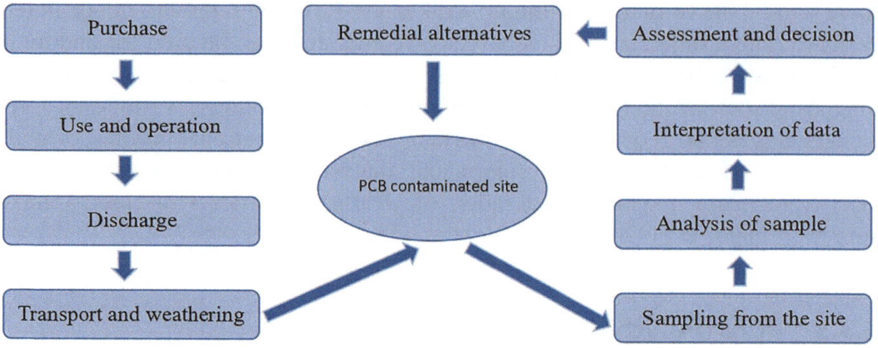

Fig. 5.3 Flowchart showing the steps of PCB site contamination and its forensic investigation (Erickson 2020)

water solubility and are highly lipophilic and have the same chemical and physical properties. In these molecules, eight possible positions were substituted by chlorine atoms, and PCDD/Fs with 2-, 3-, 7-, and 8-ring positioned chlorine congeners are considered to be more toxic in nature. They are called "the dirty 17" because they are 17 in number. For example, there are 22 isomers present in tetrachloro dibenzo-p-dioxin (TCDD) homolog group, but only 2,3,7,8-TCDD isomer is considered as toxic (Fig. 5.4).

After the combustion process, the light PCDD/Fs congeners released will be in gaseous phase and travel further from the original source. Due to high chlorination, the heavier congeners remain in the particle phase. The profiling of the PCDD/Fs patterns will be accurate for the heavier congeners because they remain in the particle phase after settling and transportation. The loss of lower chlorinated congeners should also be considered during investigation for a more predictable result. There are chances that the lighter congeners can move toward one site to another and mix with the background profile of the other site. This causes a crucial transformation of the PCDD/Fs fingerprinting patterns at the site and variability in data.

The standard analytical method used for the detection of these congeners as per US EPA includes high-resolution gas chromatography/high-resolution mass spectrometry (HRGC/HRMS) using isotope dilution (Method 1613B; US EPA 1994, Method 8290A; US EPA 2007 and Method 23; US EPA 1995). Among 201 possible congeners, 136 have four or more chlorine atoms. These 136 congeners have estimated concentration and selective ion current profile (SICP), which can be used for fingerprinting. Isomer profiling of the congeners is a recommended method for the analysis and comparison, if data is available. While comparing, the data should be from the same site and laboratory for more accurate results. Visual comparison of the major peaks will be a useful method for estimation. Immunoassays (EPA Method 4025; US EPA 2002) can also be used for the estimation of these congeners.

For the forensic investigation of the congeners, it is recommended to use multiple tools. Among them, chemical fingerprinting is a well-established technique for the differentiation and detection of PCDD/Fs. Chemical fingerprinting can include

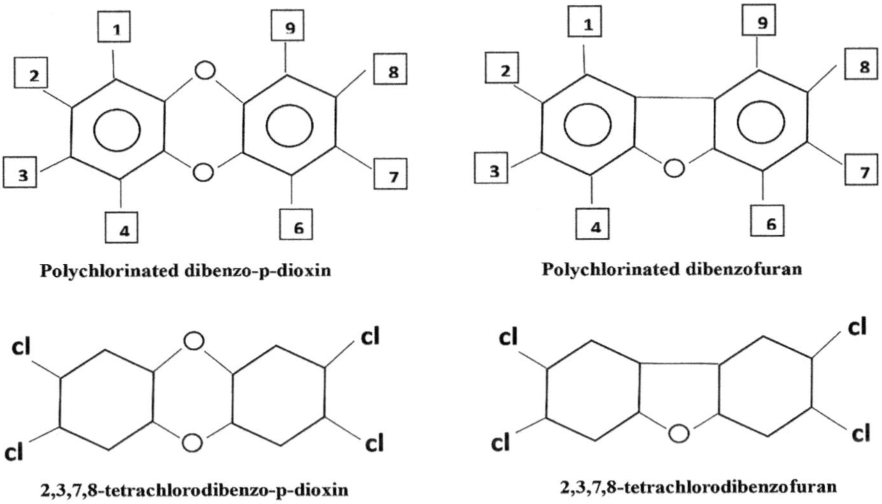

Fig. 5.4 Molecular structure and toxic isomers of PCDD/Fs congeners (Shields et al. 2015)

comparison of peaks of different individual congeners with two different compositions to complex congeners with less variability in their selective ion current profiles (SICPs). Multivariate statistical analysis, like principal component analysis (PCA) or hierarchical cluster analysis (HCA), can be a convenient tool for complex data (Shields et al. 2015).

5.3.5 PFAS (Per- and Polyfluoroalkyl Substances)

PFAS (per- and polyfluoroalkyl substances) are fluorinated anthropogenic compounds having hydrophilic and hydrophobic functionalities. The general structure is $F(CF_2)_n$-R, where n stands for number of carbons in hydrophobic alkyl chain and R stands for hydrophilic functional groups, such as neutral (alcohol), anionic (carboxylate), or cationic (ammonia). This hydrophilic end is polar in nature with the absence of a fluorine atom (Jian et al. 2018). The fluorinated contaminants are mainly used in household products, cosmetics, paints, adhesives, surfactants, textiles, pesticides, and paper (Lindstrom et al. 2011). The two methods used for the production are electrochemical fluorination (ECF) and telomerization. ECF is a method in which an electric current is passed through a solution containing substrate and hydrofluoric acid, resulting in the production of fluorinated and partially fluorinated compounds. Telomerization is a chemical chain reaction in which a fluorinated monomer reacts with a telogen (a compound that controls the chain length) to form a telomer or fluorinated oligomers (short-chain polymers) (Langberg et al. 2021). The commercial production using ECF resulted in biotransformation, forming perfluoro octane sulfonamide (PFOS), which is listed as a persistent organic pollutant (POP) in the Stockholm Convention 2009 by the United Nations

Environment Programme (UNEP 2019). Thus, PFAS produced by telomerization became a replacement method. An alternative compound for PFOS is 6:2 fluorotelomer sulfonate (FTS) and the perfluoro ether sulfonate (F-53B) (Ruan et al. 2015). Different chemical structures of PFAS are shown in Fig. 5.5.

The forensic investigation lies in the comparison of previous data from the contaminated site with the data collected afterward. The analytical method used for the estimation includes mainly LC-MS/MS and TOFMS. For example, Coggan et al. (2019) investigated the PFAS in 19 Australian wastewater treatment plants using LC-MS/MS with isotope dilution method. In this analysis, distribution coefficient, linear mixed-effect analysis, and PCA were used for the data processing. Nontargeted approach for the detection of these compounds using high-resolution mass spectrometry (HSMS) based on negative mass defect become an advanced technique for the uncharacterized PFAS. Negative mass defect means the total mass difference in the nucleus and its individual nucleons (McCord and Strynar 2019). Once the molecular features are extracted using instruments, a mass profile is formed for the comparison of upstream and downstream samples in the case of water bodies. If the downstream sample features are absent in upstream samples, then we can conclude that a point source is present between these sampling locations. Identification of carboxylic acid and sulfonic acid groups can be achieved by TOFMS using total ion

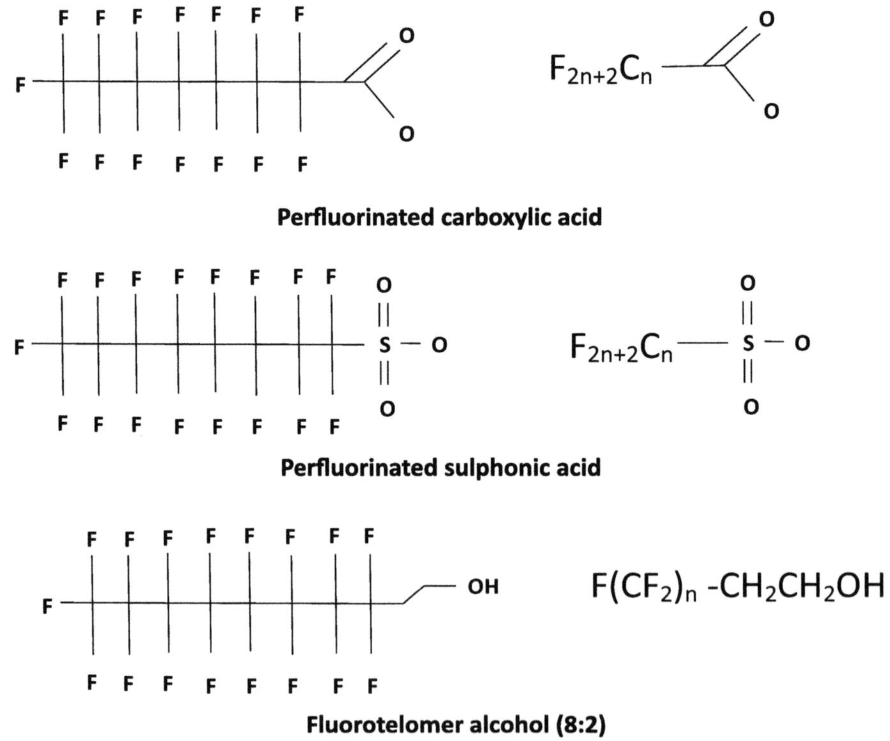

Fig. 5.5 Chemical structure of different groups of PFAS (Shields et al. 2015)

chromatogram comparison (Strynar et al. 2015). Thus, various forensic techniques are considered for the identification and estimation.

5.4 Inorganic Pollutants

Inorganic pollutants are non-biological chemical substances that are formed in the environment due to human activities such as mining, industrial discharge, smelting, battery production, etc. These pollutants are nonbiodegradable, persistent and toxic in nature, and accumulate in air, water, and soil, causing health issues in both the environment and living systems. Therefore, site-specific identification of these contaminants is necessary in order to detect the individual or organization behind this act (Hu 2002). The main inorganic pollutants are heavy metals, like arsenic, cadmium, lead, mercury, nitrites, and fluorides, and are resistant to degradation. These are produced from the sources of municipal solid waste, fossil fuel combustion, and industrial wastes (Halim et al. 2003).

5.4.1 Heavy Metals

Humans started to use different types of metals during the development of most modern societies, but these applications come at an environmental price, leaving behind the possible consequences of human health and environmental pollution. The production and use of heavy metals are still continuing but are eventually leading toward the health issues of human, aquatic, and terrestrial ecosystems (Wilson 1996). Identification of the contaminant sites and reinforcing the remedial activities at a specific site is important in environmental science. The major elements that can be encountered during these processes are arsenic (As), mercury (Hg), cadmium (Cd), lead (Pb), zinc (Zn), selenium (Se), chromium (Cr), and manganese (Mn) (Borah et al. 2020). Arsenic enters the environment during fossil fuel combustion and washing out of coal mining sites, leaching into groundwater (Herawati et al. 2000). However, it is essential for human beings, but an excess amount can lead to serious health concerns. Thus, it can reach the contaminant sites through different pollutant agents.

Forensic methods that are used for the investigation of contamination include sophisticated techniques. Among them, the most common one is chemical speciation. It is the process of identifying and quantifying the different forms of chemical elements present within the sample. For example, mercury will be having the elemental form Hg^0 and Hg^{2+} or can be encounter in its organic composition as methylmercury. This variations within the samples gives more information about different contaminant sites and their identification. For this analysis, HPLC coupled with ICP-MS is used (Zuo et al. 2018; Xing et al. 2019). Grain size analysis can also be done for the differentiation of the heavy metals (Fernández et al. 2020).

An environmental forensic investigation starts when there is a suspected contaminant site tracked, and any party, either private or government, is ready to fund the investigation. It is always recommended that the benefits of investigation should always be higher than the total cost required for the detailed study. The first step in this investigation is the site visit and preliminary survey of the contaminants using portable devices. Once the pollution is spotted, a background study is conducted by collecting the historical details of the site, location of the site, and details of the industries present nearby. After the study, proper sampling and analysis are done using various instruments and statistics.

5.4.2 Radionuclides

Radioactive substances (radioisotopes/radionuclides) are atoms that have an unstable nucleus and radiate their energy to attain a more stable state (Atwood 2013). The radioisotopes or radionuclides radiate energy in the form of alpha, beta, and gamma particles, which make them in more stable positions. This process is called radioactive decay (Lehto and Hou 2011). The contamination occurred in the environment due to radioactive wastes are extremely concerning, and identifying and tracing the impact on the environment is necessary. The production of nuclear weapons and electric power from nuclear fuel and the testing of nuclear weapons are the main sources of nuclear contamination (Hu et al. 2010). These contaminants can reach the groundwater and leach out to the surrounding areas, leading to severe environmental pollution. The radionuclides that are released to the environment from nuclear reactors are only about 2.5%, whereas the majority of them are deposited in the ocean, including ^{235}U, ^{238}U, ^{239}Pu, and ^{137}Cs (Choppin 2007).

Actinides are the radioactive elements in the periodic table starting from atomic number 89 (actinium) to 103 (lawrencium) (Freeman and Koelling 1974). The behavior of these actinides is different when their oxidative state changes. For example, in aerobic waters, PuO_2^+ is the dominant species, while in strongly acidic waters, trivalent plutonium is dominant (Fukai et al. 1987). Thus, the geochemical tracing of radionuclides is comparatively complicated. The actinides released to the environment as a byproduct of nuclear industries are mainly through the improper disposal of nuclear waste, accidental release during transportation, direct release from the enrichment process to groundwater and surface water, or during nuclear weapon testing (Runde and Neu 2010).

During forensic investigation, the source identification can be demonstrated using the isotopic ratios (Schneider et al. 2017). Yamamoto et al. (2012) demonstrated the contamination source identification of the Fukushima Daiichi Power Plant due to the nuclear reactor from the results of elevated plutonium isotopic ratios obtained from the soil samples. The determination of different isotopes from the site will be a useful method for the identification of source. This can be achieved by using alpha spectrometry, gamma spectrometry, or beta counting (Ješkovský et al. 2019). Liquid scintillation analysis (LSA) can be used as a rapid screening technique for the detection and quantification of gamma and high-energy beta nuclides with approximately 100% detection efficiency. Its complication due to the quenching effect of peaks can be rectified using multiple quenching correction approaches (Croudace et al. 2016). The inorganic mass spectrometry technique for isotopic ratio determination becomes more successful with the introduction of inductively coupled plasma (ICP) combined quadrupole mass analyzer (Houk et al. 1980). Its high accuracy and precision, along with simple design from ionizing the sample to detection using multiple types of detectors, enable its suitability to couple with different separation techniques, like gas chromatography, liquid chromatography, solid phase microextraction, and others (Becker 2012). These techniques suggest that the radioisotopic ratio is a strong forensic fingerprinting tool in differentiating contaminated sites of radioactive sources.

5.5 Biological Pollutants

The biological invasion of species into a new geographical area, which starts to colonize to form a new generation of non-native species with self-sustainability, can be termed as an introduced species (Boudouresque and Verlaque 2002). These species can harm the ecosystem, human health, or both. Some major pollutants are microorganisms, pathogens, algal blooms, biological byproducts, and others (Elliott 2003). The introduction of invasive species to a new ecosystem can alter the natural habitat and cause destruction to the quality of life, economy, and human health when they are survived and successfully breed (Olenin et al. 2011). The proper remedial measures and border control strategies are required for its prevention and mitigation. The source identification from the polluted site to determine the origin of these contaminants is important for the prevention of future invasion and rapid litigation.

5.5.1 Pathogens

Pathogen-host interaction is the basis for the environmental issues in the ecosystem. A pathogen can be defined as any microbe that causes harm or destruction to the host. Here, the host can be either a living or nonliving thing. Any kind of bacteria, virus, fungi, or parasite can be a pathogen. Detection of these microbes and preventing its spread will be the first mitigation step in order to control the growth (Casadevall and Pirofski 2002). The basic non-biological pathogen reduction processes in sewage sludge are air and heat drying, lime sanitation, heat treatment, and gamma radiation. Aerobic and anaerobic digestion, sludge lagooning, and composting are the biological stabilization processes (Dumontet et al. 1999). One of the pathogens that is considered invasive in nature is bacterial agents (Salmonella typhi, *Escherichia coli* (*E. coli*)), which can cause health issues in humans.

Invasive microorganisms can be carried from one ecosystem to another ecosystem or from one region to another region by a host. Once the host interacts with the pathogen from the external source, then they will become the carrier. They will move from one ecosystem to another through the ecotone (transition zone between two ecosystems) without detection, and the pathogen tends to spread to other hosts.

Effective mitigation strategies and early detection are crucial for its control and spread (Fig. 5.6). The tracking of the pathway of pathogen and listing the number of individuals involved in the pathway is significant to identify the source origin. The advancements in molecular and pathogen biology serve a lot of scientific methods and techniques to ascertain numerous diseases, but these advancements can be leveraged against humanity in the means of biological weapons that can cause serious harm to living things. A forensic investigation is needed when a case occurred with a pathogen as bioweapon. The source of release of these bioweapons and the motive behind such crime will be investigated in an epidemiologic investigation (Morse and Khan 2005). Since the methods adopted by culprits to deliver an infectious pathogen are simple and direct, such as anthrax letter attack (Stern and Schouten 2017), it can be encountered frequently in forensic investigations.

Methods used for the detection of microbial concentration of these pathogens from different contaminated sites are crucial for reliable results in forensic investigations. Impedance microbiology is one of the oldest rapid microbial growth detection techniques that can be used to enumerate the concentration by measuring the changes in electrical signals. Impedance is the resistance of a medium to pass electricity through it (Grossi et al. 2017). Over the last few years, formulating genetic profiles using DNA/RNA typing has become a promising technique for the identification and characterization of pathogens. But the question is whether DNA/RNA typing is able to distinguish the variants of a pathogen, since microorganisms have the ability to clone their genetic material. The limited data regarding epidemiological characteristics of microbes makes the investigation difficult to compare with known data samples (Budowle et al. 2005). The multilocus VNTR analysis (MLVA) can also be used for the detection of different strains of pathogens. In 1993, the sarin nerve gas attack in the Tokyo subway to spread anthrax was analyzed by MLVA and found that the strains are harmless from the data comparison with the existing

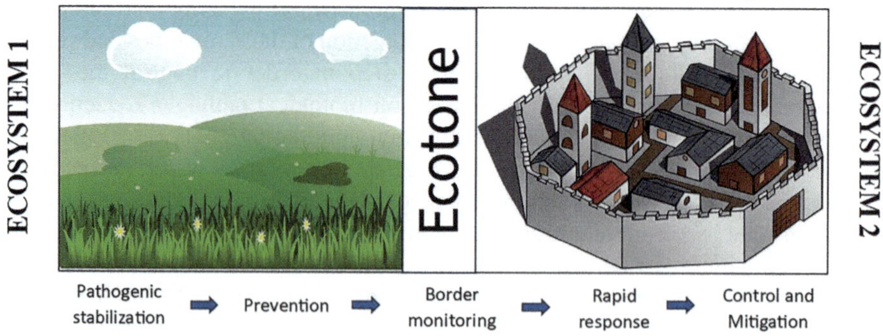

Fig. 5.6 Diagrammatic representation of steps involved in an interaction of a pathogen across the ecotone and mitigation strategies in each interaction

database. The profiling of strains from the 2001 attack in the USA using *Bacillus anthracis* spores revealed that the variants are Ames strains and are closely associated with laboratories. Thus, the source identification of laboratories were started and identified its origin (Keim et al. 2001). Genomic sequencing and bioinformatic tools can reform microbial forensics for the detection the pathogen contamination in different environments.

5.5.2 Algal Blooms

The accumulation of phytoplankton, algae, and heterotrophic protists is considerably increasing worldwide, and the concerns regarding the environmental harm to aquatic and terrestrial growth are consequential (Song et al. 2009). The aggregation of these harmful algal blooms (HABs) can decolorize the waterbodies and deplete the oxygen availability in water, which in turn affects the organisms living there. The production of hazardous toxins, which come along with the food chain, can be consumed by human beings, leading to serious health issues and death. The major toxins encountered include saxitoxins (paralytic shellfish poisons), domoic acid (amnesic shellfish poisons), okadaic acid and dinophysis toxins (diarrhetic shellfish poisons), brevetoxins (neurotoxic shellfish poisons), ciguatoxins (ciguatera fish poisoning), cyanotoxins (cyanobacteria toxin poisoning), and azaspiracid (azaspiracid shellfish poisoning) (Sellner et al. 2003). The excess anthropogenic activities and the impact of climate change synergize eutrophication, leading to the contamination of the marine and freshwater environment. Human activities cause the nutrient input in water bodies and sediment deposition, which amplifies the algal growth. The adverse results can be fish mortality, seafood contamination, and human health issues. Proper investigation should be required for the overgrowth of these algae, and rapid mitigation strategies need to be adopted for the prevention and control of harmful algal blooms (Sanseverino et al. 2016).

Sediment fingerprinting is one of the methods adopted for the detection of algal contamination in the case of eutrophication. The investigation will be focusing on the historical nutrient deposition and the release of nitrogen and phosphorus from agricultural runoff into the waterbodies, which fasten the algal growth. The responsibility will be equally given among the parties who released these pollutants. Radiometric dating can be a good technique to investigate the presence of any radioactive substances in sediments. X-ray fluorescence analyzer can be used for the analysis of major elements present in it. In order to identify specific algae, present in the contamination site, DNA/RNA profiling will be a successful tool. It can reveal the specific strain present in the particular algae. The toxicity caused by any algal toxins will be determined by HPLC, ELISA, or mass spectrometry (Wuilloud et al. 2006; Gurbuz et al. 2012).

5.6 Techniques for Pollution Source Identification

In forensic investigation, different techniques were used for the detection of pollutant source in order to identify the specific source in which it is released to the environment. These techniques can be categorized as chemical and physical analysis, spectroscopic analysis, and elemental analysis. The data that are generated from the samples collected from the contaminant site give an information of overall distribution and impact of pollutants.

5.6.1 Chemical and Physical Analysis

5.6.1.1 Chromatography

Chromatography operates on the idea that molecules in a mixture, when applied to a solid or fluid stationary phase, are separated from one another as they move with the assistance of a mobile phase. The elements influencing this separation process encompass molecular features pertinent to adsorption (liquid-solid), partition (liquid-solid), and variations in molecular weights or affinities. Due to these disparities, certain components of the mixture remain in the stationary phase for an extended duration, moving slowly inside the chromatographic system, whereas others transition swiftly into the mobile phase, exiting the system more rapidly.

This strategy comprises three components that constitute the foundation of the chromatography process.

1. Stationary Phase: This phase consistently consists of a "solid" phase or "a layer of liquid adsorbed onto the surface of a solid support."
2. Mobile Phase: This phase consists exclusively of a liquid or gaseous component.
3. Separated molecules.

The interaction among the stationary phase, mobile phase, and the chemicals in the mixture is the fundamental factor influencing the separation of molecules. Partition-based chromatography technologies are highly successful for the separation and identification of tiny molecules, such as amino acids, carbohydrates, and fatty acids. Affinity chromatographs, such as ion-exchange chromatography, are more efficient in the separation of macromolecules like nucleic acids and proteins. Paper chromatography is employed for protein separation and studies on protein synthesis; gas-liquid chromatography is utilized for the separation of alcohols, esters, lipids, and amino groups, as well as for observing enzymatic interactions, while molecular-sieve chromatography is specifically used for determining the molecular weights of proteins (Miller 2005).

The stationary phase in chromatography is either a solid phase or a liquid phase adhered to the surface of a solid phase. The mobile phase traversing the stationary phase is either a gas or a liquid. If the mobile phase is liquid, it is referred to as liquid chromatography (LC); if it is gas, it is designated as gas chromatography (GC). Gas chromatography is utilized for gases, volatile liquid mixtures, and solid substances. Liquid chromatography is particularly employed for thermally unstable and non-volatile materials (Giddings 2017).

Gas chromatography-Mass Spectrometry: Hyphenated gas chromatography (GC) and mass spectrometry (MS) are the two distinct analytical methods that make up gas chromatography mass spectrometry (GC-MS). Gas chromatography and mass spectrometry are often used in tandem as part of an analytical apparatus, with the former connected to the latter by a heated transfer line. Nevertheless, there are specialized devices that house the entire GC-MS system in a single unit, which is often a small or portable device.

Gas chromatography (GC) is a separation method for isolating and detecting individual chemical components in a sample mixture. This allows one to check whether a certain component is present and, if so, in what concentration. The information provided by GC detectors is typically two-dimensional, consisting of the analytical column retention time and the detector response; however, this is not always the case. The retention times of the peaks in a sample are compared to those in standards of recognized substances that have been evaluated using the same method for identification. The identification of unidentified, however, is beyond the scope of GC alone; here is when hyphenation to MS becomes useful. You have the option to utilize MS as the only detector or to split the column effluent between MS and GC detectors (Karasek and Clement 2012).

Molecular weight, elemental composition, and chemical structure may all be better understood with the use of mass spectrometry (MS), an analytical method that examines the mass-to-charge ratio (m/z) of the charged particles. The three-dimensional data produced by a gas chromatography-mass spectrometer (GC-MS) may be utilized for both quantitative and qualitative analysis, in addition to mass spectra for chemical identification or confirmation (in the case of unknown substances).

To identify analyte molecules, the GC separates the sample mixture before eluting it into the MS. The carrier gas carries them as it constantly circulates through the gas chromatograph (GC) and then into the mass spectrometer (MS), where it is removed using the vacuum system.

The sample is either manually or automatically put into the gas chromatography (GC) system or introduced to the carrier gas through the GC inlet. The heated GC intake turns liquid samples into vapor, which is then carried to the analytical column. The particulate matter, or "analytes," of the sample is distinguished from one another by virtue of their dissimilarities in the partitioning of the mobile phase (the gas carried by the carrier) and the stationary phase, which is either a liquid housed inside the column or, in the case of more volatile gases, a solid. A liquid stationary phase is typically used in gas chromatography-mass spectrometry examinations in conjunction with a short (10–30 m length) and narrow (0.1–0.25 mm internal diameter) column. The neutral molecules elute into the mass spectrometer through a heated transfer line after separation, which is not necessary for GC-MS analysis unless the analytes are isomers. Neutral molecules are initially ionized within the mass spectrometer, usually using electron ionization (EI). Electron injection (EI) creates a radical cation molecular ion by accelerating an electron from a filament to

70 electron volts (eV), which then knocks off an electron from the molecule. Excess energy can be lost by fragmentation, and an unstable molecule ion can be produced by this high-energy ionization. Loss of a neutral or radical molecule and subsequent chemical rearrangements are possible outcomes of bond breakage(s). A multitude of ions of varying masses are produced as a result of all this, with the molecule ion being the heaviest among them and fragment ions of lesser masses resulting from factors, such as what the chemical formula is, where the analyte's bonds have broken, and which portion of the molecule has managed to keep its charge. After that, the mass analyzer sorts the ions by mass/zoometric ratio, which is the fifth stage (Stashenko and Martínez 2014).

The huge variations in mass resolution (and, by extension, instrument expense) are a result of the wide variety of mass analyzer types. How well a mass analyzer can distinguish between ions with minute m/z variations is called mass resolution. High mass resolution (HRMS) instruments would isolate masses to four or five decimal places, in contrast to unit mass resolution instruments that can only isolate nominal masses or masses to one decimal place.

The quadrupole, the most popular unit mass instrument, is a scanning device that changes the voltage, such that only ions with a specific mass-to-charge ratio may follow a stable path between the four poles and end up in the ion detector. There are two main applications for quadrupole instruments:

When working with higher concentration analytes, developing methods, or just trying to figure out what an unknown is, full scan mode—which involves acquiring all ions within a mass range—is the way to go.

Useful for trace analysis because of increased sensitivity, but just of target analytes, is the selected ion monitoring (SIM) mode, which involves acquiring only the ions that represent the target molecule (Kopka 2006).

Ion traps are similar to two-dimensional scanning instruments, but they're three-dimensional and expel the ions in a sequential fashion according to their mass, so they may be detected. Ions are separated in time-of-flight (ToF) mass analyzers by measuring how long it takes for them to travel down the flight tube and reach the ion detector. People with lower masses will arrive first because of their greater velocity, while those with higher masses will arrive later because of their lower velocity, even though their kinetic energy is the same. Fast time-of-flight (ToF) instruments may acquire up to 1000 spectra/second, making them unit mass resolution, while slower acquisition rates are characteristic of high-resolution mass spectrometry (HRMS) ToFs. The chemical formula may be determined with the use of HRMS, although two-dimensional gas chromatography (GC × GC) applications benefit from high acquisition rates and peak widths as short as 30 ms. Hence, the application dictates the choice of ToF among the available options, which vary in speed and mass resolution; nonetheless, the GC peak width ought to correspond to the MS's acquisition rate capabilities.

Among the hyphenated HRMS devices used in GC is the magnetic sector mass analyzer, which uses electric and magnetic fields to divert ion paths for separation. In isotope ratio studies, magnetic sector GC-MS devices seem to be more prevalent.

The ion motion frequency in the HRMS orbitrap is m/z dependent and is determined by their orbit around a central spindle. The ions are then sent to the ion detector, where an electron multiplier (in low-resolution MS) or a multichannel plate (in HRMS instruments) amplifies the signal after the mass analyzer has sorted them according to their m/z. A chromatogram and mass spectra are generated for every data point by recording the signal using computer acquisition software.

Three-dimensionality is found in GC-MS data. The retention period, measured from the time of sample injection till the completion of the GC run, is shown on the x-axis. This is another way of looking at the scan number, indicating the total amount of data points that MS has collected throughout the run. The ion detector's reaction or intensity is shown on the y-axis. Over the whole mass range, the ions' m/z ratios form the z-axis.

A total ion chromatogram (TIC) is more similar to a chromatogram produced by a GC detector; it is a two-dimensional chromatogram that is created by adding up the distributions of all the ions at a single data point and graphing it against the retention time (RT)/scan number. Total ion chromatogram data points, on the other hand, are really mass spectra, and the program often allows you to examine each one in its own window (Garcia and Barbas 2011).

As the saturated hydrocarbon chain length increases, the molecular ion abundance decreases until the mass spectra show no molecular ions at all. On the other hand, because the extra energy may be more readily ingested, there is less fragmentation of unsaturated molecular ions, particularly those containing conjugated double bonds, such as aromatic molecules. The mass spectra serve as a unique identifier for each molecule and may be compared to databases of spectra acquired with the same ionization process and voltage. Electronic ion spectra generated at 70 eV are the most popular in commercial libraries. By analyzing the ion masses, isotope concentrations, and fragment ion losses, one may deduce the chemical formula and structure from the mass spectrum.

The retention time and area are typically employed for quantification and target analyte identification in gas chromatography. The peaks must exhibit strong separation via chromatography with baseline resolution in order for the quantification to be performed accurately. The mass spectrum offers an extra way to validate the target analyte with GC-MS, whether you use the complete spectrum or only a few ions' existence and relative ratios. Since there is less chance of interference from co-eluting peaks when using the area of a single, unique ion for quantitation using GC-MS data rather than the area under the TIC peak, this method is often used. The ability to spectrally resolve the peaks and perform baseline-to-baseline integration is sufficient for accurate quantification in the absence of chromatographic baseline resolution, provided that a unique ion is chosen that isn't present in the coeluting peaks.

If there is a need to analyze and understand the structure of multicomponent environmental mixtures containing various sulfur species in nanogram or picogram amounts, gas chromatography (GC), particularly high-resolution gas chromatography (HRGC) and, more recently, comprehensive two-dimensional chromatography (GC × GC), meets that need better than any other method. Note that most sulfur-containing chemical quantification and qualitative analysis need quantitative desorption from traps and selective high-resolution columns in addition to sensitive sulfur-specific detectors. In addition to reducing the likelihood of sulfur analyte rearrangements, the implemented method should lessen adsorption losses of sulfur compounds across the whole analytical system. A number of passivation methods for whole chromatographic analytical systems (including liners, tubing, cylinders, connections, etc.) were suggested in the past ten years as a means to reduce sulfur compound losses during sample, storage, transfer, analysis, etc. Complete two-dimensional gas chromatography (GC × GC) is a promising tool for improved petroleum sample characterization because of its high-resolution power and increased sensitivity. For the investigation of sulfur speciation in these types of matrices, GC × GC linked to sulfur chemiluminescent detectors (SCDs) is expected to become a crucial tool in the future. Research into sulfur analysis in a variety of environmental matrices is ongoing and showing promising results; however, many labs still do not regularly use these methods. Methods that may be implemented on the spot to take measurements in real time have to be the primary emphasis of future innovations (Koek et al. 2006).

Liquid Chromatography-Mass Spectrometry: A growing number of samples, including soil, food, body fluids, and medicines, need precise microgram and sub-microgram concentration measurements of targets, occasionally in complicated matrices. Even while this isn't an easy process, it becomes even more daunting when you consider the possibility of having to evaluate hundreds of samples rapidly while yet guaranteeing data quality.

An effective technique for analytical scientists to fulfill these demanding standards has been the combination of mass spectrometry (MS) with liquid chromatography (LC). Modern analytical laboratories are increasingly interested in liquid chromatography-mass spectrometry (LC-MS) systems because of their efficiency and adaptability. Analytical methods such as liquid chromatography-mass spectrometry (LC-MS) use physical separation of analytes (target substances) and subsequent mass-based detection. Despite its youth, this method has quickly become the gold standard for detecting analytes ranging from natural product extracts to pesticides, food adulterants, drug metabolites, and micrograms or even nanograms in amounts (Covey et al. 1986).

Analytes in a solid sample or a liquid sample can be physically separated using LC. The mobile phase consists of a solvent stream into which a small volume of sample solution is introduced. With an autosampler, you can reliably inject 0.1μL to 100μL of the sample, albeit the ideal volume depends on the experimental circumstances. The stationary phase consists of silica particles covered with another liquid and is typically housed in a column, which is a stainless-steel tube, through which the mobile phase is continually pushed. Upon reaching the column, the components of the sample solution-mobile phase mix will interact with the stationary phase in different ways, based on their physical and chemical characteristics. Classification of LC separations into several modes is based on the mechanism of analyte-stationary phase interaction. For example, one method is partition chromatography, which takes into account the analytes' varying hydrophobicity and solubility in the stationary and mobile phases. Second, analytes are separated by their ionic charges in ion-exchange chromatography. The third method is size-exclusion chromatography, which uses the analyte molecules' size differences as a separation mechanism. The fourth method is affinity chromatography, which uses the analytes' bonding abilities to the stationary phase to separate them.

Separation of analytes occurs as a result of their relative strengths of interaction with the stationary phase. The analytes that interact with the stationary phase the least will be the ones that escape from the column first. The remainder of the analytes are flushed out in a progressive manner as the mobile phase flows through the column, with the ones with the strongest interactions appearing last. Analytes are defined by the amount of time they spend in the column, which is known as their retention time (RT).

The eluent, or mobile phase, which is pumped out of the column, is analyzed by a detector that "responds" to the analytes' specific physical or chemical properties, like their light absorption or refractive index. The signal or "peak" that results from this reaction is measured in terms of the component's concentration in the sample, and its strength (peak area or peak height) is proportional to that concentration. The RT of the detector is the instant when it "sees" the analyte. If a sample's RT is compared to that of a known chemical, the identification of the sampled substance can be verified. Even while this isn't the most precise way to identify compounds, it does assist when you know a little bit about the sample beforehand (Niessen 2006).

While many other detectors have been integrated with LC to analyze various samples, the mass spectrometer has shown to be the most effective, sensitive, and versatile of them all. The mass spectrometer is designed to prevent the flow of the LC eluent containing the separated analytes, unlike other detectors. Through an interface, the mass spectrometer can function in a vacuum, while the LC system can function at ambient pressures. Applying heat evaporates the solvent as the column eluent runs into the interface, vaporizing and ionizing the analyte molecules. This is a must-do since the mass spectrometer can only pick up and measure ions in the gas phase. The process of air pressure ionization (API) is used to describe the generation of analyte ions at the interface, which is also known as the API source. Most LC-MS analyses make use of electrospray ionization (ESI) or atmospheric pressure chemical ionization (APCI) (Holčapek et al. 2012).

A mass spectrometer uses electric and/or magnetic fields to analyze the analyte ions that are pulled into the device. To separate ions according to their mass-to-charge (m/z) values, one can change the applied fields such that the ions take different flight paths. After separation, many mass detectors may be used to gather and detect the ions; the most popular of them is the electron multiplier. The dynode, an electron multiplier, releases secondary electrons when ions, which have been separated, hit its surface. A chain reaction involving many dynodes is used to multiply these secondary electrons. At each moment, the mass spectrometer's ion concentrations are connected with the amplified current produced by the secondary electron flow (De Hoffmann and Stroobant 2007).

Total ion chromatograms (TICs) display the ion abundances evaluated during LC-MS sample analysis. Analyte ion peak intensities vs RT are shown in this graph.

Mass spectra are also linked to each spot in the chromatogram. A comparison of the ion abundances with the observed m/z values is shown in the mass spectrum.

A compound's mass spectra reveal its structure via the relative abundances of the isotopic mass peaks and information on the parent compound's mass (gleaned from the m/z value of its ion). The quantification of the analyte is done by calculating its peak area.

Two modes of operation are available for the mass spectrometer: scan and selective ion monitoring (SIM). The scan mode is configured to identify all ions during a given time period, ranging from low to high m/z values. When no prior knowledge of the ions in a sample is known or when the sample's identity is uncertain, this mode is employed for analysis. Specified m/z values are measured by the mass spectrometer when it is running in SIM mode. This is the best operating mode to use if you want to accurately measure the concentration of known substances in your sample (Di Stefano et al. 2012).

5.6.1.2 Spectroscopic Analysis

Fourier Transform Infrared (FTIR) Spectroscopy: One of the most widely used techniques in modern times, Fourier transform infrared (FTIR) spectroscopy is renowned for its exceptional sensitivity, adaptability, specificity, and robustness. It has grown into one of the most popular analytical instrumental methods in the scientific community due to its versatility in handling solid, liquid, and gaseous analytes. Despite FTIR's well-documented shortcomings—for example, the technique is somewhat water-intolerant and very sensitive to the physical characteristics of the analysis matrix—it continues to enjoy widespread acclaim and widespread application in fields as varied as food and beverage, engineering, environmental science, pharmaceuticals, biomass, and clinical research. Benchtop, portable, and web-based real-time devices are now acceptable forms of instrumentation.

One key distinction between infrared and Fourier transform infrared spectroscopy is that the latter type of technique is built using an interferogram as its raw signal. In contrast to dispersive instruments, which show the light intensity as a function of wavelength, an interferometer shows the light intensity as a function of mirror position. The FT is this. For the intensity to be a function of wavenumber, the signal must first undergo a Fourier transformation (FT). Common parlance has it that Fourier transform infrared (FTIR) operates in the middle infrared spectrum. Nonetheless, FT equipment is accessible for both near-infrared and ultraviolet wavelengths. While Fourier transform infrared spectroscopy (FTIR) and Fourier transform near-infrared spectroscopy (FT-NIR) might work well together, analysts often have to choose between the two for different applications. Traditional dispersive instruments are significantly slower as compared to FTIR spectra. The FT approach outperforms IR in terms of signal-to-noise ratio and wavelength precision due to its calibration of the wavelength scale with an extremely exact reference laser (Berthomieu and Hienerwadel 2009).

The FT type is now the de facto standard for mid-IR instruments used in research and development. Albert Michelson, who studied the speed of light and won the Nobel Prize for his invention of the "interferometer" in the 1890s, is largely responsible for their development. The interferometer—a part of an FTIR instrument—uses a laser, two mirrors, a source, a beam splitter, and a detector. After the energy source is reached, the beam splitter divides the beam in half. One half is sent to a movable mirror while the other half is reflected back to a stationary one. The reaction of the calibrating laser controls the motion of the moving mirror, which is a constant velocity back and forth. The mirrors reflect the two beams, which are then merged at the beam splitter. This creates an interference pattern, that is then passed via the sample compartment (and, if applicable, the sample itself, at the absorbance-producing region) to the detector. After that, the FT function is used to create a spectrum from this signal.

The FT instrument stores all the data across all the detected wavelengths in the interference waveform, which is called an "interferogram" (to be addressed later). Nevertheless, a computationally demanding Fourier transform mathematical function must be applied to the signal before it can produce an interpretable spectrum. The "fast Fourier transform" (FFT) was created in 1966 as a consequence of the work of Coey and Tukey, which allowed for a calculating shortcut. The first commercially available FTIR, the FTS-14, was released in 1969 as a result of this and the development of the first commercial computer systems (Movasaghi et al. 2008).

A wide variety of FTIR instruments and adaptable, replaceable attachments are currently available, allowing for the analysis of gaseous, liquid, and solid samples with varying sizes and shapes using a single, fundamental equipment. Keep in mind that regular glass isn't see-through in the middle infrared spectrum; therefore, any optics used in the instruments or for the sample must be made from something that is. One of the first methods for solid sample analysis involved grinding the analyte and mixing it with an IR-transmissible substrate—typically potassium bromide (KBr)—under intense pressure in a tiny solid, transparent disk. After that, they were placed in a holder to measure transmission. It was common practice to sandwich two of these IR-transmissible discs with a little spacer to create thin films of liquid (non-water containing) samples. This approach has problems with both time and repeatability.

Most notably, the now-ubiquitous "ATR" (attenuated total reflectance) methods have seen a dramatic increase in their use throughout the past three decades. With this apparatus, a few seconds are all it takes to collect a spectrum from a little quantity of a solid or liquid sample deposited on a crystal window—no real sample preparation necessary. A top fixing clamp is used to securely push solids against the crystal window during analysis. This type of gadget is presently used in the majority of reported applications for solid samples. For more specialized uses, you can get gas sample enclosed cells or reflecting hemispheres for diffuse reflectance. To facilitate high-throughput screening with specifically designed FTIR accessory units, there are 96-position microtiter plates manufactured from gold along with other IR-compatible materials (Mohamed et al. 2017).

In a standard setup, gathering a background "blank" spectrum is the first order of business. All of the optical and atmospheric absorption values along the light beam's journey will be contained in this. Next, the sample is examined, and the spectral responses specific to the sample are obtained by subtracting the blank spectrum. In order to achieve a satisfactory signal-to-noise ratio, application-specific optimization is necessary for both the wavenumber resolution (usually 4 to 16 cm^{-1}) and the co-added scans (usually 8 to 64). In a single sample, analytical procedures on an ATR device may be completed in under 2 minutes because to the fast individual scans, which usually take less than 1 second on current instruments. Because of this, FTIR-ATR is perfect for metabolomic fingerprinting and other screening or manufacturing applications that involve hundreds or thousands of samples.

The information richness of FTIR spectra is also one of its challenges when it comes to use and interpretation. There is more than one peak in the spectrum of even the most basic, pure, one-compound samples, such as vanillin. While library matching techniques to authenticated standards may detect it in mixes including a single component, this becomes extremely challenging when dealing with complicated mixtures containing many substances. Since the majority of organic compounds include numerous combinations of carbon, hydrogen, nitrogen, and oxygen atoms bound together by either single or double bonds, the identical set of absorption peaks occurs in all of these compounds. Attempting to "eyeball" a large number of spectra in order to determine if the samples they originate from are different quickly becomes an insurmountable obstacle, even for a seasoned analyst. Combining FTIR data with statistical modeling tools like multivariate analysis (MVA), sometimes known as "chemometrics" in a chemical setting, is a typical way to tackle this challenge (Simonescu 2012).

The fundamental premise of MVA methods is to gather several spectra from each sample and combine them into a single data matrix; this makes FTIR spectral data particularly suitable for these methods. In this table, the absorbance for certain successive wavenumbers across all samples is shown in each column, and the entire spectra of individual samples are shown in each row. Using methods such as principal components analysis (PCA), it is feasible to easily examine and depict potential class-based correlations between the spectral responses of distinct sample groups through their score plots. When spectra from various classes of samples are superimposed, it becomes much easier to see the "interpretability" of the differences between them.

When concentration values or other quantitative information needs to be determined, MVA methods like partial least-squares regression can be employed to generate calibration predictions based on quantitative values for properties like chemical concentrations. This is done by utilizing information about the sample's composition that has been collected through other assay techniques. After that, the algorithm takes these values as input and uses them to find spectral characteristics that have the strongest relationship to the target external value. The second method is highly favored because, with careful preparation and validation, it can enable FTIR to efficiently substitute wet chemistry experiments for novel unknown materials (of the same kind), yielding significant cost and time savings. Making a statistical output table and sample plots from an MVA model construction helps to identify the most significant wavelength that was utilized to create the model, which is a helpful byproduct. One may usually deduce some precise chemical insights from this data.

The "mid"-FTIR form of FTIR is practically usually taken for granted. This wide range of frequencies is not achievable with non-FT mid-IR dispersive sensors due to their low power and sluggish scan rates. Contrarily, a non-FT dispersive sensor can achieve spectra comparable to mid-FTIR devices due to the significantly higher energy of NIR. Due to the increased time required, however, the resolution—the total number of detected wavenumbers—is often lower (Abidi 2021).

Raman Spectroscopy: Chemical species can be determined with great precision using Raman spectroscopy. Raman spectroscopy may identify specific light-matter interactions, much as other spectroscopic methods. Specifically, this method takes advantage of the fact that both Stokes and anti-Stokes scattering occur in order to analyze the structure of molecules. Various forms of scattering can take place when molecules interact with visible or near-infrared (NIR) light (Rostron et al. 2016).

An incident photon with energy $h\nu$ elevates the molecule from its vibrational state to one of an unlimited number of virtual states between the ground level and first electronic levels in all three kinds of scattering. The relaxation mechanism of the excited molecule determines the kind of scattering that may be seen.

How polarizable a bond is determines its Raman activity. An electric field's capacity to distort a bond can be measured in this way. How easily the bond's electrons may be shifted, creating a transient dipole, is the primary factor that determines this value. Groups or molecules have a strong Raman signal when there is an abundance of polarizability and a high concentration of weakly bound electrons in a bond. Because of this, in contrast to infrared, Raman is usually better at picking up on a molecule's overall structure rather than any one functional group. A molecule's polarity, which indicates the degree to which its electric charges are separated, is different from this. Because electrons are held so closely by electronegative atoms, polar compounds frequently exhibit extremely weak Raman signals (Smith and Dent 2019).

Raman spectroscopy lends itself to the study of organic and inorganic chemical species alike. Raman spectra are characteristic of atoms with a large number of free electrons, such as metals in coordination compounds. By analyzing the metal-ligand link, one may learn about the complexes' stability, structure, and composition by Raman spectroscopy. Metal complexes with low infrared vibrational absorption frequencies can benefit greatly from this. Organic molecular fingerprints and functional groups can also be uncovered using Raman spectroscopy. Because vibrations in molecules occur across the entire molecule and not only in discrete groups, Raman vibrations may be very identifying to a particular molecule. It is common for numerous bonds to be involved in the vibrations of the groups that do show up in Raman spectra.

A Raman spectrum is produced as a result of the examination of scattered light originating from the material. It is the portion of an incoming light beam that travels through the sample that is the most extensive (transmission). There is a tiny portion that is dispersed isotopically, which means that it is uniformly distributed in all directions, and it has the same wavelength as the beam that hits it. The Raman spectrum is a result of the fact that a minute fraction of the light that is scattered by the sample will be corresponding to a frequency that is distinct from the light that is incident onto it. The molecules undergo a transition to an excited rotational-vibrational-electronic state as a consequence of the photon that has been absorbed by them, and the photons are then emitted from the sample. Following the transition of the molecule to a higher energy state, the dispersed photons undergo a transition to a new state in order to ensure that the system's energy level remains stable. For any given pair of molecular vibrational states, there is a corresponding energy disparity between the new and old states of the scattered photon. In order for a Raman peak to be seen, the molecule's polarizability must vary throughout the transition. Two intensities, one with a perpendicular polarization and the other with a parallel polarization relative to the incident beam, are multiplied together to get the depolarization (Moon et al. 2023).

The majority of the chemicals that are utilized are aqueous solutions, and the necessary laser is activated when the proper wavelength has been chosen. After that, the spectrometer is calibrated by employing a reference sample and conducting the exposure for the right amount of time and energy. Subsequently, the sample is positioned beneath the microscope, and the focus is directed toward the layer that is being analyzed. The Raman spectrum is calculated by the monochromator by scanning a variety of wavenumbers, which ultimately results in the spectrum. Analysis is thereafter conducted by examining the data using the correct software, with the resulting spectra serving as the basis for the findings (McCreery 2005).

5.6.2 Elemental Analysis

5.6.2.1 XRF

The luminescence-based field-portable approach known as X-ray fluorescence is capable of providing elemental assessments in a reasonably affordable manner and in a relatively short amount of time. Measurements may be taken outside of a controlled laboratory environment using the battery-operated devices displayed below, which primarily function in a point-and-acquire mode. This makes them ideal for tasks such as recycling metal alloys or checking if toys or any kind of other consumer products contain heavy metals. Because it does not cause any damage to the object being examined, this technique is ideally suited for the compositional examination of precious art and antiques (Acquafredda 2019).

A more conventional laboratory spectrometer configuration, in which the material is put into the instrument, is another use for XRF equipment. These devices can also be used in this format. However, despite their potential for greater sensitivity and accuracy, this benchtop equipment often comes at a higher price.

Exciting valence electrons into an excited state is accomplished by the application of ultraviolet-visible light in the process of molecular fluorescence measurements. This process of electrons in the excited state returning to the ground state and releasing photons is what causes fluorescence to take place. Compared to the light that was used to excite the material, the wavelength that causes the most fluorescence is redshifted, which means it has a longer wavelength with a lower energy. This is considering the excited electrons return to their ground vibrational as well as rotational states before sending out a photon. The electrons in atomic fluorescence are thermally excited in a flame or plasma, and fluorescence happens as the excited electrons relax (Melquiades and Appoloni 2004).

5.6.2.2 Stable Isotope Analysis

By analyzing the stable isotope composition of the molecules of water (δ2H(H_2O) and δ18O(H_2O)), the solutes in water (δ15N(NO_3), δ18O(NO_3), δ34S(SO_4), δ18O(SO_4), δ13C(DIC)), or particulate organic matter (δ13C(POM) and δ15N(POM)), as well as the concentration of major ions in water (Na^+, Ca_2^+, Cl^-, SO_4^{2-}, NO_3 $^-$, and NH_4^+), it is possible to study hydro- and hydrogeological processes, as well as the identification of sources of pollution and pathways. Although basic hydrochemistry approaches are useful for determining pollution levels, stable isotope methods are indispensable for determining evaporation and evapo-concentration as well as for identifying sources and disentangling water and pollution budgets. Consequently, for the purpose of identifying non-particulate, non-source agricultural contaminants, techniques that integrate hydrochemical and stable isotope approaches have become indispensable instruments. The cheap cost and high success rates of these integrated strategies are their key advantages, especially when applied in areas without baseline data or long-term monitoring (Katzenberg and Waters-Rist 2018).

Water nitrate contamination sources can be identified by analyzing their stable δ15N and δ18O isotopes. Denitrification is the process by which *Pseudomonas aureofaciens* bacteria, via enzymes, transform water-soluble NO_3^- to nitrogen gas O_2. The concentration of the gaseous N_2O and the fingerprints of the δ15N and δ18O isotopes may be determined using an isotopic laser spectroscope. This information can be used as tracers to identify where the dissolved nitrate is coming from. The denitrifier technique has a number of benefits over other nitrate isotope discrimination methods. These include just a small amount of sample, the absence of harmful reagents, and the ability to utilize the approach for samples having low concentrations. Modifying the laser spectroscopy approach to handle a big sample input is possible, and it also provides robust analysis. Not only that, but this technique works just as well with saltwater as it does with freshwater, and it can be used with any water sample source that has a nitrate contamination level of 2 ppm or above (Meier-Augenstein and Kemp 2009).

Ammonium and nitrate/nitrite stable nitrogen isotope compositions (15 N/14 N) in liquid samples may be found using the micro-diffusion technique. This method involves isolating ammonium and nitrate/nitrite before analyzing 15 N/14 N using emission spectroscopy (ES) or isotope ratio mass spectrometry (IRMS). In principle, it is possible to analyze any kind of liquid; nevertheless, pre-extractions may be required for the N-pool under consideration. Soil extracts containing KCl are required for investigations on soil inorganic N. The findings will show the total of both components according to their relative quantities, as the technology cannot distinguish between nitrate and nitrite; nevertheless, in most cases, the concentration of nitrite is very low or nonexistent. The analysis of 15 N-enriched nitrate and ammonium may be used in various inorganic N-pool investigations, such as leaching, balancing, mineralization-immobilization turnover, and fertilizer usage efficiency (Meier-Augenstein 2017).

5.7 Case Study

5.7.1 Case 1

This case study talks about how a cement production plant in Coimbatore, Tamil Nadu, India, contributes to air pollution in its surroundings. Cement production is a major contributor of SPM as an industrial source that adversely affects air quality. Residents near the cement factory under study complained about air pollution, and thus, this forensic investigation was done to authenticate or refute the factory as a causative pollution source.

The investigation applied environmental forensic methodologies, including geographic information system tools and scanning electron microscopy, to identify the source of particulate matter (PM) emissions around the factory. Sampling was done at 25 locations using the Mini Laser Aerosol Spectrometer (GRIMM Model 11R) in determining PM concentrations, including PM1, PM2.5, and PM10. The study was based on some principles in the INTERPOL Pollution Crime Forensic Investigation Manual. SEM was utilized to analyze dust samples collected from local vegetation and the spectrometer's filters, which were later compared with control samples of cement particles in terms of surface texture and shape. Spatial distribution analysis of PM concentrations made use of GIS.

Among the PM10 sampling locations, most exceeded India's National Ambient Air Quality Standards. The maximum concentrations of PM10 were three times the permissible level. PM2.5 and PM1 were more uniformly distributed. SEM micrographs of dust particles on vegetation and in filters showed similar morphologies with cement particulate matter from the factory.

Spatial analysis showed high values of PM10 in the northwest direction, concordant with the dominant directions of the winds, and established the cement plant as the pollution source. This investigation conclusively established the cement factory's contribution to air pollution in its vicinity. The findings highlight the utility of integrated forensic techniques, including SEM, GIS, and air quality monitoring, in identifying pollution sources and providing evidence for regulatory and remediation efforts (Abdul Samad et al. 2020).

5.7.2 Case 2

Oil spills represent some of the most significant environmental disasters, affecting both the ecosystems and the economies in the long run. Delineation of source and composition of spilled oil is crucial to mitigate the loss and determine liabilities. This is a case study of five unique oil spill cases spread across regions with the aid of sophisticated techniques like advanced GC-MS source identification. Samples include spills from the Deepwater Horizon incident (United States), Kuwait crude oil spill (Gulf War), Prestige oil tanker spill (Spain), and others.

A multi-technique approach was employed, integrating conventional GC-MS, thermodesorption GC-MS (TD-GC-MS), and pyrolysis GC-MS (Py-GC-MS). These methods were applied to both shoreline and inland samples, including tarballs, oil-stained rocks, and sand. For instance, the Kuwait oil sample was obtained from a "dry oil lake" characterized by high TPH concentrations and a weathered tar mat. Each sample was fractionated to analyze saturates, aromatics, and asphaltenes, along with thermodesorption for assessing volatile and semi-volatile compounds.

Macondo Sample of Deepwater Horizon Incident: The samples showed high saturation of hydrocarbons, which consist of long-chain alkanes and polycyclic aromatic hydrocarbons (PAHs). Although the weathering was moderate, it had an original chemical fingerprint characteristic of Macondo well oil.

Kuwait Oil Spill: The Burgan field samples showed high sulfur levels with biodegradation. Unusual markers such as benzothiophenes and hopanes highlighted its unique geochemical signature.

Prestige Oil Tanker Spill: This heavy fuel oil, spilled off the coast of Spain, showed advanced weathering. GC-MS found a high concentration of asphaltenes and polar compounds, which indicated bioremediation efforts using oleophilic fertilizers. Advanced GC-MS methods became very useful for the identification of sources and states of weathering of oil spills to support environmental forensic investigations. They not only served to characterize a complex mixture but also showed how the combination of thermodesorption and pyrolysis might be useful. The results underpin the importance of strong analytical frameworks in responding to oil spill challenges (Kruge et al. 2018).

5.8 Conclusion

Environmental forensic investigations are crucial for the proper litigation and remedial actions. The analytical and environmental chemistry integrates to address the legal and justice-related concerns in environment with the support of forensic techniques and advancements. The growth of environmental forensics in detection of pollution sources using various methods, such as isotope analysis ratio, molecular markers and instrumentation enables accurate tracing of contaminants. The differentiation of natural and anthropogenic sources using these methods facilitates new scopes for effective environmental management and pollution control. The vital role of forensic investigation in contaminated sites shows the importance of continuous research and development of technologies in this field for a sustainable future.

References

Aarkrog A (1962) Disposal of radioactive wastes into marine and fresh waters, IAEA bibliographical series No. 5 (Vienna, 1962. 368p.$3.00). Nucl Phys 37:693

Abdul Samad MS, Mohan P, Varghese GK, Shah IK, Alappat BJ (2020) Environmental forensic investigation of the air pollution from a cement manufacturing unit. Environ Forensic 21(1):37–47

Abidi N (2021) FTIR Microspectroscopy. Springer International Publishing, Cham

Acquafredda P (2019) XRF technique. Phys Sci Rev 4(8):20180171

Akhtar N, Syakir Ishak MI, Bhawani SA, Umar K (2021) Various natural and anthropogenic factors responsible for water quality degradation: a review. Water 13(19):2660

Aldenberg T, Slob W (1991) Confidence limits for hazardous concentrations based on logistically distributed NOEC toxicity data, Report# 719102002. National Institute for Public Health and Environmental Protection

Antle PM, Zeigler CD, Wilton NM, Robbat A Jr (2013) A more accurate analysis of alkylated PAH and PASH and its implications in environmental forensics. Int J Environ Anal Chem 94(4):332–347. https://doi.org/10.1080/03067319.2013.840886

Atmadja J, Bagtzoglou AC (2001) Pollution source identification in heterogeneous porous media. Water Resour Res 37(8):2113–2125

Atwood DA (ed) (2013) Radionuclides in the environment. John Wiley & Sons

Balasubramanian A (2017) Soil erosion–causes and effects. Centre for Advanced Studies in Earth Science, University of Mysore, Mysore

Baragaño D, Ratié G, Sierra C, Chrastný V, Komárek M, Gallego JR (2022) Multiple pollution sources unravelled by environmental forensics techniques and multivariate statistics. J Hazard Mater 424:127413

Bari MA, Kindzierski WB (2018) Ambient volatile organic compounds (VOCs) in communities of the Athabasca oil sands region: sources and screening health risk assessment. Environ Pollut 235:602–614. https://doi.org/10.1016/j.envpol.2017.12.065

Bashir AS, Joshi RD (2019) Physicochemical and heavy metals analysis of effluent discharge of metal processing industries

Becker JS (2012) Inorganic mass spectrometry of radionuclides. In: Handbook of radioactivity analysis. Elsevier, pp 833–870. https://doi.org/10.1016/b978-0-12-384873-4.00013-x

Berthomieu C, Hienerwadel R (2009) Fourier transform infrared (FTIR) spectroscopy. Photosynth Res 101:157–170

Bhardwaj R, Gupta A, Garg J (2017) Evaluation of heavy metal contamination using environmetrics and indexing approach for river Yamuna, Delhi stretch, India. Water Sci 31:52–66

Borah P, Kumar M, Devi P (2020) Types of inorganic pollutants: metals/metalloids, acids, and organic forms. In: Inorganic pollutants in water. Elsevier, pp 17–31. https://doi.org/10.1016/b978-0-12-818965-8.00002-0

Boudouresque CF, Verlaque M (2002) Biological pollution in the Mediterranean Sea: invasive versus introduced macrophytes. Mar Pollut Bull 44(1):32–38. https://doi.org/10.1016/s0025-326x(01)00150-3

Budowle B, Murch R, Chakraborty R (2005) Microbial forensics: the next forensic challenge. Int J Legal Med 119(6):317–330. https://doi.org/10.1007/s00414-005-0535-y

Casadevall A, Pirofski L-A (2002) What is a pathogen? Ann Med 34(1):2–4. https://doi.org/10.1080/078538902317338580

Charbonnet JA, Rodowa AE, Joseph NT, Guelfo JL, Field JA, Jones GD et al (2021) Environmental source tracking of per-and polyfluoroalkyl substances within a forensic context: current and future techniques. Environ Sci Technol 55(11):7237–7245

Chen X, Liu M, Zuo L, Wu X, Chen M, Li X et al (2023) Environmental noise exposure and health outcomes: an umbrella review of systematic reviews and meta-analysis. Eur J Pub Health 33(4):725–731

Choppin GR (2007) Actinide speciation in the environment. J Radioanal Nucl Chem 273(3):695–703. https://doi.org/10.1007/s10967-007-0933-3
Coggan TL, Moodie D, Kolobaric A, Szabo D, Shimeta J, Crosbie ND, Lee E, Fernandes M, Clarke BO (2019) An investigation into per- and polyfluoroalkyl substances (PFAS) in nineteen Australian wastewater treatment plants (WWTPs). Heliyon 5(8):e02316. https://doi.org/10.1016/j.heliyon.2019.e02316
Cole-Dai J (2010) Volcanoes and climate. Wiley Interdiscip Rev Clim Chang 1(6):824–839
Costa HJ, Sauer TC (2005) Forensic approaches and considerations in identifying PAH background. Environ Forensic 6(1):9–16. https://doi.org/10.1080/15275920590913859
Costa HJ, White KA, Ruspantini JJ (2004) Distinguishing PAH background and MGP residues in sediments of a freshwater creek. Environ Forensic 5(3):171–182. https://doi.org/10.1080/15275920490495909
Covey TR, Lee ED, Bruins AP, Henion JD (1986) Liquid chromatography/mass spectrometry. Anal Chem 58(14):1451A–1461A
Croudace IW, Warwick PE, Reading DG, Russell BC (2016) Recent contributions to the rapid screening of radionuclides in emergency responses and nuclear forensics. TrAC Trends Anal Chem 85:120–129. https://doi.org/10.1016/j.trac.2016.05.007
Csuros M, Csuros C (2016) Environmental sampling and analysis for metals. CRC Press
Davis JM, Thompson L, Brown CJ (2002) Untreated sewage and its effects on water quality and aquatic life. J Environ Sci 15(3):235–245
Daw A, Yeo K, Karpatne A, Klein L (2022, December) Multi-task learning for source attribution and field reconstruction for methane monitoring. In: 2022 IEEE International Conference on Big Data (Big Data). IEEE, pp 4835–4841
De Hoffmann E, Stroobant V (2007) Mass spectrometry: principles and applications. John Wiley & Sons
Deolikar R, Pandit S, Jadhav J, Vyavahare G, Gurav R, Desai N, Patil R (2021) Pesticides: types, toxicity and recent updates on bioremediation strategies. In: Spatial modeling and assessment of environmental contaminants: risk assessment and remediation, pp 531–568
Di Stefano V, Avellone G, Bongiorno D, Cunsolo V, Muccilli V, Sforza S, Dossena A, Drahos L, Vékey K (2012) Applications of liquid chromatography–mass spectrometry for food analysis. J Chromatogr A 1259:74–85
Dumontet S, Dinel H, Baloda SB (1999) Pathogen reduction in sewage sludge by composting and other biological treatments: a review. Biol Agric Horticult 16(4):409–430. https://doi.org/10.1080/01448765.1999.9755243
Elliott M (2003) Biological pollutants and biological pollution—an increasing cause for concern. Mar Pollut Bull 46(3):275–280. https://doi.org/10.1016/s0025-326x(02)00423-x
Elshafei AM, Mansour R (2024) Microbial bioremediation of soils contaminated with petroleum hydrocarbons. Discover Soil 1(1):9
Erickson MD (2020) Environmental PCB forensics: processes and issues. Environ Sci Pollut Res 27:16492–16502. https://doi.org/10.1007/s11356-019-07416-9
Fernández B, Lara LM, Menéndez-Aguado JM, Ayala J, García-González N, Salgado L et al (2020) A multi-faceted, environmental forensic characterization of a paradigmatic brownfield polluted by hazardous waste containing hg, as, PAHs, and dioxins. Sci Total Environ:138546. https://doi.org/10.1016/j.scitotenv.2020.138546
Freeman AJ, Koelling DD (1974) The actinides
Fukai R, Yamato A, Thein M, Bilinski H (1987) Oxidation states of fallout plutonium in Mediterranean rain and seawater. Geochem J 21(2):51–57. https://doi.org/10.2343/geochemj.21.51
Garcia A, Barbas C (2011) Gas chromatography-mass spectrometry (GC-MS)-based metabolomics. In: Metabolic profiling: Methods and protocols, pp 191–204
Gerber PJ, Hristov AN, Henderson B, Makkar H, Oh J, Lee C et al (2013) Technical options for the mitigation of direct methane and nitrous oxide emissions from livestock: a review. Animal 7(s2):220–234

Ghorani-Azam A, Riahi-Zanjani B, Balali-Mood M (2016) Effects of air pollution on human health and practical measures for prevention in Iran. J Res Med Sci 21(1):65
Giddings JC (2017) Dynamics of chromatography: principles and theory. CRC Press
Gingerich DB, Grol E, Mauter MS (2018) Fundamental challenges and engineering opportunities in flue gas desulfurization wastewater treatment at coal fired power plants. Environ Sci 4(7):909–925
Grandjean P, Landrigan PJ (2006) Developmental neurotoxicity of industrial chemicals. Lancet 368(9553):2167–2178
Grossi M, Ricco B, Parolin C, Vitali B (2017) Bacterial concentration detection using a portable embedded sensor system for environmental monitoring. In: 2017 7th IEEE International Workshop on Advances in Sensors and Interfaces (IWASI). https://doi.org/10.1109/iwasi.2017.7974263
Gurbuz F, Metcalf JS, Codd GA, Karahan AG (2012) Evaluation of enzyme-linked immunosorbent assays (ELISAs) for the determination of microcystins in cyanobacteria. Environ Forensic 13(2):105–109
Halim M, Conte P, Piccolo A (2003) Potential availability of heavy metals to phytoextraction from contaminated soils induced by exogenous humic substances. Chemosphere 52(1):265–275. https://doi.org/10.1016/s0045-6535(03)00185-1
Halliday JR (1999) The impact of fossil fuel combustion on air quality. Environ Pollut 102(1):73–78
Harrison EZ, Oakes SR, Hysell M, Hay A (2006) Organic chemicals in sewage sludges. Sci Total Environ 367(2–3):481–497
Hatipoğlu G (2022) On the use of dimension reduction or signal separation methods for nitrogen river pollution source identification. arXiv preprint arXiv:2204.13182
Herawati N, Suzuki S, Hayashi K, Rivai IF (2000) Cadmium, copper, and zinc levels in rice and soil of Japan, Indonesia, and China by soil type. Bull Environ Contam Toxicol 64(1):33–39. https://doi.org/10.1007/s001289910006
Holčapek M, Jirásko R, Lísa M (2012) Recent developments in liquid chromatography–mass spectrometry and related techniques. J Chromatogr A 1259:3–15
Hong S, Candelone JP, Turetta C, Boutron CF (1996) Changes in natural lead, copper, zinc and cadmium concentrations in central Greenland ice from 8250 to 149,100 years ago: their association with climatic changes and resultant variations of dominant source contributions. Earth Planet Sci Lett 143(1–4):233–244
Hossini H, Karimi H, Mustafa YT, Al-Quraishi AMF (2022) Role of effective factors on soil erosion and land degradation: a review. Environmental degradation in Asia: land degradation, environmental contamination, and human activities (pp. 221–235). Springer, Cham
Houk RS, Fassel VA, Flesch GD, Svec HJ, Gray AL, Taylor CE (1980) Inductively coupled argon plasma as an ion source for mass spectrometric determination of trace elements. Anal Chem 52(14):2283–2289. https://doi.org/10.1021/ac50064a012
Hu H (2002) Human health and heavy metals exposure. In: McCally M (ed) Life support: the environment and human health. MIT Press, Cambridge, pp 65–81
Hu Q-H, Weng J-Q, Wang J-S (2010) Sources of anthropogenic radionuclides in the environment: a review. J Environ Radioact 101(6):426–437. https://doi.org/10.1016/j.jenvrad.2008.08.004
Huo H, Wu Y, Wang M (2009) Total versus urban: well-to-wheels assessment of criteria pollutant emissions from various vehicle/fuel systems. Atmos Environ 43(10):1796–1804
Ješkovský M, Kaizer J, Kontuĺ I, Lujaniené G, Müllerová M, Povinec PP (2019) Analysis of environmental radionuclides. In: Handbook of radioactivity analysis: Volume 2. Elsevier, pp 137–261. https://doi.org/10.1016/b978-0-12-814395-7.00003-9
Jian J-M, Chen D, Han F-J, Guo Y, Zeng L, Lu X, Wang F (2018) A short review on human exposure to and tissue distribution of per- and polyfluoroalkyl substances (PFASs). Sci Total Environ 636:1058–1069. https://doi.org/10.1016/j.scitotenv.2018.04.380
Johnson GW, Quensen JF III, Chiarenzelli JR, Hamilton MC (1964) Polychlorinated biphenyls. Environ Forensic:187–225. https://doi.org/10.1016/b978-012507751-4/50032-x
Karasek FW, Clement RE (2012) Basic gas chromatography-mass spectrometry: principles and techniques. Elsevier

Katzenberg MA, Waters-Rist AL (2018) Stable isotope analysis: a tool for studying past diet, demography, and life history. In: Biological anthropology of the human skeleton, pp 467–504

Keim P, Smith KL, Keys C, Takahashi H, Kurata T, Kaufmann A (2001) Molecular investigation of the Aum Shinrikyo anthrax release in Kameido, Japan. J Clin Microbiol 39(12):4566–4567. https://doi.org/10.1128/jcm.39.12.4566-4567.2001

Koek MM, Muilwijk B, van der Werf MJ, Hankemeier T (2006) Microbial metabolomics with gas chromatography/mass spectrometry. Anal Chem 78(4):1272–1281

Kooijman SALM (1987) A safety factor for LC50 values allowing for differences in sensitivity among species. Water Res 21(3):269–276

Kopka J (2006) Gas chromatography mass spectrometry. In: Plant metabolomics. Springer Berlin Heidelberg, Berlin, Heidelberg, pp 3–20

Kruge MA, Gallego JLR, Lara-Gonzalo A, Esquinas N (2018) Environmental forensics study of crude oil and petroleum product spills in coastal and oilfield settings: combined insights from conventional GC–MS, thermodesorption–GC–MS, and pyrolysis–GC–MS. In: Oil spill Environmental forensics case studies. Butterworth-Heinemann, pp 131–155

Kukuła K, Bylak A (2020) Synergistic impacts of sediment generation and hydrotechnical structures related to forestry on stream fish communities. Sci Total Environ 737:139751

Kulkarni SJ (2015) A review on studies and research on various aspects of leaching. Int J Res Rev 2(9):579–583

Kumar N, Pathera AK, Saini PARVEEN, Kumar M (2012) Harmful effects of pesticides on human health. Ann Agri-Bio Res 17(2):125–127

Landrigan PJ, Fuller R, Acosta NJ, Adeyi O, Arnold R, Baldé AB et al (2018) The Lancet Commission on pollution and health. Lancet 391(10119):462–512

Langberg HA, Arp HPH, Breedveld GD, Slinde GA, Høiseter Å, Grønning HM et al (2021) Paper product production identified as the main source of per- and polyfluoroalkyl substances (PFAS) in a Norwegian lake: source and historic emission tracking. Environ Pollut 273:116259. https://doi.org/10.1016/j.envpol.2020.116259

Lawan MS, Kumar R, Rashid J, Barakat MAEF (2023) Recent advancements in the treatment of petroleum refinery wastewater. Water 15(20):3676

Lehto J, Hou X (2011) Chemistry and analysis of radionuclides: laboratory techniques and methodology. John Wiley & Sons

Lindstrom AB, Strynar MJ, Libelo EL (2011) Polyfluorinated compounds: past, present, and future. Environ Sci Technol 45(19):7954–7961. https://doi.org/10.1021/es2011622

Lundstedt S, White PA, Lemieux CL, Lynes KD, Lambert IB, Öberg L et al (2007) Sources, fate, and toxic hazards of oxygenated polycyclic aromatic hydrocarbons (PAHs) at PAH-contaminated sites. AMBIO J Hum Environ 36(6):475–485. https://doi.org/10.1579/0044-7447(2007)36[475:SFATHO]2.0.CO;2

Maji S, Dwivedi DH, Singh N, Kishor S, Gond M (2020) Agricultural waste: its impact on environment and management approaches. In: Emerging eco-friendly green technologies for wastewater treatment, pp 329–351

Majumdar D (2003) The blue baby syndrome: nitrate poisoning in humans. Resonance 8(10):20–30

Maki AW, Bishop WE (1985) Chemical safety evaluation. In: Rand GM, Petrocelli SR (eds) Fundamentals of aquatic toxicology. Hemisphere, Washington, D.C., pp 619–635

Manisalidis I, Stavropoulou E, Stavropoulos A, Bezirtzoglou E (2020) Environmental and health impacts of air pollution: a review. Front Public Health 8:14

Mather TA, Pyle DM, Oppenheimer C (2003) Tropospheric volcanic aerosol. Geophys Monogr-American Geophysical Union 139:189–212

McCord J, Strynar M (2019) Identification of per- and polyfluoroalkyl substances in the cape fear river by high-resolution mass spectrometry and nontargeted screening. Environ Sci Technol 53(8):4717–4727. https://doi.org/10.1021/acs.est.8b06017

McCreery RL (2005) Raman spectroscopy for chemical analysis. John Wiley & Sons

Megson D, Reiner EJ, Jobst KJ, Dorman FL, Robson M, Focant J-F (2016) A review of the determination of persistent organic pollutants for environmental forensics investigations. Anal Chim Acta 941:10–25. https://doi.org/10.1016/j.aca.2016.08.027

Meier-Augenstein W (2017) Stable isotope forensics: methods and forensic applications of stable isotope analysis. John Wiley & Sons

Meier-Augenstein W, Kemp HF (2009) Stable isotope analysis: general principles and limitations. Wiley Encyclopedia of Forensic Science

Melquiades FL, Appoloni C (2004) Application of XRF and field portable XRF for environmental analysis. J Radioanal Nucl Chem 262(2):533–541

Michel J, Fingas M (2016) Oil spills: causes, consequences, prevention, and countermeasures. In: Fossil fuels: current status and future directions, pp 159–201

Miller JM (2005) Chromatography: concepts and contrasts. John Wiley and Sons

Mohamed MA, Jaafar J, Ismail AF, Othman MHD, Rahman MA (2017) Fourier transform infrared (FTIR) spectroscopy. In: Membrane characterization. Elsevier, pp 3–29

Moon J, Li M, Ramirez-Cuesta AJ, Wu Z (2023) Raman spectroscopy. In: Springer handbook of advanced catalyst characterization. Springer International Publishing, Cham, pp 75–110

Morrison RD (2000) Critical review of environmental forensic techniques: part I. Environ Forensic 1(3):157–173. https://doi.org/10.1006/enfo.2000.0017

Morse SA, Khan AS (2005) Epidemiologic investigation for public health, biodefense, and forensic microbiology. Microbial Forensics:157–171. https://doi.org/10.1016/b978-012088483-4/50011-1

Movasaghi Z, Rehman S, ur Rehman DI (2008) Fourier transform infrared (FTIR) spectroscopy of biological tissues. Appl Spectrosc Rev 43(2):134–179

Munshed M, Thé JVG, Fraser R, Matthews B, Ramadan A (2023) Extending multi-pathway human health risk assessment from regional to country-wide—a case study on Kuwait. Atmosphere 14(8):1247. https://doi.org/10.3390/atmos14081247

Murphy BL, Morrison RD (eds) (2014) Introduction to environmental forensics

Nearing MA, Xie Y, Liu B, Ye Y (2017) Natural and anthropogenic rates of soil erosion. Int Soil Water Conserv Res 5(2):77–84

Niessen WM (2006) Liquid chromatography-mass spectrometry. CRC Press

Odubo TC, Kosoe EA (2024) Sources of air pollutants: impacts and solutions. In: Air pollutants in the context of one health: fundamentals, sources, and impacts. Cham, Springer Nature Switzerland, pp 75–121

Oppenheimer C (2003) Ice core and palaeoclimatic evidence for the timing and nature of the great mid-13th century volcanic eruption. Int J Climatol: J Royal Meteorol Soc 23(4):417–426

Olenin S, Elliott M, Bysveen I, Culverhouse PF, Daunys D, Dubelaar GBJ et al (2011) Recommendations on methods for the detection and control of biological pollution in marine coastal waters. Mar Pollut Bull 62(12):2598–2604. https://doi.org/10.1016/j.marpolbul.2011.08.011

PAHO (1998) Natural disaster mitigation in drinking water and sewerage systems. World Health Organization, Washington, DC

Parris K (2011) Impact of agriculture on water pollution in OECD countries: recent trends and future prospects. Int J Water Resour Dev 27:33–52

Pędziwiatr A, Potysz A, Kaczmarczyk I, Sulej J, Kwasowski W, Uzarowicz Ł (2025) Combined approach using soil and Fly ash analysis to understand the Environmental consequences of coal combustion in thermal power stations in the city. Water Air Soil Pollut 236(3):146

Pericherla MN, Kumar S, Reddy MS (2020a) Organic matter, nutrients, and pathogens carried into aquatic systems by runoff from cattle operations. Environ Sci Pollut Res 27(12):14567–14578

Pericherla S, Karnena MK, Vara S (2020b) A review on impacts of agricultural runoff on freshwater resources. Int J Emerg Technol 11:829–833

Pozzer A, Tsimpidi AP, Karydis VA, de Meij A, Lelieveld J (2017) Impact of agricultural emission reductions on fine-particulate matter and public health. Atmos Chem Phys 17(20):12813–12826

Rejah BK, Alameer NKA, Kadim WH, Murad STM (2018) Estimate level of radon concentration for drinking water in some regions of Baghdad City. Arab J Sci Eng 43:3831–3835

Rostron P, Gaber S, Gaber D (2016) Raman spectroscopy, review. Laser 21:24

Ruan T, Lin Y, Wang T, Liu R, Jiang G (2015) Identification of novel polyfluorinated ether sulfonates as PFOS alternatives in municipal sewage sludge in China. Environ Sci Technol 49(11):6519–6527. https://doi.org/10.1021/acs.est.5b01010
Rubin K (1997) Degassing of metals and metalloids from erupting seamount and mid-ocean ridge volcanoes: observations and predictions. Geochim Cosmochim Acta 61(17):3525–3542
Runde W, Neu MP (2010) Actinides in the geosphere. In: The chemistry of the actinide and transactinide elements. Springer, pp 3475–3593. https://doi.org/10.1007/978-94-007-0211-0_32
Sagasta JM, Zadeh SM, Turral H (2017) Water pollution from agriculture: a global review. Food and Agriculture Organization of the United Nations (FAO) and the International Water Management Institute (IWMI), Colombo
Sanseverino I, Conduto ADS, Pozzoli L, Dobricic S, Lettieri T (2016) Algal bloom and its economic impact
Save the Water (2022) Eutrophication: An overview of pollution effects and removal technologies
Schneider S, Bister S, Christl M, Hori M, Shozugawa K, Synal H-A et al (2017) Radionuclide pollution inside the Fukushima Daiichi exclusion zone, part 2: forensic search for the "forgotten" contaminants Uranium-236 and plutonium. Appl Geochem 85:194–200. https://doi.org/10.1016/j.apgeochem.2017.05.022
Sellner KG, Doucette GJ, Kirkpatrick GJ (2003) Harmful algal blooms: causes, impacts and detection. J Ind Microbiol Biotechnol 30(7):383–406. https://doi.org/10.1007/s10295-003-0074-9
Shah MP (ed) (2021) Biological treatment of industrial wastewater. Royal Society of Chemistry
Shahabudin, M. M., & Musa, S. (2018). Occurrence of surface water contaminations: an overview. In IOP Conference Series: Earth and Environmental Science (Vol. 140, 1, 012058). IOP Publishing
Sharma S, Bhattacharya A (2017) Drinking water contamination and treatment techniques. Appl Water Sci 7(3):1043–1067
Shields WJ, Saba T, Boehm PD, Pietari J (2015) Congeners. In: Murphy BL, Morrison RD (eds) Introduction to environmental forensics, 3rd edn. Elsevier, pp 347–393. https://doi.org/10.1016/b978-0-12-404696-2.00010-2
Simonescu, C. M. (2012). Application of FTIR spectroscopy in environmental studies. Advanced aspects of spectroscopy, 29(1), 77–86
Smith E, Dent G (2019) Modern Raman spectroscopy: a practical approach. John Wiley & Sons
Song X, Huang L, Zhang J, Huang H, Li T, Su Q (2009) Harmful algal blooms (HABs) in Daya Bay, China: an in-situ study of primary production and environmental impacts. Mar Pollut Bull 58(9):1310–1318. https://doi.org/10.1016/j.marpolbul.2009.04.030
Stashenko E, Martínez JR (2014) Gas chromatography-mass spectrometry. In: Advances in gas chromatography, pp 1–38
Stern J, Schouten R (2017) Chapter 3: Lessons from the anthrax letters. In: Insider threats. https://doi.org/10.7591/9781501705946-006
Stockholm Convention Secretariat (2001) UNEP. http://chm.pops.int/. Accessed Sept 2015
Strynar M, Dagnino S, McMahen R, Liang S, Lindstrom A, Andersen E et al (2015) Identification of novel perfluoroalkyl ether carboxylic acids (PFECAs) and sulfonic acids (PFESAs) in natural waters using accurate mass time-of-flight mass spectrometry (TOFMS). Environ Sci Technol 49(19):11622–11630. https://doi.org/10.1021/acs.est.5b01215
Textor C, Graf HF, Timmreck C, Robock A (2004) Emissions from volcanoes. In: Emissions of atmospheric trace compounds. Springer Netherlands, Dordrecht, pp 269–303
Tripathy S, Kar OP, Pradhan A (2025) Challenges and innovations in industrial wastewater treatment: safeguarding water resources and promoting sustainable practices. Water Air Soil Pollut 236(92)
United Nations Environment Programme (2019) All POPs listed in the Stockholm Convention. Retrieved August 21, 2020, from http://chm.pops.int/TheConvention/ThePOPs/AllPOPs/tabid/2509/Default.aspx
United States Environmental Protection Agency (2008, January) Polycyclic aromatic hydrocarbons (PAHs). Office of Solid Waste, Washington, DC

US EPA (1994) Method 1613: tetra- through octa-chlorinated dioxins and furans by isotope dilution HRGC/HRMS. In: US Environmental Protection Agency, Office of Water, Engineering, and Analysis Division, Washington, D.C.
US EPA (1995) Method 23: determination of polychlorinated dibenzo-p-dioxins and polychlorinated dibenzofurans from municipal waste combustors. US Environmental Protection Agency
US EPA (2002) Method 4025: screening for polychlorinated dibenzodioxins and polychlorinated dibenzofurans (PCDD/Fs) by immunoassay. US Environmental Protection Agency, Washington, D.C.
US EPA (2007) Method 8290A: polychlorinated dibenzo-to-dioxins (PCDDs) and polychlorinated dibenzofurans (PCDEs) by high-resolution gas chromatography/high resolution mass spectrometry (HRGC/HRMS), SW-846 Test Methods for Evaluating Solid Waste. Physical/Chemical Methods, US Environmental Protection Agency
USEPA (2005) Protecting water quality from agricultural runoff; united state Environmental Protection Agency (USEPA), Washington, DC
van Straalen NM, Denneman CA (1989) Ecotoxicological evaluation of soil quality criteria. Ecotoxicol Environ Safety 18:241–251
Varol M, Sen B (2012) Assessment of nutrient and heavy metal contamination in surface water and sediments of the upper Tigris River, Turkey. Catena 92:1–10
Wagner C, Løkke H (1991) Estimation of ecotoxicological protection levels from NOEC toxicity data. Water Res 25(10):1237–1242
Wang Z, Fingas M, Page DS (1999) Oil spill identification. J Chromatogr A 843(1–2):369–411. https://doi.org/10.1016/s0021-9673(99)00120-x
Wilson AJ (1996) The living rock: the story of metals since earliest time and their impact on civilization. Whitehurst & Clark, p 272
Wu X, Nawaz S, Li Y, Zhang H (2024) Environmental health hazards of untreated livestock wastewater: potential risks and future perspectives. Environ Sci Pollut Res 31(17):24745–24767
Wuilloud RG, Altamirano JC, Smichowski PN, Heitkemper DT (2006) Investigation of arsenic speciation in algae of the Antarctic region by HPLC-ICP-MS and HPLC-ESI-ion trap MS. J Anal At Spectrom 21(11):1214–1223
Xing G, Sardar MR, Lin B, Lin J-M (2019) Analysis of trace metals in water samples using NOBIAS chelate resins by HPLC and ICP-MS. Talanta 204:792798. https://doi.org/10.1016/j.talanta.2019.05.041
Yamamoto M, Takada T, Nagao S, Koike T, Shimada K, Hoshi M et al (2012) An early survey of the radioactive contamination of soil due to the Fukushima Dai-ichi Nuclear Power Plant accident, with emphasis on plutonium analysis. Geochem J 46(4):341–353. https://doi.org/10.2343/geochemj.2.0215
YCoelho LM, Rezende HC, Coelho LM, Sousa PAR, Melo DFO, Coelho NMM (2015) Bioremediation of polluted waters using microorganisms. In: Shiomi N (ed) Advances in bioremediation of wastewater and polluted soil. IntechOpen, London, pp 1–22
Zuo TT, Li YL, Jin HY et al (2018) HPLC–ICP-MS speciation analysis and risk assessment of arsenic in *Cordyceps sinensis*. Chin Med 13(19). https://doi.org/10.1186/s13020-018-0178-9

Ecological Crime Scene Investigation

6

Akash Kumar, Chhote Raja Patle, Supriya Tomar,
Murali Manohar Yadav, and Girraj Sharma

Abstract

Operational practices between forensic science and environmental crime investigation make up ecological crime scene investigation (ECSI), which unites multiple disciplines to document and identify ecosystem violations for prosecution. These environmental offenses include the illicit disposal of hazardous substances along with forest clearance and protected animal poaching and trafficking, discharge of pollutant and emerging contaminats into air, water, and earth surface (soil) aside from leaving behind intricate physical evidence. Moreover, reconstruction of ecological crimes and tracking of their sources requires integration between forensic ecology, environmental forensics, forensic toxicology, and criminal justice principles and techniques through ECSI. Environmental forensic scientist detect identify and quantify pollutants, as well as emerging contaminants while tracing their sources through techniques, which include water and soil collection as well as satellite analysis and DNA identification and chemical substance tracing. Ecological investigations handle harmed ecosystems and species using specialized techniques and legal frameworks because they do not solely conduct investigations on human victims like traditional forensics do. The ECSI studies extended harm resulting from damage to the environment that causes damage to biodiversity and impacts human health and climate systems.

A. Kumar · M. M. Yadav
Dr. A.P.J. Abdul Kalam Institute of Forensic Science and Criminology, Bundelkhand University, Jhansi, Uttar Pradesh, India

C. R. Patle · S. Tomar
Department of Forensic Science, Medi-Caps University, Indore, India

G. Sharma (✉)
Jindal Institute of Behavioural Sciences, O.P. Jindal Global University, Sonipat, Haryana, India

M. S. Sankhla et al. (eds.), *Introducing the Synergy Between Forensic Science and Environmental Analysis*, https://doi.org/10.1007/978-981-95-0199-1_6

The advanced scale of environmental crimes requires ECSI to serve multiple functions that advance both environmental protection and fairness. Law enforcement depends on this field, together with policy development and public awareness, because it translates intricate ecological harm into evidence for legal proceedings. Additionally, ECSI maintains its rising importance because societies worldwide are focusing on safeguarding environmental integrity. This chapter also supports United Nations Sustainable Development Goals; Climate action (SDG 13), Life on land (SDG 15), Life Below Water (SDG 14), Peace, Justice and Strong Institutions (SDG 16).

Keywords

Forensic science · Ecology · Investigation · Crime · Crime scene · Ecological crime · SDG-13 · SDG-15 · SDG-14 · SDG-16

6.1 Introduction to Ecological Crime

Ecological crime or environmental crime refers to a wide range of illegal or unlawful anthropogenic activities that negatively impact our environment (Barclay and Bartel 2015). These offences not only violate national and international laws and regulations of the environment but also threaten the solidity of ecosystems and biodiversity (Barclay and Bartel 2015). Some of the common examples of ecological crimes include poaching and trafficking of illegal logging, unregulated fishing, dumping of emerging contaminants and pollutants, and habitat destruction (Lirëza and Koçi 2023).

These activities are not only harmful to the ecosystem but also associated with organized crime, exploitation and poor governance (White 2013). It has become more crucial than ever to address ecological crime in recent years because of deteriorating global climate conditions, the speedy loss of biodiversity, and increasing human-wildlife battles (P. Wiltshire 2010). Ecological crimes threaten livelihoods, reduce natural resources and challenge public health and safety. For example, poaching endangered species damages ecosystem services like pollination and seed dissemination in addition to dropping population levels (Shepherd et al. 2020).

Divergent legal systems and cultural perspectives further complicate the categorization of ecological crimes (Barclay and Bartel 2015). Some environmentally destructive practices may be accepted or even legal in many nations despite their negative effects on the environment (Kotzé and Adelman 2023). This gap between legality and harm raises questions about the sufficiency of present legal definitions and enforcement methods. Green criminologists believe that concentrating only on breaking the law is insufficient and that actions that hurt the environment, even if they are lawful, should be examined from a wider ethical and ecological perspective (Lynch and Stretesky 2016). For example, widespread deforestation that is authorized by the government may not be illegal, but because of its detrimental effects on biodiversity and indigenous populations, it might nevertheless be regarded as an ecological crime. Several authorities enforce rules and regulations, including the UNODC, which also classifies ecological crime. This organization's response to crimes that affect the environment extends from the crime scene to the courtroom, strengthening its capacity to detect, intercept, investigate, and prosecute crimes such as:

- Wildlife poaching & trafficking (SDG 16)
- Forest crimes (SDG 16)
- Illegal mining (SDG 16)
- Crimes in the fishing sector (SDG 16), Discharging of emerging contaminants & pollutants (SDG 16). Environmental forensics efforts and methods are closely aligned with the United Nations SDGs, particularly Climate Action, Life Below Water, Life on Land, and Peace, Justice, and Strong Institutions, which collectively emphasize the need to protect ecosystems, ensure environmental justice, and promote sustainable use of natural resources.

6.1.1 The Major Ecological Crimes

Most ecological crimes deal with devastation and the implications of environmental rules and regulations. Hence, the type of nature and conduct obtain various types of e-crimes (ecological crimes), which contain separate types of acts to conduct. In Table 6.1 the various types of ecological crimes are reported, which shows the

Table 6.1 The categories of ecological crimes and their impact on the environment

Ecological crime	Description	Environmental impact	Key example/reference
Oil spills	Deliberate or accidental release of petroleum into marine environments	Long-term damage to marine life, habitat destruction, disrupted biological dynamics	Deepwater Horizon spill released ~4.9 million gallons (Freudenburg and Gramling 2010)
Deforestation	Large-scale clearing of forests for agriculture, mining, or infrastructure	Loss of biodiversity, climate change, habitat destruction, increased CO_2 emissions	Illegal deforestation in conserved areas (White 2013)
Poaching	Illegal killing or capturing of wildlife, often for commercial gain	Endangerment and extinction of species, disruption of ecosystems	Poaching of rhinos, elephants, Malayan tigers, pangolins (Clements et al. 2010)
Chemical dumping	Unauthorized release of toxic substances into land or water	Soil and water contamination, threats to human and ecological health	Toxic runoff leading to environmental degradation (Lastumäki et al. 2020)
Overfishing	Excessive and illegal fishing reducing fish stocks and disrupting marine food chains	Collapse of fisheries, unbalanced marine ecosystems, food insecurity	IUU fishing in poorly regulated regions (Agnew et al. 2009)

inevitable concern in this new domain of crime scenes that follows (SDG 13, 14, 15, & 16).

Oil Spills (SDG 14): It is the deliberate or accidental spilling of petroleum into marine areas, leading to the destruction of habitat and wildlife. Oil spills rank among the most catastrophic environmental crimes because of their long-term effects on marine organisms. The deepwater horizon disaster is a main example of a corporate ecological crime that caused irreversible marine damage (Freudenburg and Gramling 2010). Nearly 4.9 million gallons of petroleum-based products leaked into the Gulf of Mexico in 2010 because of the oil spill that occurred on the Deepwater Horizon, destroying marine life, destroying coastlines, and changing biological dynamics (Jernelöv 2010).

Deforestation (SDG 13): This is the extensive destruction or clearing of forests, usually for the purpose of mining, urbanization, building new infrastructure, or agriculture. This process can be legal or illegal, but when it is carried out in conserved areas or without the required approval, it is considered an ecological crime. The consequences of deforestation are the loss of biodiversity, disruptive changes in global climate, the emission of greenhouse gases, and the destruction of native wildlife habitat (White 2013).

Poaching (SDG 15): This is the illegal or unlawful capturing, killing, or hunting of wildlife, often for commercial purposes. As a result of this, wildlife populations are falling, and species are becoming endangered, e.g., poaching of rhinos for horns and elephants for ivory. Significant population reductions are brought on by the illegal trade in animals, including Malayan tigers, elephants, and pangolins (Cardoso et al. 2021).

Chemical Dumping (SDG 14 & 15): The unauthorized release of toxic substances into land or water bodies causes contamination of soil and water bodies. These contaminants can threaten human health and cause environmental catastrophes. The harmful constituents of releasing material can degrade both the water and soil quality and can lead to toxic surroundings for fauna and flora (Ankit et al. 2021).

Overfishing (SDG 15): This is disturbing marine food webs, reducing fish stocks, and capturing fish at excessive rates. Illegal, unreported, and unregulated (IUU) fishing weakened the fisheries management system and marine food webs. This problem is common in areas with lax regulatory supervision or fraudulent officials (Agnew et al. 2009).

6.1.2 Impact of Ecological Crimes

Environmental crime has a negative effect on the one health approach. Even, it can have detrimental consequences on the economy as well as the security of a country. For humanity, local communities, and indigenous people, it may impact widely the public health, livelihoods, and lower property values, as well as impacting on non-human species, nature itself, as well as future generations (Malakouti and Hazrati 2025).

Biodiversity Loss: Ecological habitat destruction leads to species extinction, and this loss is heavily caused by environmental crimes (Mozer and Prost 2023).

Human Health Risks: Pollutants and degraded ecosystems can lead to waterborne diseases, respiratory conditions, and exposure to heavy metals. The outcome of these effects is the increased rates of morbidity in humans (Shetty et al. 2023).

Global Environmental Consequences: Loss of ecosystem services, degradation of soil, and climate change are all consequences of cumulative harm to the environment (Eekhout and De Vente 2022). The repercussions of these impacts significantly worsen the health of Earth.

6.1.3 Forensic Ecology

Forensic ecology has innovated relatively recently, and came at 40 years ago, it was not listened of as an aid to criminal or civil investigation. In recent years, it has gained popularity and became an important field in the environmental forensic. Forensic ecology is the application of ecological theory and practice to criminal or civil examination for the criteria of obtaining intelligence for cases, or for obtaining probative evidence that may be presented in a judicial ground (P. E. J. Wiltshire 2019).

Several organisms, or parts of organisms, are competent of being retrieved from crime scenes and other relevant locations, as forensic exhibits. They can, therefore, provide precious material evidence. Few organisms are highly characterized in their ecological requirements and may have specific ecological and geographical distributions, while others are less demanding and have very wide distributions. This does not mean that they have no value in forensic investigation (P. E. J. Wiltshire 2009).

Forensic science frequently encounters analyses such as follows: DNA, fibers, footmarks, gunshot residue, fingerprints, and others. Though these techniques relatively common also have significant value in ecological investigation. Forensic entomology is another routinely used field likewise, the forensic botany, palynology (the examination of tiny microscopic entity (palynomorph), which has been scattered away from its source), phycology (the study of algae, including diatoms), mycology (fungal science), paedology (soil science), and geology are however not applied so frequently (P. E. J. Wiltshire 2009).

6.2 The Crime Scene Investigation

Ecological crime scene investigation is an organized and scientific method for comprehending, recording, and dealing with environmental crimes. The following are some of main procedure steps.

Ecological Crime Detection and Identification: The main goal of this objective is to identify and detect unlawful acts that negatively impact the environment. This includes activities such as poaching, overfishing, deforestation, and illegal garbage disposal. Investigators can identify ecological infractions by using techniques such as field surveys, water and soil testing, and satellite photography (Eliason 2003).

Preservation and Documentation of Evidence: Ecological crime scene investigation emphasizes proper preservation and documentation of the crime scene and evidence associated with it. Proper investigation ensures that the evidence remains uncontaminated and legally admissible. Court arguments and legal claims are supported when evidence is handled properly (White 2022).

Evaluation of Environmental Harm: Proper investigation assesses the type, extent and severity of environmental harm. This investigation aims to detect, identify and measure emerging contaminants & pollutants concentrations in the environment and evaluate changes in the ecosystem. It is possible to estimate the level of damage using sophisticated techniques like ecological indicators and Geographic Information Systems (GIS) (Cao et al. 2016).

Assistance with Legal Structures: Investigation findings are used to support environmental legal procedures. These initiatives guarantee that offenders suffer the penalties while policies are improved for more effective enforcement in the future (White 2013).

Fostering Advocacy and Awareness among People: The results of investigations can be shared with the public to encourage their involvement in conservation and increase environmental awareness. Local communities can be empowered to take on the role of environmental stewards through reports, media briefings, and community engagement (Lirëza and Koçi 2023).

Remediation and Recovery Planning: Investigation contributes to the creation of plans to remedy environmental damage. Recommendations may include habitat restoration, bioremediation, or replanting to ensure that ecosystems recover and resume their functions (Ashraf 2017).

Supporting Scientific Investigations: Information acquired during this investigation can support academic and applied research in the fields of biology, toxicology, wildlife, etc. These revelations assist in improving future research methods and guiding sustainable policy (Gouda et al. 2020).

Preventing Future Environmental Crimes: Investigation ultimately seeks to lessen the occurrence of ecological crimes by addressing the underlying causes, enforcement flaws, and policy gaps. A proactive strategy promotes a culture of environmental stewardship and aids in the development of resilient ecosystems (White 2013).

6.2.1 Identification of Nature of Crime

Imagine a river with discolored water, a foul chemical smell, and floating dead fish. A once-thriving woodland nearby is now home to exotic plant species and exhibits withering vegetation (P. Wiltshire 2010). These locations are representative of common ecological crime sites and serve as starting points for inquiry.

Delineating the area, identifying salient features, and gathering preliminary observations are the first steps used by field investigators. Establishing baseline evidence for further analysis depends heavily on this early reporting (P. Kumar et al. 2021). Although systematic fieldwork is essential for successful ECSI. Important methods include (SDG 16):

Field Journals: These are used to record environmental observations, such as weather, species sightings, date, time, and geographic location. Logging regularly promotes traceability and authenticity (Blount et al. 2022).
Sketching: The scene's organizational layouts, sources of pollution, or impacted areas are highlighted in visual diagrams (Donnermeyer 2015).
Photography: It is one of the essential methods for documenting a crime scene. Evidence like dead animals, soil degradation, contaminated water, and invading species is captured by high-resolution cameras. The scientific value is increased by including reference scales (Lee et al. 2001).

6.2.2 Essence of Analyses

Emphasis is given here to the importance of aspects of forensic botany, forensic ecology, and forensic mycology. These have contributed positively to investigations involving:

Unexpected and Sudden Wildlife Deaths: Dead birds, mass fish kills, or the extinction of other species without a clear natural (Rana and Kumar 2023). Sign of illicit chemical runoff, pesticide toxicity, or toxic dumping (Rana and Kumar 2023). For example, Minamata Bay's mercury pollution caused neurological conditions and the mass death of marine life (Harada 1995).
Odor or discoloration in Soil or Water: Changes in the color of the water, such as red tides or an oiled or muddy look (Dutta et al. 2024). Odors of strange chemicals or unpleasant scents (Dutta et al. 2024). Indicates the presence of sewage, industrial waste, or illegal discharge (Meghwal et al. 2020).
Destroyed Vegetation: Patchy or burnt vegetation but shows no signs of wildfire (Bytnerowicz 2009). Abnormal development patterns in the rings of trees, which are bioindicators of prolonged exposure (Lev-Yadun et al. 2023). It could be the consequence of deforestation, chemical dumping, or acid rain.
Biodiversity Disruption: A sudden decline or extinction of keystone species (Cubas et al. 2022). The spread of invading species in an ecosystem that was once in equilibrium (X. Wang et al. 2013). Frequently connected to pollution, illegal trafficking, or habitat damage (Adla et al. 2022).
The Presence of Foreign Substances: Finding heavy metals, artificial substances, or microplastics in soil, water, or tissue samples (F. Wang et al. 2024). May be found through toxicological studies or bioassays.
Evidence of Human Activity: The exhibits of illegal camps or dumpsites (Triassi et al. 2015). Tire tracks or machine markings (Cooper and Cooper 2013). Illegal mining, logging, or construction sites inside protected areas (Cooper and Cooper 2013).

Abnormal Animal Behavior: Animals who act aggressively, leave their habitats, or behave lethargically may be exhibiting symptoms of environmental imbalance or chemical exposure (Cooper and Cooper 2013).

6.3 Collecting of Evidence

The foundation of ecological crime scene investigation is the accurate gathering and recording of physical evidence. In contrast to conventional crime scenes, ecological crime scenes are dynamic, frequently encompassing considerable regions and incorporating a variety of environmental media, including soil, water, air, and living things.

This chapter highlights the different types of evidence that are frequently found in ecological investigations, the instruments that are used to gather them, and the procedures that are used to document and handle the data that is produced (Hall and Byrd 2012).

6.4 Types of Evidence and Analyses

There are several types of evidence that participate in the ecological context of forensic science and assist courts or judicial proceedings. Due to traceability of the evidence, the process of collection and preservation requires a sensitive approach. Table 6.2 represents the plant-based ecological evidence and their collection and preservation (SDG 16).

These ecological analysis techniques in crime scene investigation use of environmental science principles to reconstruct events, identify sources, and estimate the time of events. Techniques include diatom analysis, forensic ecogenomics, and environmental forensics, which help trace evidence, understand the movement of objects, and determine the provenance of evidence (Bock and Norris 2016).

Diatom Analysis: Diatoms, microscopic algae, can be found in various environments (soil, water, even human tissue) and can be used to determine the environmental origin of a crime scene or evidence. By identifying specific diatom species, investigators can gain insights into the potential location of a crime scene or where an object may have come from (Bock and Norris 2016).

Forensic Ecogenomics: This technique uses the study of microbial communities to analyze environmental samples and reconstruct events. For example, microbial analysis can be used to estimate the postmortem interval (PMI) or the time since a body was deposited. It can also help determine the origin of soil samples or other environmental evidence, aiding in the identification of clandestine grave locations or the reconstruction of events (Ralebitso-Senior 2018).

Table 6.2 Various types of ecological evidence and their collection and examination procedure

Type of evidence	Collection procedure	Examination procedure
Algae (e.g., diatoms, chlorophyta)	Collect water, sediment, or debris samples from the scene (e.g., clothing, hair, surfaces). Preserve in ethanol or formalin	Microscopic examination (light/SEM), identification via morphology of frustules (for diatoms); may use DNA for chlorophytes; compare species with environmental records (Pandey et al. 2016)
Testate amoebae	Extract from soil, sediment, or peat samples using flotation or chemical treatments	Microscopic analysis post-acid treatment; identify tests (shells) composed of silica/$CaCO_3$; compare morphology with reference slides (Charman et al. 2000)
Miscellaneous organisms (e.g., mites, nematodes, mollusks)	Recover via sieving soil/debris, flotation, or palynological prep	Microscopic identification; taxonomic classification to infer environmental/temporal context; relevant for linking bodies to sites (Campbell et al. 2011)
Fungi and spores	Swab surfaces, scrape material from remains, clothing, or environment. Preserve with minimal moisture to prevent overgrowth	Identification via spore morphology (light microscopy/SEM); DNA analysis for species-level ID; assess habitat/geographic specificity; used for time-since-death estimation (P. Wiltshire 2010)
Non-seed plants (bryophytes, ferns)	Collect plant fragments or spores from clothing, footwear, or surrounding area; sample soil for spores	Identify via microscopy (surface sculpture); evaluate habitat specificity; useful in situational provenance and environmental reconstruction (Margesin and Schinner 2005)
Seed plants (gymnosperms/angiosperms)	Collect seeds, pollen, or plant parts from scene, suspects, or objects; use forceps/gloves to avoid contamination	Use light microscopy for pollen grains; SEM for detailed wall patterning; botanical identification to species/genus/family using floras or databases; possible DNA testing (Albach et al. 2021)
Wood (dendrochronology/dendrochemistry)	Collect wood fragments with clean tools. Record context (e.g., embedded in object, near remains)	Analyze growth rings under magnification for age/dating; dendrochemistry (e.g., EDXRF) for pollution history and environmental exposure (Ballikaya et al. 2022)
Molecular plant evidence	Extract DNA from plant tissues, seeds, or fragments; preserve with silica gel or freeze	PCR-based DNA identification for known species; limited use due to lack of databases; suitable for Cannabis or tomato in specific contexts (De Boer et al. 2021)

(continued)

Table 6.2 (continued)

Type of evidence	Collection procedure	Examination procedure
Pollen and spores (palynology)	Collect soil, dust, or swabs from suspect/scene/items. Use tape lifts or rinse collections	Light microscopy, SEM; identify via sporopollenin patterns, exine sculpture; match against local flora databases or palynological references (Jones and Bryant 2007)
Phytoliths, trichomes, acritarchs, fossil spores	Extract from soils/ sediments via heavy-liquid separation; preserve dry	Examine using SEM or polarized light microscopy; match micromorphological features with known databases; used for provenance and activity reconstruction (Mazuy et al. 2024)

Environmental Forensics: This broader discipline uses chemical, physical, and statistical techniques to investigate contaminants and other environmental evidence. For instance, chemical fingerprinting can help determine the source of a chemical spill or the origin of trace evidence (Introduction to Environmental Forensics 2007; Environmental Forensics 2005).

6.4.1 Water Samples

A lot of the time, toxins enter an environment through water. Researchers examine water to determine the amount and presence of toxic materials and to gauge the general health of the ecosystem (Boelee et al. 2019).

Key parameters include:

pH: It is an indicator of alkalinity or acidity. Variations from typical ranges may be a sign of runoff from farms, acid rain, or industrial pollutants (Dey et al. 2021).

Turbidity: Usually associated with soil erosion, algae blooms, or building activity, turbidity is the cloudiness of water brought on by suspended solids (Davies-Colley and Smith 2001).

Nitrates and Phosphates: Excess amounts of these constituents cause eutrophication, which depletes oxygen and causes large die-offs in aquatic life (Badamasi et al. 2019).

Collection Techniques: Samples must be collected in sterile, chemically inert, and clean containers (such as HDPE bottles or borosilicate glass) (Wątor and Dobrzyński 2022).

Samples should ideally be taken at various depths and locations to account for spatial variability (Haining 2001).

Chemical and biological components can change quickly for that water must be stored at 4 °C and examined promptly (Lloyd et al. 2022).

Temperature, weather, water flow, and visual observations (such as color and odor) should all be included in field notes (Zhang and Wen 2023).

6.4.2 Soil Samples

The long-term contamination, particularly from industrial spills, excessive pesticide use, or heavy metal deposition, might be inferred from soil data. Soil can affect nearby ecosystems as a source and sink of pollution through leaching or dust (Tan 2005).

Collection Technique:
Investigators take different soil samples from predetermined depths using soil augers or corers, frequently in a grid or transect pattern.

The samples are kept in nonreactive, airtight containers such glass jars or polyethylene bags (Tan 2005). It is important to sieve and air dry the soil sample prior to analysis (Tan 2005). Care must be given to prevent cross-contamination between locations (Tan 2005). Measuring pH, organic matter content, and the levels of pollutants like lead, arsenic, and hydrocarbons are all part of the analytical focus (Carter and Gregorich 2008).

6.4.3 Biological Samples

Ecological stress, which can act as a bioindicator, can be reflected by biota throughout time. Living or dead species, such as fish, invertebrates, plant tissues, or the carcasses of impacted wildlife, are examples of biological evidence. These samples aid in the establishment of biological effects such as illness, infertility, or death (Oosthuizen et al. 2022).

Collection Techniques: To prevent contamination, samples are handled with sterile gloves and equipment (Singh 2021).

Tissues and animal remains are kept in airtight containers or frozen, while plant material is frequently wrapped in paper envelopes to keep mold from growing (Funk et al. 2017).

Chain-of-custody records are essential, especially if samples are going to be used in court.

Toxic substances like dioxins, pesticides, or endocrine-disrupting chemicals can be found in laboratories by genetic testing, bioaccumulation investigations, or histopathology (Diamanti-Kandarakis et al. 2009).

6.5 Tools Used in Evidence Collection

The quality and appropriateness of the tools or instruments used in the field have a direct impact on the accuracy of environmental data. Investigators are outfitted with several specialized tools to gather, measure, and preserve evidence under a range of environmental circumstances(Goray et al. 2024).

6.5.1 Thermometer

Typical instruments include thermometers, which are used to measure the ambient temperature of soil, water, and air. Both biological activity and chemical reactions are influenced by temperature (Yolcubal et al. 2004).

6.5.2 Field Test Kits

These portable kits provide quick on-site testing for a variety of parameters, including nitrates, phosphates, dissolved oxygen, and pH (Morrison et al. 2018).

6.5.3 Hand Lenses or Magnifying Glasses

These tools facilitate the close inspection of plant structures, micro debris, and invertebrates (Lirëza and Koçi 2023).

6.5.4 Collection Containers

These containers may include vials, glass containers, centrifuge tubes, and sample bags, depending on the type of material and the needs of the analysis (Malakouti and Hazrati 2025).

To make sure measurements are accurate and to keep contaminants from getting into samples, each tool must be properly adjusted and cleaned.

6.6 Data Recording

Documentation of field investigations must be comprehensive and well-organized. Evidences credibility depends on how it is recorded, preserved, and presented in addition to how it is gathered (Sachowski 2016).

Some of the data recording practices includes:

6.6.1 Tables

Used to record statistics, including sample amounts, pollutant levels, and temperature readings.

6.6.2 Graphs and Charts

Useful for showing patterns over time or across locations, like variations in nitrate or pH levels (Sachowski 2016).

6.6.3 Maps

Give sample locations, impacted areas, pollution sources, and ecological boundaries a spatial context. GPS and GIS technology can be used to combine geospatial data (Sachowski 2016).

6.6.4 Field Logs

Documentation of reports, such as weather, odd smells or colors, and the behavior of wildlife (Sachowski 2016).

The adoption of the required procedure potentially impacts the investigation and improves the outcomes. Hence, in crime scene investigation, the role of ecological evidence extends the limitation beyond a shadow of doubt. Table 6.3 represents the key benefits of ecological analysis in crime scene investigation.

6.7 Possible Ecological Offenders

Finding possible culprits is crucial to figuring out the underlying causes of environmental damage in ecological crime scene investigations. Ecological crimes frequently involve businesses, procedures, or systematic actions that disturb ecosystems, in contrast to traditional criminal investigations, where suspects are frequently people or organizations. The main types of ecological offenders are examined in this chapter, along with how evidence might be linked to them and profiles of common ecological suspects (SDG-13, 14, 15 & 16).

6.7.1 Industrial Pollution

Industries are frequently the main culprits in ecological degradation because they release pollutants into the air, water, and soil. Hazardous materials may be released by some of the chemical plants, factories, and manufacturing facilities either lawfully under lax regulations or illegally to save money on disposal. For example, releasing heavy metals like lead, cadmium, and mercury into streams can have a major negative impact on human and wildlife health as well as cause long-term ecological harm (Das et al. 2023). Frequently linked to industrial operations, persistent organic pollutants (POPs) bioaccumulate in the food chain and interfere with both human and animal endocrine systems (Martínez Valenzuela et al. 2022).

Table 6.3 Significance of ecological analysis in crime scene investigation

Aspect	Description	Application/benefit
Tracing evidence	Analysis of ecological materials such as soil, pollen, or water	Helps link suspects to crime scenes by matching trace materials (Alotaibi et al. 2020)
Reconstructing events	Examining environmental conditions and disturbances	Aids in determining the sequence and timing of events at the crime scene (A. Kumar et al. 2022)
Estimating PMI/ PMSI	Use of ecological data (e.g., insect activity, water conditions)	Assists in estimating post-mortem interval or submersion time (Y. Wang et al. 2022)
Identifying grave locations	Analysis of soil chemistry, microbial activity, and vegetation changes	Useful in detecting clandestine graves or disturbed burial sites (Watson et al. 2021)
Upstream/ downstream analysis	Sampling water bodies to trace pollution origins	Pinpoints sources of contamination (factories, runoff), crucial in water-related crime scenes (Chakraborti 2021)
Satellite imagery	High-resolution images to detect land-use changes	Detects illegal activities such as deforestation or dumping by comparing past and present images (Merugu and Jain 2013)
GIS mapping	Spatial analysis of contamination and ecological features	Visualizes pollutant spread or ecological damage across a region (Chen et al. 2022)
Weather and wind pattern analysis	Correlating weather data with pollutant movement	Traces air pollution, smoke, or odor sources; supports arson and environmental crime investigations (Bolan et al. 2024; Rahman et al. 2022)
Timeline determination	Sediment analysis, chemical degradation, historical data comparison	Establishes when environmental crimes occurred, aiding in legal proceedings and environmental remediation (Mudge 2008)
Environmental forensics	Integration of ecology, chemistry, and data science	Provides scientific evidence in court for ecological damage and identifies responsible parties (Kolobaric et al. 2023)

6.7.2 Agricultural Runoff

Agriculture, especially industrial-scale operations, contributes heavily to nonpoint source pollution through runoff, including fertilizers, pesticides, and animal waste. In aquatic environments, added nutrients from nitrogen and phosphorus compounds can cause eutrophication, which can result in hypoxic "dead zones" and algal blooms (Smith and Schindler 2009). These regions lose oxygen, endangering fisheries and aquatic biodiversity. Furthermore, the ecological impact may be further complicated by pesticide residues that linger in the environment and damage soil organisms as well as food webs (Tilman et al. 2002).

6.7.3 Illegal Logging

Deforestation and forest degradation are largely caused by illegal logging, particularly in tropical areas with abundant biodiversity. In addition to lowering the capacity for storing carbon, this activity damages habitats, which results in species extinction and ecological imbalance (Gunawan et al. 2024). Illegal logging operations, which are frequently associated with organized crime, circumvent legal frameworks, fabricate permits, and take advantage of remote forest locations where enforcement is scarce. Satellite photography and remote sensing technologies are being utilized more and more to identify such operations and link them to accountable parties. (Hansen et al. 2013).

6.7.4 Poaching or Overhunting

Many species are at risk from poaching and unsustainable hunting methods, particularly those that are already endangered or vulnerable. Poachers, who are driven by survival, tradition, or financial gain, frequently employ advanced techniques to avoid detection, including snares, guns, and sophisticated tracking systems. The illicit wildlife trade, which is estimated to cost billions of dollars a year, not only causes population decline but also modifies ecosystem dynamics by displacing keystone species and upsetting predator-prey relationships (Rana and Kumar 2023). Among the methods used to track down poached animals and identify the people or networks responsible are forensic ballistics, satellite tracking, and DNA analysis.

6.8 Clue Relationships: Correlating Suspect Behavior with Evidence

Gathering and examining a range of data, such as chemical residues, wildlife corpses, tree stumps, satellite imaging, and eyewitness reports, is necessary to establish the connection between ecological damage and possible culprits. For example, water samples carrying certain contaminants, such as PCBs or nitrates, might be linked to industrial or agricultural sources depending on local land use patterns and discharge records (Diamond 1999).

When illegal logging occurs, the origin of confiscated timber can be verified by comparing the growth ring patterns or DNA of the wood with stumps discovered at the site. Similarly, movement disturbances or odd deaths suggestive of poaching might be found using GPS tracking of tagged animals. By generating believable cause-and-effect links, these hints assist investigators in delineating the actions and intentions of ecological offenders.

6.8.1 Profile Creation

Ecological crime scene investigators create thorough profiles of suspected criminals to support investigations and enforcement efforts. These profiles give a thorough picture of the suspect's ecological footprint by combining information from public sources, forensic analysis, and field observations.

Factory X: Factory X is situated upstream from a river system that is rich in biodiversity, previously cited for exceeding waste discharge limitations. Near its exit, recent samples reveal high amounts of cadmium and phenol, indicating continuous pollution. Witness reports and drone footage also show unregistered pipes avoiding treatment facilities.

Farm Y: A significant farming operation that grows soybeans and maize. Fertilizer application schedules align with elevated nitrate and glyphosate levels seen in neighboring wetlands. Direct flow into nearby water bodies during the rainy season is shown by drainage schemes and soil erosion patterns.

Logging Group Z: Using fictitious permissions to operate in a restricted forest area. Illegal access roads and inconsistent clearing patterns are seen in satellite images. While residents report increasing truck movement at night, tree stump research reveals that protected species are being harvested.

Poaching Ring A: Several tagged elephants in a wildlife reserve went missing as a result of poaching by Ring A. Snares, severed tusks, and bullet casings found in the region are consistent with techniques employed by poaching rings that have already been discovered. Suspects are present at the scene when the killings occur, thanks to mobile phone data triangulation.

These profiles are crucial for guiding policy responses, public awareness campaigns, and ecological restoration initiatives, in addition to being used in court.

6.9 Notable Cases in Forensic Ecology

Forensic ecology has played a critical role in solving several high-profile criminal and environmental cases. In one notable instance, soil analysis linked a suspect to a remote crime scene by matching rare pollen types found on his shoes. In another landmark case, satellite imagery and GIS mapping were used to expose illegal deforestation activities tied to corporate land encroachment. Forensic ecologists have also successfully traced industrial chemical spills to their sources, leading to prosecutions for environmental violations. These cases highlight the growing impact of ecological evidence in both criminal justice and environmental law enforcement.

6.9.1 BP Deepwater Horizon Oil Spill

One of the worst environmental disasters in history was the BP Deepwater Horizon oil spill in the year 2010. Local businesses, marine ecosystems, and coastal habitats were all severely damaged when an estimated 4.9 million barrels of oil spilled into the Gulf of Mexico over a period of 87 days (National Commission on the BP Deepwater Horizon Oil Spill and Offshore (Makocha et al. 2019).

Ecological crime scene investigators employed chemical fingerprinting to track the oil's movement and detect long-term environmental harm. Although comprehensive, cleanup operations were criticized for using chemical dispersants that caused harm to marine life (Peterson et al. 2003). To prevent future similar catastrophes, this occurrence made clear the necessity of quick environmental assessments and more robust regulatory frameworks. Even after spending billions on rehabilitation, almost ten years later, some deep-sea communities are still severely affected.

6.9.2 Amazon Rainforest Illegal Logging

One chronic environmental crime that greatly contributes to global deforestation and biodiversity loss is illegal logging in the Amazon rainforest. Investigating these crimes has relied heavily on forensic wood examination, satellite images, and drone footage (Brancalion et al. 2018). Ecological investigators frequently come across obstacles, including violent opposition from illicit activities and dishonest law enforcement.

Degradation of the Amazon impacts indigenous cultures, many species, and the stability of the global climate. Even while global pressure and certification programs like the Forest Stewardship Council (FSC) have reduced certain illicit activity, political unrest, and enforcement shortcomings still stand in the way of advancement (Barlow et al. 2016).

6.9.3 Chernobyl Wildlife Recovery

After the Chernobyl nuclear accident in 1986, huge areas of land were rendered unusable by people. The quarantine zone has surprisingly become an accidental wildlife haven. The region has been utilized as a "living laboratory" by ecologists to investigate the long-term effects of radiation on ecosystems (Deryabina et al. 2015).

The recovery of wildlife populations, such as wolves, boars, and lynx, indicates that, under certain situations, the ecological consequences of radiation are outweighed by the lack of human interference. However, in other species, genetic changes and population decreases have also been reported by researchers (Møller and Mousseau 2011). Ecosystems can be resilient, as demonstrated by the Chernobyl case, but it also emphasizes how complicated the long-term effects of contamination can be.

6.10 Conclusive Learning: What Worked, What Didn't

The recognized ecological crime cases have shown insightful analyses of both effective tactics and routine obstacles in environmental forensic investigations.

What was effective, successes have come cooperation, public involvement, and the accurate use of forensic technology. Recent chemical analyses, for example, helped to identify the source of pollution in the Deepwater Horizon oil disaster, hence guiding focused cleanup actions. Continuous satellite monitoring has also shown unlawful deforestation in the Amazon basin, hence providing real-time proof to back legal actions.

Many steps, nevertheless, have been hampered by poor investigation systems, tardy reactions, and improper long-term planning. Traditional environmental studies following the Chernobyl catastrophe overstated radiation hazards, which resulted in false public policy. Moreover, the lack of continuous ecological monitoring has hampered a better knowledge of ecosystem recovery even if animals in the exclusion zone were clearly returning. These instances underline the need of constant control, openness, and long-term environmental management to guarantee ecological justice.

The youth pave new ways. Particularly in environmental advocacy and ecological responsibility, new generation activism has developed as a transforming force. As Greta Thunberg's Fridays for Future campaign shows, the potential of young-led objectives in changing the conversation on climate action. Her focus on moral duty and scientific integrity has motivated millions to call for change from governments and companies both (Thunberg 2019).

Youth participation via eco-clubs and local groups has had real environmental effects at the grassroots level. Young activists are not only raising awareness but also producing useful data for scientific research by planning reforestation campaigns to carrying out biodiversity surveys in regions impacted by illegal logging or oil spills (Roberts 2009). These youth-led initiatives highlight the possibility of community science in supporting official ecological research.

Young environmental leaders' dedication and tenacity show that the younger generation can propel significant and lasting change in ecological preservation and justice when armed with information, tools, and a platform.

6.11 Challenge to Investigate Ecological Crime Scene

The most difficult concern of ecological crime scene examination, which we have learnt, is controlling the immense complexity and massive size of these crimes. The concern is very high, not just for the environment but for the civilization of humans as well, whether it be pollution, illicit logging, wildlife trafficking, or ecological loss. The experts have the difficult challenge of negotiating a labyrinth of legal, environmental, and ethical issues, as well as naming the offenders. Important Issues in Ecological Crime Scene Investigation are mentioned below.

6.11.1 Explanation of "Ecocide"

To what extent does the environmental degradation turn into crime? Still developing is the process of defining ecocide—extreme and widespread environmental destruction—as a legal offence. It calls for not only scientific clarity but also worldwide agreement on what constitutes permanent harm.

6.11.2 The Concern Over: Protection Versus Development

Presently, the world is at an endless debate over a sensitive topic to balance environmental protection with economic growth. Though safeguarding ecosystems is important, on the other hand, the industrial activities of "ecocide" might impede economic development. This conflict may sometimes put lawmakers and probes in a tight position.

6.11.3 Environmental Crimes Don't Affect Scenery

They wasted communities. The loss is unbearable from degradation of environment to extinction of species. These offences endanger the cultural legacy of indigenous people, water supplies, and food security as well.

Unlike many ongoing crimes, the outcomes of environmental crimes are typically gradual and long-reaching. Deliberate deforestation may change climatic patterns for centuries, while one unlawful garbage dump now might cause cancer epidemics decades later.

6.11.4 A Global Issue Too Large for One Country Alone

Environmental offences are seldom limited by boundaries. Cross-border pollution, animal trafficking, and illegal garbage exports need for coordinated global response. Regrettably, uneven legislation and insufficient collaboration might let criminals fall in the gaps.

Ecological crime scenes are seas, woods, and rivers, not cordoned-off rooms. Collecting and preserving evidence is made very challenging by remote places, severe conditions, and natural deterioration. Key clues may be washed away by a rainfall. A changing tide may wipe out a path for all time.

Knowledge across disciplines is crucial: No one specialist can fix an ecological crime. Environmental scientists, forensic analysts, legal professionals, and conservation biologists working together produces results. Missing vital insights would be impossible without this mix of expertise.

6.11.5 Increasing Awareness and Fortifying Justice

At last, we must confront a sad reality: Many still do not grasp the gravity of environmental offences. Public awareness, legal system improvement, and more enforcement are vital first measures not just for looking into these offences but also for their prevention.

6.12 Conclusion

As the world steps into an era defined by environmental protection urgency, ecological crime scene investigation is no longer a niche field; it is a global need. The alarming increase in crimes that destroy our ecosystems, threaten biodiversity, and compromise people's health demands a shift in understanding justice, science, and responsibility. When looking ahead, the trend of ecological crime investigation must be rooted in collaboration, innovation, and accountability. Innovation in forensic ecogenomics, remote sensing, artificial intelligence, and large data analytics is tended to revolutionize the detection, analysis, and prosecute environmental crimes. Just as important is the integration of traditional ecological knowledge from indigenous communities, whose insights can deepen our understanding of ecosystem change. Moreover, technological development individually will not be sufficient. Strict international laws defining ecocide, separate environmental courts, and international cooperation are essential to closing legal loop holes and delivering meaningful justice. Education and youth involvement will also play a critical role, ensuring that future generations not only inherit a liveable planet but are also equipped to protect it. These cordially joint efforts can directly aid the United Nations Sustainable Development Goals—SDG 13 (Climate Action), SDG 14 (Life Below Water), SDG 15 (Life on Land), and SDG 16 (Peace, Justice, and Strong Institutions)—which collectively advocate for environmental preservation, justice, and sustainable ecosystems.

Conclusively, ecological crime scene investigation confronts and evolves into a pillar of environmental governance. It must be proactive rather than reactive, guided by the principle that protecting nature is not just a scientific or legal duty, it is a moral imperative. In doing so, we lay the foundation for a world where justice is not limited to humanity but extended to the living planet we all share.

References

Adla K, Dejan K, Neira D, Dragana Š (2022) Degradation of ecosystems and loss of ecosystem services. In: One health. Elsevier, pp 281–327. https://doi.org/10.1016/B978-0-12-822794-7.00008-3

Agnew DJ, Pearce J, Pramod G, Peatman T, Watson R, Beddington JR, Pitcher TJ (2009) Estimating the worldwide extent of illegal fishing. PLoS One 4(2):e4570. https://doi.org/10.1371/journal.pone.0004570

Albach DC, Tsymbalyuk ZM, Mosyakin SL (2021) Pollen morphology of Ellisiophyllum and Sibthorpia (Plantaginaceae, tribe Sibthorpieae) and phylogenetics of the tribe. Plant Syst Evol 307(6):66. https://doi.org/10.1007/s00606-021-01786-9

Alotaibi SS, Sayed SM, Alosaimi M, Alharthi R, Banjar A, Abdulqader N, Alhamed R (2020) Pollen molecular biology: applications in the forensic palynology and future prospects: a review. Saudi J Biol Sci 27(5):1185–1190. https://doi.org/10.1016/j.sjbs.2020.02.019

Ankit SL, Kumar V, Tiwari J, Sweta RS, Singh J, Bauddh K (2021) Electronic waste and their leachates impact on human health and environment: global ecological threat and management. Environ Technol Innov 24:102049. https://doi.org/10.1016/j.eti.2021.102049

Ashraf MA (2017) Persistent organic pollutants (POPs): a global issue, a global challenge. Environ Sci Pollut Res 24(5):4223–4227. https://doi.org/10.1007/s11356-015-5225-9

Badamasi H, Yaro MN, Ibrahim A, Aliyu Bashir I (2019) Impacts of phosphates on water quality and aquatic life. Chem Res 4:124–133

Ballikaya P, Marshall J, Cherubini P (2022) Can tree-ring chemistry be used to monitor atmospheric nanoparticle contamination over time? Atmos Environ 268:118781. https://doi.org/10.1016/j.atmosenv.2021.118781

Barclay E, Bartel R (2015) Defining environmental crime: the perspective of farmers. J Rural Stud 39:188–198. https://doi.org/10.1016/j.jrurstud.2015.01.007

Barlow J, Lennox GD, Ferreira J, Berenguer E, Lees AC, Nally RM, Thomson JR, Ferraz SFDB, Louzada J, Oliveira VHF, Parry L, De Castro R, Solar R, Vieira ICG, Aragão LEOC, Begotti RA, Braga RF, Cardoso TM, De Oliveira RC, Souza CM Jr et al (2016) Anthropogenic disturbance in tropical forests can double biodiversity loss from deforestation. Nature 535(7610):144–147. https://doi.org/10.1038/nature18326

Blount D, Gero S, Van Oast J, Parham J, Kingen C, Scheiner B, Stere T, Fisher M, Minton G, Khan C, Dulau V, Thompson J, Moskvyak O, Berger-Wolf T, Stewart CV, Holmberg J, Levenson JJ (2022) Flukebook: An open-source AI platform for cetacean photo identification. Mamm Biol 102(3):1005–1023. https://doi.org/10.1007/s42991-021-00221-3

Bock JH, Norris DO (2016) Forensic plant science. Elsevier

Boelee E, Geerling G, Van Der Zaan B, Blauw A, Vethaak AD (2019) Water and health: from environmental pressures to integrated responses. Acta Trop 193:217–226. https://doi.org/10.1016/j.actatropica.2019.03.011

Bolan S, Padhye LP, Jasemizad T, Govarthanan M, Karmegam N, Wijesekara H, Amarasiri D, Hou D, Zhou P, Biswal BK, Balasubramanian R, Wang H, Siddique KHM, Rinklebe J, Kirkham MB, Bolan N (2024) Impacts of climate change on the fate of contaminants through extreme weather events. Sci Total Environ 909:168388. https://doi.org/10.1016/j.scitotenv.2023.168388

Brancalion PHS, De Almeida DRA, Vidal E, Molin PG, Sontag VE, Souza SEXF, Schulze MD (2018) Fake legal logging in the Brazilian Amazon. Sci Adv 4(8):eaat1192. https://doi.org/10.1126/sciadv.aat1192

Bytnerowicz A (2009) Wildland fires and air pollution. Elsevier

Campbell G, Moffett L, Straker V (2011) Environmental archaeology: a guide to the theory and practice of methods from sampling and recovery to post-excavation

Cao C, Yang B, Xu M, Li X, Singh RP, Zhao X, Chen W (2016) Evaluation and analysis of post-seismic restoration of ecological security in Wenchuan using remote sensing and GIS. Geomat Nat Haz Risk 7(6):1919–1936. https://doi.org/10.1080/19475705.2015.1084952

Cardoso P, Amponsah-Mensah K, Barreiros JP, Bouhuys J, Cheung H, Davies A, Kumschick S, Longhorn SJ, Martínez-Muñoz CA, Morcatty TQ, Peters G, Ripple WJ, Rivera-Téllez E, Stringham OC, Toomes A, Tricorache P, Fukushima CS (2021) Scientists' warning to humanity on illegal or unsustainable wildlife trade. Biol Conserv 263:109341. https://doi.org/10.1016/j.biocon.2021.109341

Carter MR, Gregorich EG (eds) (2008) Soil sampling and methods of analysis, 2nd edn. Canadian Society of Soil Science/CRC Press

Chakraborti L (2021) Impact of upstream plant level pollution on downstream water quality: evidence from the clean water act. J Environ Plan Manag 64(3):517–535. https://doi.org/10.1080/09640568.2020.1776227

Charman D, Hendon D, Woodland WA (2000) The identification of testate amoebae (Protozoa: Rhizopoda) in peats, Technical Guide No. 9. Quaternary Science Reviews, 9
Chen L, Mao Y, Zhao R (2022) GIS application in environmental monitoring and risk assessment. In: 2022 3rd International Conference on Geology, Mapping and Remote Sensing (ICGMRS), pp 908–917. https://doi.org/10.1109/ICGMRS55602.2022.9849269
Clements R, Rayan DM, Ahmad Zafir AW, Venkataraman A, Alfred R, Payne J, Ambu L, Sharma DSK (2010) Trio under threat: can we secure the future of rhinos, elephants and tigers in Malaysia? Biodivers Conserv 19(4):1115–1136. https://doi.org/10.1007/s10531-009-9775-3
Cooper JE, Cooper ME (2013) Wildlife forensic investigation: principles and practice. CRC Press
Cubas J, Martín-Esquivel JL, Marrero-Gómez M, Díaz JRD, Rodríguez F, González-Mancebo JM (2022) Climate change causes rapid collapse of a keystone shrub from insular alpine ecosystems. J Nat Conserv 69:126263. https://doi.org/10.1016/j.jnc.2022.126263
Das S, Sultana KW, Ndhlala AR, Mondal M, Chandra I (2023) Heavy metal pollution in the environment and its impact on health: exploring green Technology for Remediation. Environ Health Insights 17:11786302231201259. https://doi.org/10.1177/11786302231201259
Davies-Colley RJ, Smith DG (2001) Turbidity suspenied sediment, and water clarity: a review. JAWRA J Am Water Res Assoc 37(5):1085–1101. https://doi.org/10.1111/j.1752-1688.2001.tb03624.x
De Boer H, Rydmark MO, Verstraete B, Gravendeel B (2021) Molecular identification of plants: from sequence to species. Pensoft Publishers. https://doi.org/10.3897/ab.e98875
Deryabina TG, Kuchmel SV, Nagorskaya LL, Hinton TG, Beasley JC, Lerebours A, Smith JT (2015) Long-term census data reveal abundant wildlife populations at Chernobyl. Curr Biol 25(19):R824–R826. https://doi.org/10.1016/j.cub.2015.08.017
Dey S, Botta S, Kallam R, Angadala R, Andugala J (2021) Seasonal variation in water quality parameters of Gudlavalleru Engineering College pond. Curr Res Green Sustain Chem 4:100058. https://doi.org/10.1016/j.crgsc.2021.100058
Diamanti-Kandarakis E, Bourguignon J-P, Giudice LC, Hauser R, Prins GS, Soto AM, Zoeller RT, Gore AC (2009) Endocrine-disrupting chemicals: An Endocrine Society scientific statement. Endocr Rev 30(4):293–342. https://doi.org/10.1210/er.2009-0002
Diamond J (1999) Restoring life in running waters: better biological monitoring. JR Karr, EW Chu. J N Am Benthol Soc 18(2):297–298. https://doi.org/10.2307/1468472
Donnermeyer JF (2015) Crime in the rural context. In: International encyclopedia of the Social & Behavioral Sciences. Elsevier, pp 158–163. https://doi.org/10.1016/B978-0-08-097086-8.45049-X
Dutta S, Adhikary S, Bhattacharya S, Roy D, Chatterjee S, Chakraborty A, Banerjee D, Ganguly A, Nanda S, Rajak P (2024) Contamination of textile dyes in aquatic environment: adverse impacts on aquatic ecosystem and human health, and its management using bioremediation. J Environ Manag 353:120103. https://doi.org/10.1016/j.jenvman.2024.120103
Eekhout JPC, De Vente J (2022) Global impact of climate change on soil erosion and potential for adaptation through soil conservation. Earth Sci Rev 226:103921. https://doi.org/10.1016/j.earscirev.2022.103921
Eliason S (2003) Illegal hunting and angling: the neutralization of Wildlife Law Violations. Soc Anim 11(3):225–243. https://doi.org/10.1163/156853003322773032
Environmental forensics (2005) Elsevier. https://doi.org/10.1016/B978-0-12-507751-4.X5021-6
Freudenburg WR, Gramling R (2010) Blowout in the Gulf: the BP oil spill disaster and the future of energy in America. The MIT Press. https://doi.org/10.7551/mitpress/9141.001.0001
Funk V, Gostel M, Devine A, Kelloff C, Wurdack K, Tuccinardi C, Radosavljevic A, Peters M, Coddington J (2017) Guidelines for collecting vouchers and tissues intended for genomic work (Smithsonian Institution): Botany Best Practices. Biodiver Data J 5:e11625. https://doi.org/10.3897/BDJ.5.e11625
Goray M, Taylor D, Bibbo E, Fantinato C, Fonneløp AE, Gill P, Van Oorschot RAH (2024) Emerging use of air eDNA and its application to forensic investigations—a review. Electrophoresis 45(9–10):916–932. https://doi.org/10.1002/elps.202300228

Gouda S, Kerry RG, Das A, Chauhan NS (2020) Wildlife forensics: a boon for species identification and conservation implications. Forensic Sci Int 317:110530. https://doi.org/10.1016/j.forsciint.2020.110530

Gunawan H, Setyawati T, Atmoko T, Kwatrina RT, Yeny I, Yuwati TW, Effendy R, Abdullah L, Lastini T, Arini DI, Sari UK (2024) A review of forest fragmentation in Indonesia under the DPSIR framework for biodiversity conservation strategies. Glob Ecol Conserv 51:e02918. https://doi.org/10.1016/j.gecco.2024.e02918

Haining RP (2001) Spatial sampling. In: International encyclopedia of the social & behavioral sciences. Elsevier, pp 14822–14827. https://doi.org/10.1016/B0-08-043076-7/02510-9

Hall DW, Byrd JH (2012) Forensic botany: a practical guide. John Wiley & Sons. https://doi.org/10.1002/9781119945734

Hansen MC, Potapov PV, Moore R, Hancher M, Turubanova SA, Tyukavina A, Thau D, Stehman SV, Goetz SJ, Loveland TR, Kommareddy A, Egorov A, Chini L, Justice CO, Townshend JRG (2013) High-resolution global maps of 21st-century forest cover change. Science 342(6160):850–853. https://doi.org/10.1126/science.1244693

Harada M (1995) Minamata disease: methylmercury poisoning in Japan caused by environmental pollution. Crit Rev Toxicol 25(1):1–24. https://doi.org/10.3109/10408449509089885

Introduction to environmental forensics (2007) Elsevier. https://doi.org/10.1016/B978-0-12-369522-2.X5000-3

Jernelöv A (2010) The threats from oil spills: now, then, and in the future. Ambio 39(5–6):353–366. https://doi.org/10.1007/s13280-010-0085-5

Jones GD, Bryant VM (2007) A comparison of pollen counts: light versus scanning electron microscopy. Grana 46(1):20–33. https://doi.org/10.1080/00173130601173897

Kolobaric A, Cozzolino D, Chapman J (2023) Environmental forensics. In: Sharma V, Sauzier G, Lewis SW (eds) Chemometric methods in forensic science. Royal Society of Chemistry, pp 254–293. https://doi.org/10.1039/BK9781839166099-00254

Kotzé LJ, Adelman S (2023) Environmental law and the unsustainability of sustainable development: a tale of disenchantment and of hope. Law Critique 34(2):227–248. https://doi.org/10.1007/s10978-022-09323-4

Kumar P, Debele SE, Sahani J, Rawat N, Marti-Cardona B, Alfieri SM, Basu B, Basu AS, Bowyer P, Charizopoulos N, Jaakko J, Loupis M, Menenti M, Mickovski SB, Pfeiffer J, Pilla F, Pröll J, Pulvirenti B, Rutzinger M et al (2021) An overview of monitoring methods for assessing the performance of nature-based solutions against natural hazards. Earth Sci Rev 217:103603. https://doi.org/10.1016/j.earscirev.2021.103603

Kumar A, Sharma V, Kumar S (2022) Tools for crime scene reconstruction-a review study. Int J Med Toxicol Legal Med 25(3 and 4):268–272. https://doi.org/10.5958/0974-4614.2022.00091.2

Lastumäki A, Turja R, Brenner M, Vanninen P, Niemikoski H, Butrimavičienė L, Stankevičiūtė M, Lehtonen KK (2020) Biological effects of dumped chemical weapons in the Baltic Sea: a multi-biomarker study using caged mussels at the Bornholm main dumping site. Mar Environ Res 161:105036. https://doi.org/10.1016/j.marenvres.2020.105036

Lee HC, Palmbach T, Miller MT (2001) Henry Lee's crime scene handbook. Academic

Lev-Yadun S, Račko V, Kardošová M, Ďurkovič J (2023) Indented growth-rings (hazel wood) deserve more attention. Dendrochronologia 79:126073. https://doi.org/10.1016/j.dendro.2023.126073

Lirëza L, Koçi G (2023) Environmental crimes: their nature, scope, and problems in identification. Interdisc J Res Dev 10(1 S1):237. https://doi.org/10.56345/ijrdv10n1s135

Lloyd CEM, Johnes PJ, Pemberton JA, Yates CA, Jones D, Evershed RP (2022) Sampling, storage and laboratory approaches for dissolved organic matter characterisation in freshwaters: moving from nutrient fraction to molecular-scale characterisation. Sci Total Environ 827:154105. https://doi.org/10.1016/j.scitotenv.2022.154105

Lynch MJ, Stretesky PB (2016) Exploring green criminology, 0th edn. Routledge. https://doi.org/10.4324/9781315581644

Makocha I, Ete T, Saini G (2019) Deepwater horizon oil spill: a review. Int J Tech Innov Mod Eng Sci 5:65–71

Malakouti Z, Hazrati M (2025) Interconnected challenges: examining the nexus of environmental crime and money laundering in the context of energy transition. J Econ Criminol 8:100151. https://doi.org/10.1016/j.jeconc.2025.100151

Margesin R, Schinner F (2005) Monitoring and assessing soil bioremediation. Springer Berlin Heidelberg. https://doi.org/10.1007/3-540-28904-6

Martínez Valenzuela C, Gavilán García A, Conde Avila V, Barrientos Alemán D, Apodaca Avalos M, Luna Valdez JG, Castro Carranza G, Masías Ambríz LO (2022) Applying the global monitoring plan and analysis of POPs results in atmospheric air in Mexico (2017–2018). Chemosphere 303:135154. https://doi.org/10.1016/j.chemosphere.2022.135154

Mazuy A, Ferrara V, Ekblom A, Delhon C (2024) A rapid and simple method for the extraction of biogenic silica (BSi) in phytolith-poor sediments and soils. MethodsX 12:102634. https://doi.org/10.1016/j.mex.2024.102634

Meghwal K, Kumawat S, Ameta C, Jangid NK (2020) Effect of dyes on water chemistry, soil quality, and biological properties of water. In: Wani KA, Jangid NK, Bhat AR (eds) Advances in human services and public health. IGI Global, pp 90–114. https://doi.org/10.4018/978-1-7998-0311-9.ch005

Merugu S, Jain K (2013) Change detection and estimation of illegal mining using satellite images

Møller AP, Mousseau TA (2011) Conservation consequences of Chernobyl and other nuclear accidents. Biol Conserv 144(12):2787–2798. https://doi.org/10.1016/j.biocon.2011.08.009

Morrison J, Watts G, Hobbs G, Dawnay N (2018) Field-based detection of biological samples for forensic analysis: established techniques, novel tools, and future innovations. Forensic Sci Int 285:147–160. https://doi.org/10.1016/j.forsciint.2018.02.002

Mozer A, Prost S (2023) An introduction to illegal wildlife trade and its effects on biodiversity and society. Forensic Sci Int 3:100064. https://doi.org/10.1016/j.fsiae.2023.100064

Mudge SM (2008) Environmental forensics and the importance of source identification. In: Hester RE, Harrison RM (eds) Environmental forensics, 1st edn. The Royal Society of Chemistry, pp 1–16. https://doi.org/10.1039/9781847558343-00001

Oosthuizen T, Howes LM, White R (2022) Forensic science and environmental offences: litter, DNA analysis and surveillance. Forensic Sci Int 2:100042. https://doi.org/10.1016/j.fsiae.2022.100042

Pandey L, Ojha K, Singh P, Singh C, Dwivedi S, Bergey E (2016) Diatoms image database of India (DIDI): a research tool. Environ Technol Innov 5:148–160

Peterson CH, Rice SD, Short JW, Esler D, Bodkin JL, Ballachey BE, Irons DB (2003) Long-term ecosystem response to the Exxon Valdez oil spill. Science 302(5653):2082–2086. https://doi.org/10.1126/science.1084282

Rahman MM, Shuo W, Zhao W, Xu X, Zhang W, Arshad A (2022) Investigating the relationship between air pollutants and meteorological parameters using satellite data over Bangladesh. Remote Sens 14(12):2757. https://doi.org/10.3390/rs14122757

Ralebitso-Senior TK (ed) (2018) Forensic ecogenomics: the application of microbial ecology analyses in forensic contexts. Academic Press

Rana AK, Kumar N (2023) Current wildlife crime (Indian scenario): major challenges and prevention approaches. Biodivers Conserv 32(5):1473–1491. https://doi.org/10.1007/s10531-023-02577-z

Roberts NS (2009) Impacts of the National Green Corps Program (Eco-Clubs) on students in India and their participation in environmental education activities. Environ Educ Res 15(4):443–464. https://doi.org/10.1080/13504620902994127

Sachowski J (2016) Maintain evidence-based reporting. In: Implementing digital forensic readiness. Elsevier, pp 137–142. https://doi.org/10.1016/B978-0-12-804454-4.00013-7

Shepherd CR, Leupen BTC, Siriwat P, Nijman V (2020) International wildlife trade, avian influenza, organised crime and the effectiveness of CITES: the Chinese hwamei as a case study. Glob Ecol Conserv 23:e01185. https://doi.org/10.1016/j.gecco.2020.e01185

Shetty SS, Sonkusare S, Naik PB, Kumari NS, Madhyastha H (2023) Environmental pollutants and their effects on human health. Heliyon 9(9):e19496. https://doi.org/10.1016/j.heliyon.2023.e19496

Singh HN (2021) Collection, preservation and transportation of biological evidence for forensic DNA. Analysis 9:1123–1130

Smith VH, Schindler DW (2009) Eutrophication science: where do we go from here? Trends Ecol Evol 24(4):201–207. https://doi.org/10.1016/j.tree.2008.11.009

Tan KH (2005) Soil sampling, preparation, and analysis, 0th edn. CRC Press. https://doi.org/10.1201/9781482274769

Thunberg G (2019) No one is too small to make a difference (First published 2019, reprinted with new material 2019). Penguin Books

Tilman D, Cassman KG, Matson PA, Naylor R, Polasky S (2002) Agricultural sustainability and intensive production practices. Nature 418(6898):671–677. https://doi.org/10.1038/nature01014

Triassi M, Alfano R, Illario M, Nardone A, Caporale O, Montuori P (2015) Environmental pollution from illegal waste disposal and health effects: a review on the "Triangle of death". Int J Environ Res Public Health 12(2):1216–1236. https://doi.org/10.3390/ijerph120201216

Wang X, Wang Y, Wang Y (2013) Use of exotic species during ecological restoration can produce effects that resemble vegetation invasions and other unintended consequences. Ecol Eng 52:247–251. https://doi.org/10.1016/j.ecoleng.2012.11.007

Wang Y, Wang M, Xu W, Wang Y, Zhang Y, Wang J (2022) Estimating the postmortem interval of carcasses in the water using the carrion insect, brain tissue RNA, bacterial biofilm, and algae. Front Microbiol 12:774276. https://doi.org/10.3389/fmicb.2021.774276

Wang F, Xiang L, Sze-Yin Leung K, Elsner M, Zhang Y, Guo Y, Pan B, Sun H, An T, Ying G, Brooks BW, Hou D, Helbling DE, Sun J, Qiu H, Vogel TM, Zhang W, Gao Y, Simpson MJ et al (2024) Emerging contaminants: a one health perspective. Innovation 5(4):100612. https://doi.org/10.1016/j.xinn.2024.100612

Wątor K, Dobrzyński D (2022) Towards a better practice in water sampling: case studies on used in practice geothermal waters. Chemosphere 303:134913. https://doi.org/10.1016/j.chemosphere.2022.134913

Watson CJ, Ueland M, Schotsmans EMJ, Sterenberg J, Forbes SL, Blau S (2021) Detecting grave sites from surface anomalies: a longitudinal study in an Australian woodland. J Forensic Sci 66(2):479–490. https://doi.org/10.1111/1556-4029.14626

White R (2013) Environmental harm: An eco-justice perspective. Policy Press. https://doi.org/10.51952/9781447300427

White R (2022) Green criminology, environmental harms and eco-justice. In: Nelken D, Hamilton C (eds) Research handbook of comparative criminal justice. Edward Elgar Publishing, pp 315–331. https://doi.org/10.4337/9781839106385.00030

Wiltshire PEJ (2009) Forensic ecology, botany, and palynology: some aspects of their role in criminal investigation. In: Ritz K, Dawson L, Miller D (eds) Criminal and environmental soil forensics. Springer, Cham, pp 129–149. https://doi.org/10.1007/978-1-4020-9204-6_9

Wiltshire P (2010) Forensic ecology. In: Crime scene to court: The essentials of forensic science. Royal Society of Chemistry

Wiltshire PEJ (2019) Introduction to the special issue for forensic ecology. Forensic Sci Int 299:238. https://doi.org/10.1016/j.forsciint.2019.01.024

Yolcubal I, Brusseau ML, Artiola JF, Wierenga P, Wilson LG (2004) Environmental physical properties and processes. In: Environmental monitoring and characterization. Elsevier, pp 207–239. https://doi.org/10.1016/B978-012064477-3/50014-X

Zhang W, Wen L (2023) Research on the evaluation system of "vegetation-atmosphere-water" environmental coordination development under ecological management. Heliyon 9(11):e21359. https://doi.org/10.1016/j.heliyon.2023.e21359

Wildlife Forensics

7

Badal Mavry, Sneha Lohar, Kumud Kant Awasthi, Mahipal Singh Sankhla, Anuj Sharma, Archana Gautam, Rajeev Kumar, Garima Awasthi, and Anoop M. Divakaran

Abstract

Animals are increasingly being used as raw materials for human goods, ranging from food to jewelry and cosmetics, raising concerns about the extinction of many species and the need to preserve and safeguard wildlife for that a relatively advanced field of study called wildlife forensics uses forensic methodologies to handle situations involving animal species and their habitats. This chapter examines the function of forensics and its methods in crimes involving wildlife. Since crime is evolving and becoming more sophisticated, it is necessary for forensics to use increasingly improved methods in order to track down and solve cases involving wildlife, such as poaching, trafficking, hunting, and any kind of destruction of wildlife habitats and species. Forensic pathology, DNA barcoding, and the use of isotopes for species and geographic identification are some examples of these techniques. Additionally, a group of highly qualified and specialized scientists work in the field of wildlife forensics to address various instances and circumstances that may arise in connection with animal species.

B. Mavry · S. Lohar
Department of Forensic Science, School of Bioscience and Technologies, Galgotias University, Greater Noida, Uttar Pradesh, India

K. K. Awasthi · G. Awasthi (✉) · M. S. Sankhla
School of Basic and Applied Sciences, K.R. Mangalam University, Gurugram, Haryana, India

A. Sharma · A. Gautam
Department of Forensic Science, SOS, ITM University, Gwalior, Madhya Pradesh, India

R. Kumar
Department of Forensic Science, IILM University, Greater Noida, Uttar Pradesh, India

A. M. Divakaran
Department of Science and Technology, Toyota Technological Institute, Nagoya, Japan

M. S. Sankhla et al. (eds.), *Introducing the Synergy Between Forensic Science and Environmental Analysis*, https://doi.org/10.1007/978-981-95-0199-1_7

Keywords

Wildlife · Forensic · DNA · Investigation

7.1 Introduction

Numerous animal and plant species are being exploited unsustainably and their habitats are being destroyed as a result of the Illegal Wildlife Trade (IWT). In addition to having detrimental socioeconomic effects on a local and global level, this eventually causes species extinction, which results in a loss of biodiversity and ecological services (Johnson et al. 2014). Any legal system that respects biodiversity and holds organized syndicates accountable for their financial benefit from this illicit trade must prevent and investigate suspected wildlife crimes (Nishant et al. 2017). Compared to human identification, wildlife forensic science is a broad field that may take many different forms based on the specifics of the accusation. One significant distinction is that there is sometimes no "victim" to offer information about the inquiry in cases of suspected crimes against animals (Linacre and Tobe 2011). Furthermore, compared to the single species examined in human identification, the list of species seen in animal forensic science is larger (Huffman and Wallace 2011). There are many different types of wildlife crime, such as trafficking in live specimens, hunting out of season, animal abuse, habitat damage, poaching for meat, poaching for trophies, poaching for the use of animal parts in medicines, horns, and tusks for jewelry and decorations, and more (Ogden and Linacre 2015). India has a robust legal and policy framework in place to control and limit the trade in wildlife. However, given the current state of affairs in India, robust policies and their execution are imperative. It is illegal to sell wild animals, plants, and their derivatives under the Wildlife Protection Act (WPA) of 1972. After weapons and drugs, wildlife goods are the third most illegally traded item. The area covered by wildlife crime is unclear, although it is growing quickly (Gouda et al. 2020). Every year, thousands of wild animals, birds, insects, and fish are traded worldwide as part of the highly organized illicit wildlife trafficking in mammals, birds, insects, and reptiles (WWF IND 2014). Numerous projects have been started in India to conserve wild creatures, such as sea turtles, crocodiles, tigers, elephants, one-horned rhinos, and hangul (red deer) (Cooper and Cooper 2013). Seahorses and pipefish are illegally traded from Tamil Nadu, Kerala, Maharashtra, and Karnataka in India. It is mostly exploited in antiaging medications, asthma, excessive cholesterol, goiter, renal disease, skin conditions, and infertility. Diagnostic indicators for species identification in wildlife forensics worldwide include anatomical investigations, microscopic examinations of body parts, morphological studies of the species, footprint analysis at crime scenes, and serological techniques. Hair, long bones, tooth morphology particular to a species, and other related features serve as the basis for this. In wildlife forensics, the evidence used to identify species includes dried and cooked foods, dried shark fins, egg shells, animal hairs, bone, ivory, rhinoceros' horns, turtle shells, feathers, and fish scales (Bowen et al. 2005; Wallace and Ross 2012).

Expansion of evolutionary applications has also helped wildlife forensic sciences. Ballistics, veterinary sciences, morphology, chemistry, and—most importantly—molecular applications are among the scientific disciplines that wildlife officers use to support their investigations into illegal wildlife activities, such as poaching and local and international trade. Species identification has become a standard analysis in animal forensics as it deals with a range of species-specific rules and regulations based on local, regional, national, and international legislation (Kanthaswamy 2024).

7.2 Applications of Wildlife Forensics

In order to work on various instances and bring about justice for animal species, wildlife forensics uses a variety of applications. Forensic science applications are used to identify species, examine seized samples, and determine the nature of the sample and crime (Huffman and Wallace 2011; Nishant et al. 2017). These programs may also help identify the individual or group's method of operation in order to prevent future crimes.

7.2.1 Morphological Analysis

The easiest and least costly forensic investigation strategy for identifying evidence in animals is to use morphological or physical traits. Important hints based on outward appearance are provided by wildlife flora and fauna morphological traits, which are used to identify species (Trail 2021). Generally speaking, different species have unique physical characteristics, such as the color of their skin coat, their coloring patterns, eyes, pinnas, tails, ivory, etc. Ivory, a kind of dentine that is illegally trafficked from Asian or African elephants, has a distinctive pattern called Schreger. Additionally, it possesses an angled pattern that serves as the foundation for identification, even within species (Sankhla et al. 2016; Ueland et al. 2020). In biology, morphological analysis follows the taxonomic ladder from order to family to genus to species, noting common derived characteristics (synapomorphies) at each level. For instance, a complete trophy mount of a polar bear (Ursus maritimus) may be classified as belonging to the family Ursidae by its size, body form, and plantigrade feet; to the order Carnivora by its carnassial teeth and strong claws; and to the species by its all-white pelage and well-furred foot pads (Burnham-Curtis et al. 2021).

Some animals or species have unique morphological characteristics that can be used for identification, such as paw or footprints, which can be used for both gender and species identification. On the other hand, hairs can be used to identify species because each species has different hair segments. Skin can also be used for identification because fish and reptiles have epidermic and dermic scale patterns, and birds can be identified by their feathers (Johnson 2011; Linacre and Ciavaglia 2017). This leads to the conclusion that morphological analysis of species and their use in wildlife forensics are crucial.

7.2.2 Microscopic Examination

Given that mammals make up one of the largest groups of poached species and that many wildlife crime cases require the identification of species from hair, microscopy is one of the most helpful tools in wildlife forensics, particularly when dealing with hair evidence. The range of physical evidence types in wildlife forensics is extremely broad, as are the techniques used (Sarma et al. 2014; Sahajpal et al. 2021). The biological sciences have also made extensive use of microscopic hair features to analyze dietary patterns, prey-predator interactions, and animals living in trees or dens. Mathiak created a guide to identify the hairs of Southern Michigan animals in 1938. Williams created a key to distinguish between shrew and mole hairs that same year (Sahajpal and Goyal 2009). Based on hair evidence, animals can be identified using a scanning electron microscope (SEM). Higher magnification and linked Energy Dispersive Spectra (EDS) are two advantages of scanning electron microscopy that enable the identification of geographical regions through elemental analysis, such as that of sodium, potassium, calcium, and sulfur (Coghlan 2014). It is simple to identify distinct animal species since different animals have varied hair shapes, scale patterns, medullary index, and hair appearances (Cooper 2021).

7.2.3 Footprints Analysis

Wildlife species' paw or footprints are crucial pieces of evidence for forensic species analysis. Animal footprints are the impressions left by an animal's foot on surfaces where it walks or in captivity (Dharmik et al. 2024). The size and pattern of the footprints are used to determine the species' age and kind. In order to aid with gender identification, male and female paw prints differ; typically, the male paw print is wider and shorter, while the female paw print is greater in length and narrower (Mores et al. 2025). This piece of evidence aids in field investigation; it typically doesn't require laboratory examination, allowing an investigator to examine the majority of the animal or case details in the field, which can expedite case resolution, but undetectable footprints on hard surfaces and occasionally polluted footprint sites by the presence of other animals are the main issues with using footprint imprints to identify species (Mayer 2019).

7.2.4 Serological Techniques

Similar to the lock and key metaphor, serological methods rely on the interaction between the antibody and its cognate antigen. The biological evidence's species of origin is identified using species-specific antibodies (Kanthaswamy 2024). The lack of species-specific antibodies in our lab is one of the troubling issues with serological methods. There is also enough evidence that antibody cross-reactions in nontarget species can occasionally have favorable outcomes. Consequently, in forensic analysis, it is only regarded as a presumptive test (Scaglione et al. 2018).

7.2.5 Pathological Studies

Pathological techniques aid in determining the nature of the crime, cause of death, and method of operation of any crime. There are several ways for a species to die, including being injured by a person or another animal, since different creatures in natural environment have the ability to kill other animals. Finding any lesions, wounds, bruises, abscesses, or scars can be crucial to figuring out how and when a wild animal passed away or how healthy it was before (Viner and Kagan 2018). Postmortem examinations should, whenever possible, be carried out by an experienced comparative pathologist. Some investigations in the field aim not only to identify animal carcasses and tissues but also to determine the circumstances of death. Some significant examples, particularly relevant to endangered species, are provided below (Byard and Boardman 2011; Gudmannsson and Berge 2019). In certain crimes, like poisoning, routine sample collections should include liver, kidney, urine, blood, and stomach content. However, it is always helpful to have a general idea of what substances should be screened based on the history, location, or lesions found in the animal (Munro and Munro 2013). Complementary analysis, or toxicology, is necessary to identify the substance responsible for the death or injury in conjunction with routine necrotic examination. This is crucial because poisoning may not have any pathognomonic gross or histopathologic findings (Sriraman 2021). Although postmortem change (autolysis) can make it more difficult to research carcasses, it can also yield important details about the animal's time of death and the surrounding conditions at that time if it is thoroughly examined and analyzed (Cooper 2002; Sahajpal et al. 2021). Scavenging and disruption by wild, feral, and domesticated animals can impede the collecting of bodies and complicate counting in investigations when carcasses must be gathered for inspection (Viner et al. 2016).

7.2.6 Molecular Biology Techniques

The development of analytical techniques to provide solid DNA-based evidence to help conservation law enforcement—also referred to as "wildlife DNA forensics"—is a recognized area of conservation genetics that is currently receiving more attention (Butler 2008; Robinson 2013). Compared to other approaches including anatomy, morphology, serology, and microscopic analysis, molecular techniques have been shown to be more trustworthy. Random Amplified Polymorphic DNA (RAPD), Amplified Fragment Length Polymorphism (AFLP), and Random

Fragment Length Polymorphism (RFLP) were the first DNA-based techniques used (Gunn 2019; Pérez-Espona and Consortium 2021). The identification of seized material to determine the species, individual identity or connection, and source population of the sample is a component of wildlife forensics. Wildlife forensics is not without its difficulties though. DNA forensics has emerged as a key investigative technique to reduce wildlife crime, even in the face of national and international legislation aimed at preserving declining habitat, biological species variety, and ensuring the long-term survival of species (Sankhla et al. 2016). As the complexity of species identification, the DNA markers employed in wildlife forensic research are more diverse than those utilized in the forensic human identification community. For putative individualization, pedigree assignment, and the assignment of an unknown sample to a population, microsatellite markers (also known as Short Tandem Repeat, or STR, loci) and increasingly Single Nucleotide Polymorphisms (commonly known as SNPs) are employed. Most frequently utilized for species identification, mitochondrial DNA (mtDNA) can also be used for putative individualization when a study of the female lineage is adequate (Harrison 2023). According to reports, the Natural Resources DNA Profiling and Forensic Centre (NRDFC) was created for the extraction and DNA profiling of wildlife species when forensics and biological research in Canada joined in order to conserve genetic biodiversity. However, such an institution must be set up for the purpose of profiling and creating a database for wildlife species in a country like India (Liu et al. 2023). A recently developed technique for population identification uses biopolymer-like proteins, which are relatively more stable than DNA polymers. In order to identify various populations, protein-based markers must be optimized instead of relying just on DNA. As a result, there is a great need to concentrate on screening for a distinct protein marker that offers a same level of discriminating as DNA-based markers (Frankham et al. 2025).

7.2.7 Stable Isotopes Analysis (SIA)

Stable isotopes are atoms of the same element that do not decay over time and have various atomic weights (caused by varying quantities of neutrons). Almost 80% of the dry mass of plant and animal tissues is composed of the most prevalent elements in living things: carbon (C), nitrogen (N), sulfur (S), oxygen (O), and hydrogen (H) (Prigge et al. 2024). In wildlife forensics, stable isotopes are being utilized more and more to pinpoint the origin and migration of animals. The label that is absorbed into animal tissue through feeding is provided by the large and systematic global variations in the heavy isotope content of precipitated water and snow (dDp, d18Op). These isotopes may thus provide excellent geographic origin tracers (Andersson et al. 2021; Sahajpal et al. 2021). The proportion of naturally occurring light stable isotopes (D/H, 13C/12C, 15N/14N, and 34S/32S) in food items may change predictably throughout biomes that organisms inhabit, which is exploited by this novel method. Numerous complex and interconnected environmental elements, including as temperature, altitude, soil and geological composition, food chain length, tissue turnover, and the duration of an individual's residence, influence the overall isotopic makeup of people in food webs of a given biome (Hopkins III et al. 2022). Many additional elements, such as trace elements and heavy metals, have measurable isotopes in addition to the study of conventional isotopes. The limited quantity and added difficulties in obtaining and studying these isotopes have made analysis of these unconventional elements less popular. Furthermore, it is typically more costly to test the isotopes of heavy metals and trace elements since they are more susceptible to contamination (Bartelink et al. 2014, 2016). The five primary phases of the SIA workflow are weighing, mass spectrometry, sample collection and preparation, data interpretation, and applications (Guilfoyle 2023). Extra stages can be needed, depending on the samples utilized and the goals of the inquiry. Samples for SIA processing are often prepared simply and frequently don't need any extra specialized equipment. But occasionally, specialized tools could be needed, such as a centrifuge to extract collagen or lipids or a Soxhlet extractor to defatten or degrease materials (Bodey et al. 2011). SIA is not widely used and its application in wildlife forensics is not well documented; it has only been used seldom. SIA has been used in wildlife trade investigations in three primary areas: species identification, determining regional origin, and distinguishing between wild and captive-bred/cultivated individuals (Brandis et al. 2018).

7.3 Recent Advances and Case Studies in Wildlife Forensics

A major threat to the survival of wildlife worldwide is the growing trend of illegal wildlife trade, poaching, hunting, and killing. Since ancient times, people have used and abused wild animals for human benefit. In addition to live animal trafficking, the illegal wildlife product trade accounts for a sizable amount of wildlife seizure records. This illicit activity is a complex and profitable industry that threatens both flora and fauna.

For morphological studies, identifying a species from the data gathered from wildlife crimes is typically a difficult task. Since materials like dried or processed skin are challenging to evaluate, the recent advancements in DNA-based methods for species identification are more appropriate and legally recognized as unassailable proof. In wildlife forensics, DNA barcoding technology is being used as a possible tool. Short segments of mitochondrial DNA (mtDNA) are amplified and read to barcode or identify the species with the highest specificity and differentiation from other species.

7.3.1 Case Study1

To identify the species, officials from the Tamil Nadu Forest Department forwarded the six ($n = 6$) dried genital organs that were confiscated (packs of three, each containing two samples) to the Advanced Institute for Wildlife Conservation in Tamil

Nadu, India (https://www.aiwc.res.in). The sample arrived properly sealed in a zip-lock cover.

For the analysis of samples, DNA extraction, amplification, and its sequencing using sanger sequencer were done. According to the results, DNA collected from genital organs using a modified procedure shown excellent amplification for the cytb gene. The lengths of the sequences obtained from the PCR amplicons were reported to be 434 bp for sample A, 430 bp for sample B, and 425 bp for sample C. The results showed that sample A, sample B, and sample C had 99.38%, 99.20%, and 99.00% similarities with the Bengal monitor lizard's cytb gene, respectively, according to the BLAST similarity search. Additionally, the contig sequence from the confiscated dried genital organs (sample A–C), an in-house repository cytb sequence for the Bengal monitor lizard, and additional Varanus spp. cytb sequences downloaded from the NCBI database were used to construct a maximum likelihood phylogenetic tree. According to reports, the dendrogram revealed that the dried sexual organs (samples A–C) that were confiscated were exactly like those of the Bengal monitor lizard.

TRAFFIC India brought attention to the poaching and sale of lizards for their skin, flesh, blood, and oil. Their burrows target them, and noose traps and snares are frequently used to catch them. Locals view the lizard's flesh as a delicacy, and the oil extracted from its skin and tail is marketed as a home cure for a number of illnesses, including joint discomfort. Several media outlets have reported on lizard poaching in India; for instance, the Odisha Forest Department confiscated an Indian monitor from merchants (Singh 2017).

The World Wide Fund for Nature (WWF-India) and TRAFFIC India are intensifying their efforts to stop the trade in monitor lizards in order to rescue the species. The campaign's main theme was "Buying is Stealing" and it highlighted the present danger to monitor lizards. People are cautioned in this message not to buy medical supplies made from monitor lizards and other animal species that are protected.

7.3.2 Case Study 2

In Tamil Nadu, authorities from the State Forest Department seized two separate products. The team conducting the investigation had gathered evidence of a suspected musk pod and a canid head from the crime scenes. The Advanced Institute for Wildlife Conservation Institute (AIWC), Vandalur, Chennai, received the same from the Centre for Wildlife Forensic Sciences. In order to identify the species, the cases were recorded for genetic study. "Seizure 1" was assigned to the canid head sample, and "Seizure 2" to the putative musk pod. After the collection and preservation of evidences, the samples were subjected for DNA extraction, amplification using PCR, and sequencing.

The sample's extracted DNA was amplified for the mitochondrial regions of 12S rRNA and 16S rRNA. With regard to each gene, the sequences containing three hits for each species that were comparable or nearly related were chosen. Sequences of the golden jackal (Canis aureus), domestic dog (Canis lupus familiaris), gray wolf

(Canis lupus), and Japanese wolf (Canis lupus hodophilax) were taken from the NCBI database for investigation of the 16S rRNA region. Similarly, golden jackal (Canis aureus), coyote/American jackal (Canis latrans), gray wolf (Canis lupus), and red wolf (Canis rufus) are the species selected for 12S rRNA regions. The top hits or entries found when the query sequences are compared to the NCBI database determine which species are chosen for 16S rRNA and 12S rRNA.

Both of the seizures' mitochondrial genes (16S rRNA and 12S rRNA) were effectively amplified and sequenced. Seizures 1 and 2's 12S rRNA assembled sequences were identical to those of the Indian golden jackal (Canis aureus) in every way. Seizures 1 and 2's 16S rRNA sequences also had 99.54% of their similarities with those of the Indian Golden jackal (Canis aureus).

The Golden jackal (Canis aureus), which is usually disregarded due to its "common status" and Least Concern IUCN category in India, emphasizes the need for more attention to regulate its illicit trafficking.

Less attention is paid to the illegal sale of these less charismatic species, which contributes to their decline or extinction. The general lack of awareness about jackal hunting gives poachers and traffickers an advantage when it comes to practicing religious or superstitious beliefs concerning animal items like jackal horns, skin, and skull. The market for false pods made of jackal skin is driven by the need for musk pods, especially in the black magic community (Jose et al. 2024).

7.4 Challenges and Future Directions

The fact that species identification is only conceivable if there is a "hit" or match between the forensic sample and a previously described species stored in the databases further exacerbates the issue with species testing (Budowle and van Daal 2009). The absence of established procedures in the wildlife forensic community is even more distressing (Jota Baptista et al. 2022). The lack of consistency may have been exacerbated by the fact that many heavily trafficked species are found in small quantities over restricted regions (Keskin et al. 2023). More reliance on scientific evidence from animal sources may result from sound animal forensic science employing standard operating procedures (SOPs) and adherence to quality assurance/quality control (QA/QC) standards, including DNA analysis using standardized, validated, and accredited procedures and protocols. These difficulties include a lack of funds, databases, proven techniques, standardized procedures, laboratory accreditation, personnel expertise, and other QA/QC needs.

By lowering sample maintenance expenses and evaluating complicated mixes and deteriorated material, the advancement of NGS technology promises upcoming new opportunities in wildlife forensic (Barbaro 2024). In conservation biology, next-generation sequencing methods have proven very useful for producing phylogenic data, life histories, and demographic data for endangered species.

Applying NGS and AI-assisted techniques promises to enhance wildlife forensic investigation by helping with complex mixture and low copy number issues that arise in the analysis of bio-specimens from wildlife crimes as well as by quickly and affordably screening for forensically relevant genomic variations from publicly accessible databases.

7.5 Conclusion

Protecting wild animals is a top priority because they are a weak link in the coexistence of humans and nature. Worldwide, conservationists and wildlife specialists recognize the significance of wildlife monitoring and protection in the face of increased fraud and the trafficking in wild animal products. Since many new species are found each year and many more are added to the list of endangered species, wildlife forensics' applications are expected to grow in order to facilitate the quick and precise identification of wildlife species. In order to identify species, establish origins, and connect suspects to crime scenes, this specialized area employs a multidisciplinary approach, studying a broad variety of biological evidence, from animal tissues and bloodstains to bones and animal products. Some specialized techniques determine their geographic origin, and even differentiate between wild and domestic individuals has significantly improved with the advent of sophisticated tools, especially DNA analysis, mitochondrial microsatellite analysis, and DNA barcoding.

These methods will not only support appropriate legal proceedings and conviction procedures, but they will also resolve a number of outstanding issues pertaining to taxonomy, identification, evolution, diversity, and environmental effect, among other topics. Wildlife forensics and its relevance have been further reinforced by the emergence of interdisciplinary methodologies that combine fundamental fields such as biological sciences, bioinformatics, genetics, and other related omics techniques. Artificial intelligence and/or big data analytics, which have previously been created and used in other fields, may also be used in this field to help create a systematic and well-organized chain of detection, evidence gathering, detention public health surveillance, and communication. Multidisciplinary teams with different professional backgrounds are also crucial for gathering and analyzing data from different angles.

Even with the extensive use of wildlife forensics, mainstream wildlife research has yet to adopt this new field. For the protection and survival of wildlife on Earth, a global effort involving the international community, national policymakers, funding agencies, forensic experts, and—above all—encouragement to advance wildlife forensics to a level comparable to human forensic science is desperately needed.

References

Andersson A, Gibson L, Baker D, Cybulski J, Wang S, Leung B, Chu L, Dingle C (2021) Stable isotope analysis as a tool to detect illegal trade in critically endangered cockatoos. Anim Conserv 24:1021–1031

Barbaro A (2024) Overview of NGS platforms and technological advancements for forensic applications. In: Next Generation Sequencing (NGS) Technology in DNA Analysis (pp. 35–58). Academic Press.

Bartelink E, Berry R, Chesson (2014) Stable isotopes and human provenancing. In: Advances in forensic human identification (pp. 165–192). CRC Press
Bartelink EJ, Mackinnon AT, Prince-Buitenhuys JR, Tipple BJ, Chesson LA (2016) Stable isotope forensics as an investigative tool in missing persons investigations. In: Handbook of missing persons, pp 443–462
Bodey T, Bearhop S, McDonald RA (2011) Invasions and stable isotope analysis–informing ecology and management. In: Island invasives: eradication and management, vol 148. IUCN, Gland, p 151
Bowen GJ, Wassenaar LI, Hobson KA (2005) Global application of stable hydrogen and oxygen isotopes to wildlife forensics. Oecologia 143:337–348
Brandis KJ, Meagher PJ, Tong LJ, Shaw M, Mazumder D, Gadd P, Ramp D (2018) Novel detection of provenance in the illegal wildlife trade using elemental data. Sci Rep 8:15380
Budowle B, van Daal A (2009) Extracting evidence from forensic DNA analyses: future molecular biology directions. BioTechniques 46:339–350
Burnham-Curtis MK, Straughan DJ, Hamlin BC, Draheim HM, Gray Partin TK, Wostenberg DJ (2021) Wildlife forensic genetics and biodiversity conservation: the intersection of science, species management, and the law. In: Wildlife Biodiversity Conservation: Multidisciplinary and Forensic Approaches (pp. 163–191). Cham: Springer International Publishing
Butler JM (2008) Molecular Forensics. JSOTR 114–115
Byard RW, Boardman W (2011) The potential role of forensic pathologists in veterinary forensic medicine. Forensic Sci Med Pathol 7(3):231–232. Springer
Coghlan ML (2014) Development of new molecular techniques for wildlife forensic applications (Doctoral dissertation, Murdoch University)
Cooper J (2002) Diagnostic pathology of selected diseases in wildlife. Revue scientifique et technique-Office international des épizooties 21:77–90
Cooper JE (2021) Wildlife forensic pathology. In: Wildlife Biodiversity conservation: multidisciplinary and forensic approaches. Springer, Cham, pp 211–286
Cooper JE, Cooper ME (2013) Wildlife forensic investigation: principles and practice. CRC Press, Boca Raton
Dharmik RC, Bhardwaj A, Bisen L, Bhisikar G, Baghele C, Nayak H (2024) Identification of animals through footprint analysis. In: 2024 International Conference on Artificial Intelligence and Quantum Computation-Based Sensor Application (ICAIQSA). IEEE, pp 1–6
Frankham GJ, Ogden R, Baker BW, Ewart KM, Johnson RN, Kuiper I, Lindquist CD, Moore MK, Ndiaye A, Webster LM (2025) Standards in wildlife forensic science, with a focus on non-human DNA analysis. Anim Genet 56:e70005
Gouda S, Kerry RG, Das A, Chauhan NS (2020) Wildlife forensics: a boon for species identification and conservation implications. Forensic Sci Int 317:110530
Gudmannsson P, Berge J (2019) The forensic pathology of fatal attacks by the large mammals inhabiting the Nordic wilderness—a literature review. J Forensic Sci 64:976–981
Guilfoyle K (2023) Refining the use of stable isotope analysis in detecting wildlife laundering. University of Central Florida, Orlando
Gunn A (2019) Essential forensic biology. Wiley, Hoboken
Harrison S (2023) Developing strategies to improve educational programs for wildlife forensic laboratory careers. University of Central Oklahoma, Edmond
Hopkins JB III, Frederick CA, Yorks D, Pollock E, Chatfield MW (2022) Forensic application of stable isotopes to distinguish between wild and captive turtles. Biology 11:1728
Huffman JE, Wallace JR (2011) Wildlife forensics: methods and applications. Wiley, Hoboken
Johnson RN (2011) Conservation genetics and wildlife forensics of birds. In: Wildlife forensics: methods and applications. Wiley, Hoboken, pp 293–325
Johnson RN, Wilson-Wilde L, Linacre A (2014) Current and future directions of DNA in wildlife forensic science. Forensic Sci Int Genet 10:1–11
Jose SA, Amarnath CB, Thiyagarajan KB, Sekar S (2024) Molecular species identification from seized dried genital organs: a case study of suspected Bengal monitor lizard (Varanus bengalensis). Discover Animals 1:22

Jota Baptista C, Seixas F, Gonzalo-Orden JM, Oliveira PA (2022) Wildlife forensic sciences: a tool to nature conservation towards a one health approach. Forensic Sci 2:808–817
Kanthaswamy S (2024) Wildlife forensic genetics—biological evidence, DNA markers, analytical approaches, and challenges. Anim Genet 55:177–192
Keskin BB, Griffin EC, Prell JO, Dilkina B, Ferber A, MacDonald J, Hilend R, Griffis S, Gore ML (2023) Quantitative investigation of wildlife trafficking supply chains: a review. Omega 115:102780
Linacre A, Ciavaglia SA (2017) Wildlife forensic science. In: Handbook of forensic genetics: biodiversity and heredity in civil and criminal investigation (pp. 449–471). World Scientific Publishing
Linacre A, Tobe SS (2011) An overview to the investigative approach to species testing in wildlife forensic science. Investig Genet 2:1–9
Liu X, Du W, Wang C, Wu Y, Chen W, Zheng Y, Wang M, Liu H, Yang Q, Qian S (2023) A multilocus DNA mini-barcode assay to identify twenty vertebrate wildlife species. Iscience 26:108275
Mayer C-H (2019) Combating wildlife crime in South Africa: using gelatine lifters for forensic trace recovery. Springer, Cham
Mores CM, de Castro MB, de Melo CB (2025) Forensic examinations in the investigations of crimes involving animals: a review. Ciência Rural 55:e20240071
Munro R, Munro H (2013) Some challenges in forensic veterinary pathology: a review. J Comp Pathol 149:57–73
Nishant K, Vrijesh K, Ajay K (2017) Wildlife forensic: current techniques and their limitations. J Forensic Sci Criminol 5:402
Ogden R, Linacre A (2015) Wildlife forensic science: a review of genetic geographic origin assignment. Forensic Sci Int Genet 18:152–159
Pérez-Espona S, Consortium C (2021) Conservation-focused biobanks: a valuable resource for wildlife DNA forensics. Forensic Sci Int Anim Environ 1:100017
Prigge TL, Andersson AA, Hatten CE, Leung EY, Baker DM, Bonebrake TC, Dingle C (2024) Wildlife trade investigations benefit from multivariate stable isotope analyses. Biol Rev 100:1083
Robinson LA (2013) Genetic methodologies in wildlife crime investigations. In: Wildlife forensic investigation. CRC Press, Boca Raton, pp 396–409
Sahajpal V, Goyal S (2009) Microscopic examinations in wildlife investigations. In: Forensic science in wildlife investigations. CRC Press, Boca Raton, pp 33–74
Sahajpal V, Mishra S, Bhandari D (2021) Forensic analysis in wildlife crime cases: microscopy, DNA profiling and isotope analysis. In: Forensic analysis-scientific and medical techniques and evidence under the microscope. IntechOpen, London
Sankhla MS, Kumari M, Nandan M, Kumar R, Agrawal P (2016) The role of wild life DNA forensics in identification of endangered species. Imp J Interdiscip Res (IJIR) 24:27
Sarma K, Bhattacharjee P, Dey S (2014) Microscopical analysis of guard hair of Tiger, Panthera tigris, with reference to wildlife forensic applications. J Adv Microsc Res 9:199–205
Scaglione FE, Pregel P, Bertolino S, Gabba L, Sereno A, Evangelista R, Starvaggi Cucuzza L, Peano A, Cannizzo FT, Bollo E (2018) Preliminary investigations in forensic pathology of eastern gray squirrel's (Sciurus carolinensis) skin lesions. In: Atti del 72° Convegno SISVet 2018, p 244. Soc. It. delle Scienze Veterinarie
Singh SK (2017) Road traffic accidents in India: issues and challenges. Transp Res Procedia 25:4708–4719
Sriraman P (2021) Wildlife necropsy and forensics. CRC Press, Boca Raton
Trail PW (2021) Morphological analysis: a powerful tool in wildlife forensic biology. Forensic Sci Int Anim Environ 1:100025
Ueland M, Brown A, Bartos C, Frankham GJ, Johnson RN, Forbes SL (2020) Profiling volatilomes: a novel forensic method for identification of confiscated illegal wildlife items. Separations 7:5
Viner TC, Kagan RA (2018) Forensic wildlife pathology. In: Pathology of wildlife and zoo animals. Elsevier, Amsterdam, pp 21–40

Viner T, Hamlin B, McClure P, Yates B (2016) Integrating the forensic sciences in wildlife case investigations: a case report of pentobarbital and phenytoin toxicosis in a bald eagle (Haliaeetus leucocephalus). Vet Pathol 53:1103–1106

Wallace JR, Ross JC (2012) The application of forensic science to wildlife evidence. In: Wildlife forensics: methods and applications. Wiley, Hoboken

Case Studies in Environmental Crime Investigations

8

Anjali Dubey, Varsha Rani Patel, Kavita Sharma, and Baljeet Yadav

Abstract

This chapter analyses the procedural, legal, and practical dimensions of engaging with environmental crime investigators who identify violations of national environmental statutes and regulations. The initial section of the chapter examines the repercussions and results of environmental offences, encompassing illicit mining, deforestation, and breaches of pollution laws related to hazardous waste. It suggests that individuals are increasingly aware of environmental violations, perceiving them as significant dangers to national security, the environment, and public health. The chapter presents its findings utilising both legislation and governmental statistics. It also illustrates progress in the surveillance and reaction to environmental offences, recognising both the achievements and obstacles faced in attaining compliance. An analysis of the investigative process encompasses the registration of crimes, evidence collection and examination, scene management, and the execution of legal procedures, employing a comprehensive marketing model that incorporates methods utilised both prior to and subsequent to the emergence of an issue. Priority is assigned to organic and scientific disciplines globally. In cases of environmental crime, it is necessary to have the necessary information, collaborate with other agencies, and follow the law, which

A. Dubey (✉)
Medicaps University, Indore, Madhya Pradesh, India
e-mail: Anjali.dubey@medicaps.ac.in

V. R. Patel
MATS University, Raipur, Chhattisgarh, India

K. Sharma
Shri Vaishnav Vidyapeeth Vishwavidyalaya, Indore, Madhya Pradesh, India

B. Yadav
School of Basic and Applied Sciences, K. R. Mangalam University, Gurugram, Haryana, India

M. S. Sankhla et al. (eds.), *Introducing the Synergy Between Forensic Science and Environmental Analysis*, https://doi.org/10.1007/978-981-95-0199-1_8

includes keeping the chain of custody intact. This chapter discusses enforcement measures, including active listening during investigations. Investigators ought to ground their analysis in interviewing, documentation, enquiry, and scene observation. The subsequent chapter emphasises the necessity of robust institutions, stringent rules, and local community engagement to address environmental issues and achieve sustainable justice.

Keywords

Environmental crime · Investigation model · Criminal investigation · Chain of custody · Environmental justice

8.1 Introduction

Environmental crime includes not just illegal animal trafficking but also violations of fisheries and forestry laws, illegal hazardous waste dumping, ozone-depleting compound smuggling, and illegal mining operations (Tiwari and Prasad 2020). According to Tiwari and Prasad (2020), illegal mining in particular causes serious environmental harm, such as ecosystem devastation, pollution, radiation hazards, mercury contamination from artisanal gold mining, and the loss of crops and agricultural land. Environmental crime can be broadly defined as financial crimes involving natural resources like oil, timber, and marine life, including tax evasion, money laundering, and illegal trafficking of hazardous materials, all of which disrupt ecosystems and sociopolitical structures (Sharma 2019a, b).

All living things are supported by nature, a basic and service-oriented entity. However, when human activity taints or compromises it, it becomes a source of destruction (Verma 2018a, b). According to Sharma (2019a, b), human activities have caused serious environmental issues such acid rain, climate change, biodiversity loss, and overall ecological disturbance. As a result of unchecked environmental resource extraction, uncommon species are going extinct, habitats are in danger, and human and animal health problems like allergies, cancer, and birth defects are becoming more common (Verma 2018a, b). It is becoming more well recognised that environmental crime is one of the criminal industries that is growing the fastest, having a substantial effect on the economy, future generations' welfare, and national security (Tiwari and Prasad 2020).

Ineffective enforcement, regulatory flaws, inadequate infrastructure, and political stagnation all contribute to air pollution, which is primarily caused by CO_2 emissions from industrial and transportation activities, agricultural waste incineration, deforestation, and poor waste management (Singh 2021). India faces serious environmental challenges, including the consequences of climate change, hazardous waste management, and pollution of the air, water, and noise (Singh 2021). According to the World Health Organization, India, which is ranked 177th out of

180 countries in terms of air quality measures, is home to 14 of the 20 most polluted cities in the world (WHO 2022).

In 2022, environmental crime in India decreased by 18% compared to 2021. There were indications of an increase in the violation of significant environmental statutes (National Crime). The National Crime Records Bureau of India includes crimes against minors in its 2022 report. The Forest Act of 1927, the Forest Conservation Act of 1980, the Wildlife Protection Act of 1972, the Environmental (Protection) Act of 1986, the Air and Water (Prevention and Control of Pollution) Act, the Cigarette and Other Tobacco Products Act of 2003, the Noise Pollution (Regulation and Control) Rules of 2000, and the National Green Tribunal Act of 2010 are all considered environmental offences. The NCRB reported that these offences were recorded in 2022. India's comprehensive environmental protection framework encompasses several critical laws, including the Forest Act (1927), the Forest Conservation Act (1980), and wildlife legislation. The Protection Act (1972), the Environment (Protection) Act (1986), the Air (Prevention and Control of Pollution Act), the Water (Prevention and Control of Pollution) Act, the Cigarettes and Other Tobacco Products Act (2003), the Noise Pollution (Regulation and Control) Rules (2000), and the National Green Tribunal Act (2010) were enacted (Tiwari and Prasad 2020).

Especially, cases that come up under four statutes such as the National Green Tribunal Act, the Air Act, the Water Act and the Environment (Protection) Act all pointed towards the increase in pattern in 2021 and 2022; the number of cases involving these domains went up, as the NCRB reported it in 2022 (NCRB 2022). Table 8.1 shows the number of offences registered under key acts in 2021 and 2022.

Table 8.1 Number of environmental offences registered under key acts in 2021 and 2022 with percentage change

Environment-related acts	No. of offences registered in 2022	No. of offences registered in 2021	Percentage change
Forest Act and The Forest Conservation Act	1921	2292	−16.19
Wildlife Protection Act	535	615	−13.01
Environmental (Protection) Act	329	252	30.56
Air and Water (Prevention and Control of Pollution) Act	78	55	41.82
Noise Pollution Acts	7378	7217	2.23
National Green Tribunal Act	37	16	131.25
Cigarette and Other Tobacco Products Act	42,642	54,024	−21.07

Courtesy by: National Crime Records Bureau (2022)

8.1.1 Investigation of Environmental Crimes

Enforcement agencies have created defined processes for investigating environmental crimes. It offers guidance on the most efficient methods for police agencies to engage with specific communities. Certain environmental infractions pertain to enhance agencies performance by adhering to these regulations and conforming to environmental requirements to protect public health and the ecology.

8.1.2 Environmental Crime Investigation Model

Environmental crime investigations often use either a reactive or proactive tactic. In reactive investigations, issues are identified, an initial evaluation is conducted, the appropriate location is examined, and pertinent evidence and legal documents are collected. These enquiries are started when certain incidents, such as complaints or difficult situations, are reported. Their creation attempts to guarantee adherence and deal with urgent issues. On the other hand, proactive investigations help find misbehaviour and try to stop it before it happens. These results were discussed by Tiwari and Prasad in 2020. Focusing on interagency relationships, analysing the data gathered, relying on confidential informants, and evaluating any risks are all crucial. As work moves on, routine inspections and preventative measures are carried out.

8.1.3 Registering the Crime

Regular inspections conducted by agencies like the Central Pollution Control Board (CPCB), State Pollution Control Boards (SPCBs), and local intelligence can also be reported by individuals or groups not affiliated with the government. Show-cause notifications are typically sent by SPCBs to request an explanation during investigations involving industrial security violations. The water supply or electrical connections are cut off entirely if the reaction is deemed inadequate. While SPCBs are occupied with industrial inspections, local authorities are now responsible for ensuring that enterprises and properties adhere to environmental regulations. Every time someone files a complaint regarding noise pollution, a formal complaint must be submitted because it is a violation of the Bharatiya Nagarik Suraksha Sanhita (BNSS). This report should include all supporting documentation, a list of the accused, and the name of the offence.

8.1.4 Search and Seizure

Professionals have the right to inspect the property and conduct activity checks under the Water Act, Section 23; the Air Act, Section 24; and the Environment Protection Act, Section 10. The teams will also seize objects or records that show misconduct. The appropriate BNSS authorities must be involved for searches and seizures to be permitted. They must be as soft as they can be and when necessary, they should appropriately, safely seize the pertinent materials and accurately document them. SPCB officials should contact the police if there appears to be a conflict or a safety risk. The principles and processes for gathering environmental data are described in the Environment Act and are provided in Part 26 of the Air Act, Section 21 of the Water Act, and Section 11 of the Environmental Act. Air, water, and emission samples are a few examples. Later on, the following queries are posed, which include as follows: Can the evidence be used legally? Is the sample stored in perfect condition? To preserve the chain of custody and the integrity of the evidence, it should only be handled by the investigators and occasionally by outside parties. All standard laws and regulations should be known to officials in order for the enquiry to proceed unhindered.

8.1.5 Scene of Crime (SOC) Investigation: Dos and Don'ts

Preserving the crime scene is essential during an environmental investigation. Access to the location should be restricted to permitted personnel, which must be meticulously secured and enclosed. Photographs of the scene's comprehensive structure and any further information should be captured—illustrations and recordings. Appropriate devices and containers must be utilised to collect samples of soil, water, air, and other components to prevent contamination. Preventing unauthorised access is crucial for maintaining the integrity of evidence collection (Sharma 2019a, b).

8.1.6 Investigation and Prosecution of Offenders: Dos and Don'ts

Meticulous evidence handling is crucial for effective investigation and prosecution of environmental crimes. All actions taken for courtroom evidence presentation must comply with legal standards and criteria. In robust cases, collaboration with public prosecutors is essential. To guarantee the expeditious progression of processes, delays must be averted. Defending sensitive information is crucial. To maintain privacy and the integrity of the investigation, unauthorised access will be prohibited.

8.1.7 Evidence Collection, Preservation, and Chain of Custody of Evidence

The acquired items should be preserved as physical samples and records to avoid deterioration. Digital information must always be kept in appropriately regulated settings. A chain of evidence might be considered particularly sturdy if it remains intact. All transactions and the analytical specifics of every evidence transfer are included. The timing of every transfer must be recorded for monitoring purposes. Signatures and dates are also necessary for documentation (Verma 2018a; b).

8.2 Skilled Investigation

Future environmental infractions are deterred by thorough investigations that result in successful convictions. Clear and effective investigations encourage community cooperation and public trust, which encourages the reporting of environmental violations. A comprehensive strategy to environmental crime is made possible by competent investigators who collaborate effectively with the public, governmental entities, and non-governmental organisations (Tiwari and Prasad 2020). The core investigative competencies required for environmental crime investigations are outlined in Fig. 8.1, which highlights essential skills and knowledge areas for effective enforcement. The Core Investigative Competencies in Environmental Crime Investigations are as follows:

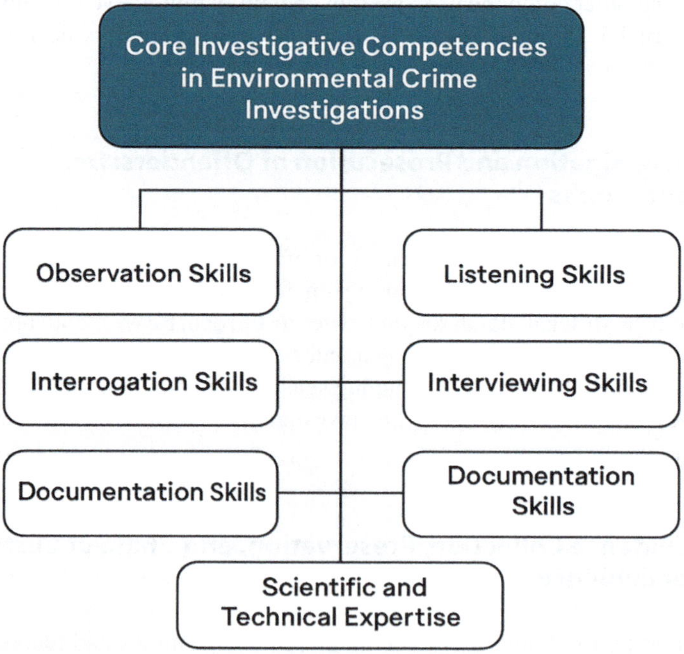

Fig. 8.1 Core investigative competencies required for environmental crime investigations

8.2.1 Observation Skills

Detecting environmental infractions requires keen observation. Investigators are required to spot irregularities, such as unusual smells, discoloured water, or signs of illegal dumping, and to carefully document their findings.

8.2.2 Listening Skills

Detectives can establish connections with communities, witnesses, and suspects by listening well. Through empathetic, culturally aware communication, active listening builds trust and makes it easier to acquire accurate information.

8.2.3 Interviewing Skills

Finding trustworthy information from multiple sources is what conducting interviews requires. Effective interviewers use open-ended questions, maintain objectivity, and create a relaxed atmosphere to elicit thorough answers.

8.2.4 Interrogation Skills

The purpose of interrogation tactics is to get crucial information or admissions from suspects. These strategies, which strike a balance between firmness and compassion, must abide by legal and ethical requirements. Despite the fact that most law enforcement officers are trained in these methods, employees of civic and environmental organisations might not be; therefore, police involvement in environmental crime investigations is essential.

8.2.5 Documentation Skills

A reliable investigation is built on accurate and comprehensive documentation. This means keeping records, writing thorough reports, and making sure all documents are clear for judicial use and legally valid.

8.2.6 Scientific and Technical Expertise

Scientific and technological methods are often used in environmental investigations today. This includes surveillance technologies, forensic sample testing, and digital analysis. Forensic science departments may be consulted by investigators, and its employees may testify as experts in court. Proficiency in these domains is necessary for appropriate evidence gathering and analysis.

8.3 Case Study

8.3.1 Illegal Sand Mining in Uttar Pradesh

In Uttar Pradesh, illegal sand mining has become widespread, particularly along the National River Ganga and its main tributaries, the Yamuna, Chambal, Ken, Betwa, and Son rivers. Unrestricted riverbed extraction has occurred in several rivers during the past 10 years, causing considerable ecological degradation and endangering aquatic biodiversity (SANDRP 2023). Mining puts infrastructure and local livelihoods at risk, and it has a negative impact on ecosystems. The Ganga River Basin's geographical layout is crucial for understanding its ecological dynamics. Figure 8.2 illustrates the main tributaries, populated cities, and sample locations.

8.3.1.1 Judicial and Investigative Response

The National Green Tribunal (NGT) has been instrumental in enforcing the elimination of illicit sand mining operations. These days, satellite imaging and drones are essential parts of the procedure. According to NGT (2020), technologies and GPS systems used to monitor mining operations, including machinery, must be reported in almost real time. The industry has stressed how much accountability is involved. Local authorities must grant approval, and it must be in compliance with existing environmental laws.

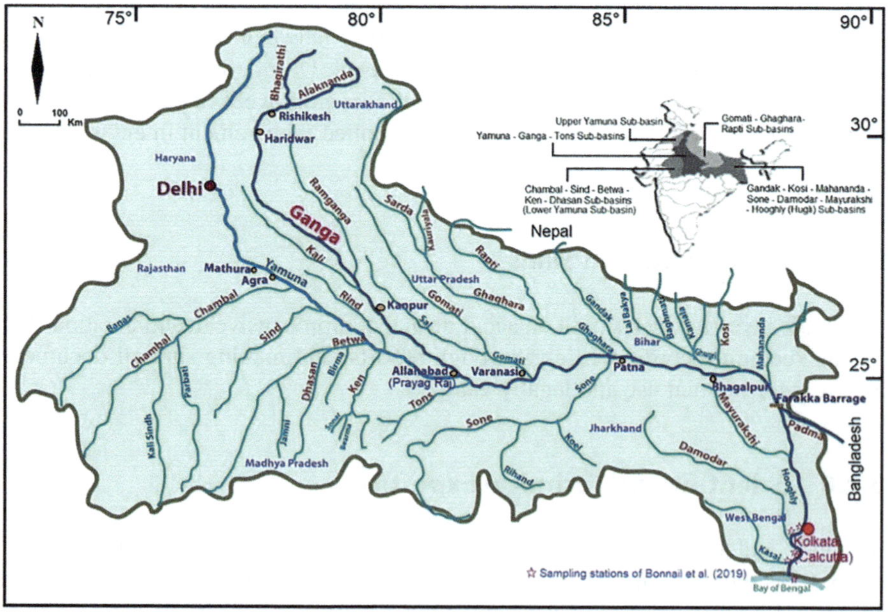

Fig. 8.2 The Ganga River Basin's main tributaries, populated cities, and sample locations are outlined

The Central Bureau of Investigation (CBI) launched an investigation into illicit sand mining in the Betwa River area of Hamirpur in the latter days of 2018. The district magistrate, the mining minister, and numerous other authorities were present during the trial, which revealed a significant fraud case (Indian Express 2019). Mechanised sand mining activities are prominently featured in Fig. 8.3, which showcases excavators and dump trucks operating in a riverbed. This figure provides a visual representation of the scale and machinery involved in such activities.

Although the CBI has conducted raids, made arrests, and seized documents and digital evidence, progress has been slow. Advancement has been hampered by bureaucratic roadblocks and the frequent relocation of investigation professionals. According to analysts, the delay could be politically motivated and meant to protect influential people from prosecution (Wikipedia 2024).

8.3.1.2 Investigation Process

A number of complaints from NGOs and environmental campaigners prompted the CBI to launch its investigation. Several crucial procedures were carried out:

Evidence Acquisition: Teams collected soil samples, GPS data, documents, and equipment from affected areas.

Witness Testimonies: Statements from individuals, members of non-governmental organisations, and lower-level officials who voiced concerns about the acts were recorded.

Fig. 8.3 Mechanised sand mining activity in a riverbed using excavators and dump trucks. (Courtesy: ©SANDRP (South Asia Network on Dams, Rivers and People)

Search and Seizure Operations: Transaction records, unauthorised equipment, and mobile data were retrieved during raids at the homes and workplaces of key suspects. Interrogation and Forensic Analysis: In addition to questioning suspects, forensic examination of computers, mobile devices, and banking records made it easier to track down illegal financial transactions.

Coordination with Local Agencies: To get mining permits and environmental clearance documents, the CBI worked in tandem with the State Pollution Control Board and Mining Department.

Investigators faced challenges despite these efforts, including as missing documentation, uncooperative local authorities, and witness intimidation. Unexpected reassignments of several officers hampered development (SANDRP 2020).

8.3.1.3 Offenders and Modus Operandi

Most of the criminals were members of local mafia groups that had political backing. In order to avoid detection, they used big machinery and unregistered cars for nocturnal mining, avoiding legal permission requirements. Sand was transported by trucks via uncontrolled country roads, and illegal earnings were hidden through cash transactions and shell companies (SANDRP 2023).

8.3.1.4 Environmental and Social Impacts

Significant environmental damage has resulted from illegal sand mining, including severe erosion of riverbanks, loss of habitat for aquatic animals including turtles and Ganga dolphins, and disruption of groundwater recharge (Wildlife Conservation Trust 2024). Social consequences include increased violence, evictions of farmers, and the degradation of rural infrastructure, especially bridges and embankments.

8.3.1.5 Conclusion

This instance illustrates how political corruption affects environmental governance and how regulatory frameworks are inherently inadequate. Although the lack of further action highlights the urgent need for independent monitoring, judicial oversight, and more community involvement to address environmental crimes of this magnitude, the CBI's investigation found serious weaknesses (Wikipedia 2024).

8.3.2 Vedanta Sterlite Copper Plant Pollution, Tamil Nadu

The Vedanta Sterlite Copper plant in Thoothukudi, Tamil Nadu, has had serious negative effects on the environment and human health, making it one of the most controversial industrial operations in India. Established in 1996, the plant produces copper and other metals, which have been accused of causing significant environmental damage despite their significant economic benefit. Due to the facility's primary activity of smelting copper, significant amounts of sulphur dioxide (SO_2) and other dangerous pollutants are released. In addition to health concerns like

Fig. 8.4 Protest against the Sterlite Copper plant in Thoothukudi, Tamil Nadu. (Courtesy: *The Indian Express* (2018). Retrieved from https://indianexpress.com/...)

respiratory disorders and dermatological diseases, residents have voiced concerns about contamination of the air, water, and soil (The News Minute 2018). The community's opposition to the Sterlite Copper plant in Thoothukudi, Tamil Nadu, is evident in Fig. 8.4, which captures a protest scene.

8.3.2.1 Environmental Violations

The pollution linked to the Vedanta Sterlite Copper plant has caused a great deal of concern. The following are the main violations:

Air Pollution: High levels of sulphur dioxide, a dangerous chemical that causes acid rain and presents serious health hazards, including respiratory problems, are produced during the copper smelting process. The facility's sulphur dioxide emissions, according to the local communities, regularly exceeded permissible limits, causing persistent air pollution (The News Minute 2018).

Water Pollution: The factory is accused of discharging untreated wastewater into the nearby Thamirabarani River, which provides the locals with drinkable water. These effluents pose serious risks to human health and aquatic ecosystems because they include harmful chemicals including heavy metals (Times of India 2018).

Soil Contamination: The discharge of hazardous material has contaminated the earth near the factory in addition to polluting the air and water. Food security is under risk due to this contamination's detrimental effects on local vegetation and agriculture (Ej Atlas n.d.).

8.3.2.2 Investigation and Legal Proceedings

Numerous environmental authorities, including the Tamil Nadu Pollution Control Board (TNPCB), have regularly evaluated the Sterlite Copper plant's environmental performance. Legal actions against the company have not advanced quickly, despite allegations of environmental law violations.

Initial Legal Actions and Closures: In 2018, the Tamil Nadu government ordered the factory to close after protests by locals and environmental groups. The ruling came after months of protests during which demonstrators and police clashed violently. During these protests, police fatally shot over 13 people (Al Jazeera 2018).

NGT and Environmental Impact Assessment (EIA): The Ministry of Environment, Forests, and Climate Change (MoEF&CC) and the National Green Tribunal (NGT) have been working to assess the environmental damage caused by the Sterlite plant. Vedanta had already been fined by the NGT for violating environmental laws. However, the company regularly challenged these rulings, which caused further regulatory actions to be delayed (Bar and Bench 2018).

Supreme Court Orders: The Supreme Court of India has stepped in and suspended Vedanta's operating licence several times for violations. However, the management of the facility continued to challenge these orders in court, often citing the factory's economic importance and the jobs it provides for local workers (Reuters 2024).

8.3.2.3 Health and Social Impact

The pollution originating from the Sterlite Copper plant has severely affected neighbouring populations:

Health Consequences: People in Thoothukudi, especially those living close to the factory, have complained of a high frequency of dermatological conditions, eye pain, and respiratory conditions. Chronic health problems including asthma, bronchitis, and other respiratory ailments are believed to have been caused by prolonged exposure to sulphur dioxide and other pollutants (The News Minute 2018).

Social Unrest: Numerous protests and public discontent have been sparked by environmental and health concerns. Violent protests broke out in 2018 because the facility continued to operate in spite of environmental rules. Ultimately, law enforcement killed several protesters during these demonstrations (Al Jazeera 2018).

Economic Impact: Despite employing thousands of locals, the company's negative health effects and environmental damage have significantly reduced the standard of living in the community. Many residents argue that the costs of pollution outweigh the economic benefits, especially when considering the long-term medical costs associated with pollution-related diseases (OpIndia 2024).

8.3.2.4 Environmental and Regulatory Measures

Regulatory Oversight Many regulatory solutions have been proposed in response to the growing pollution problems. To reduce sulphur dioxide emissions and improve waste management practices, the National Green Tribunal (NGT) has pushed for the use of cutting-edge technologies (The New Indian Express 2018). The Tamil Nadu Pollution Control Board (TNPCB) has been monitoring trash disposal procedures and enforcing stricter emission regulations.

Corporate Accountability Vedanta is forced to raise its environmental standards in the face of growing protests and environmental lawsuits. According to environmental campaigners, Vedanta should take more action to reduce the negative impacts and its contributions, which include the development of cleaner technologies, giving environmental restoration initiatives more consideration (Foil Vedanta 2020).

8.3.2.5 Conclusion

The conflict between environmental preservation and economic growth is demonstrated by this example. They do, however, make significant contributions to the modern economy. Although the industry boosts the economy, people are put off by the environmental harm because of the potential harm to society and the environment. Protests with legal concerns have been sparked by this destruction. The case demonstrates the necessity for increased enforcement. To assess current practices for tracking unlawful activity and corporate responsibility to reduce manufacturing pollution, environmental legislation is required. It is well known that the factory's closure has been universally praised. The full theological implications for the area are still unknown despite the fact that environmental justice organisations have commended it.

8.3.3 Poaching and Wildlife Trafficking: The Sansar Chand Case

There have been some terrible reports of wildlife poaching and trade in India; Sansar Chand, for instance, was a well-known participant in the wildlife trafficking industry. Known as the "King of Wildlife Smugglers", Sansar Chand was a key player in the illegal trafficking and extensive poaching of endangered species. He pursued his objectives both inside and outside of numerous Indian states. In the 1990s and early 2000s, Chand played a key role in wildlife trafficking, mostly dealing in the sale of animal hides, bones, and other body parts used for medical and decorative uses (Bahuguna 2014). It wasn't until the middle of the 2000s that he was caught and his network was broken up despite the fact that police had been aware of his illegal activities for years. Figure 8.5 starkly illustrates the illicit wildlife trade, showcasing a display of illegal wildlife products at a market, highlighting the concerning reality of wildlife trafficking.

8.3.3.1 Poaching Activities

Sansar Chand's poaching activities were executed with exceptional subtlety and organisation. His endeavours predominantly focused on:

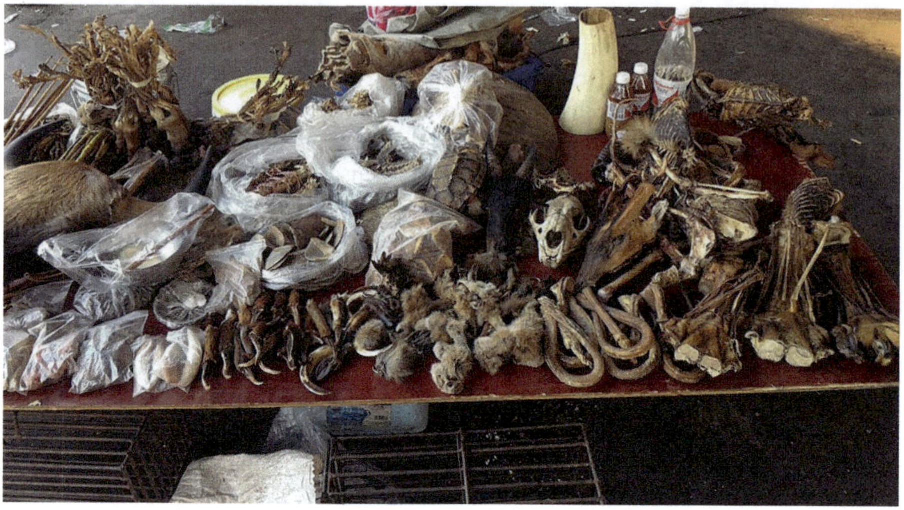

Fig. 8.5 Display of illegal wildlife products at a market. (Courtesy: Nikki van Schyndel, via Wikimedia Commons, licensed under CC BY-SA 4.0)

Tigers and Leopards: Their pelts, skeletal remnants, and various other anatomical parts were highly sought after in illegal markets both locally and abroad, leading to the extinction of these animals. Although the Wildlife Protection Act of 1972 forbids the illegal hunting of tigers, the demand for tiger bones and skins is often linked to ornamental and traditional Chinese medicinal purposes (Wildlife Trust of India 2024).

Elephants: The illegal hunting of elephants for their tusks was linked to Chand's network. The illegal ivory trade, a multibillion-dollar global enterprise, uses elephant tusks. India's elephant herds suffered greatly as a result of the extinction of these amazing animals (Wildlife Trust of India 2024).

Other Species: Crocodiles; other bird species, including the Indian peafowl; and rhinoceroses were also the focus of Chand's actions (Wildlife Trust of India 2024).

His operations were of alarming scale, including a network of wildlife criminals ranging from domestic poachers to international traffickers. Where national parks and wildlife sanctuaries were not well protected, Chand's business flourished, giving him easy access to endangered animals (Wildlife Trust of India 2024).

8.3.3.2 The Network and Operations

Across multiple Indian states, including Madhya Pradesh, Rajasthan, Uttarakhand, and Uttar Pradesh, Sansar Chand oversaw a vast network. According to sources, the Chinese culture uses particular animal parts, and their products finally made their way to China and other countries in Asia through a trafficking network that the Wildlife Trust of India (2024) has discovered links to in Southeast Asia and Nepal. People knew his circle of associates were vicious; they paid poachers to kill protected animals using illegal weapons, and they received instructions from him on how to do it.

Through his secret operations, Chand managed to get several animal parts to their target without anyone noticing. Local politicians received financial contributions from the trafficking network. Here, legally permitted commodities were moved to conceal illicit wildlife trade items and to conceal money laundering through legitimate commercial operations. Market businesses have also been examined in this study (Wildlife Trust of India 2024).

Both snares and traps were used to catch animals illegally, as well as technology advanced enough to provide tranquiliser darts to catch the animals alive, instead of just killing animals for their body parts (WTI 2024). Figure 8.6 provides an overview of the ongoing cases against notorious poacher Sansar Chand across India, detailing the scope and extent of his alleged involvement in wildlife crimes.

8.3.3.3 Legal Investigations and Arrest

The Wildlife Crime Control Bureau and the Central Bureau of Investigation are both incorporated. Two Indian agencies tasked with environmental protection have

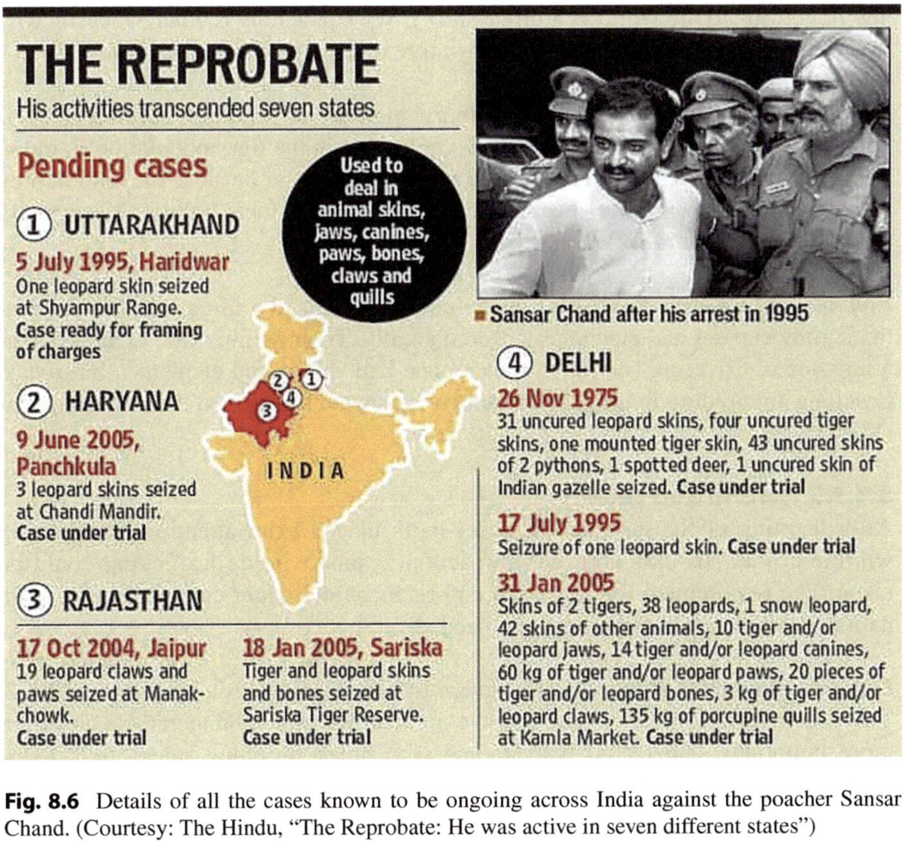

Fig. 8.6 Details of all the cases known to be ongoing across India against the poacher Sansar Chand. (Courtesy: The Hindu, "The Reprobate: He was active in seven different states")

commenced efforts to address the increase in poaching and the trafficking of animal parts. As law enforcement intensified their emphasis throughout the years, numerous individuals were apprehended following raids targeting Chand's enterprises.

Initial Investigation and Arrest First, the crew found and detained Buoying, a notorious poacher and trafficker. Chand was able to evade capture by the authorities for many years, thanks to his extensive network as he was gaining strong backing and authority from politicians. Many police forces, departments, and agencies together with the CBI and WCCB arrested him in 2005 (Wildlife Trust of India 2024).

Court Trials and Conviction The court found him guilty in a groundbreaking case in 2010 in cases related to wildlife trafficking and poaching. He participated in hunting both tigers and other animals. He injured endangered animals; he was sentenced to life in prison (India Kanoon 2010). Through thorough efforts, the case brought a major success in combatting wildlife trafficking and poaching.

8.3.3.4 Impact on Wildlife and Conservation

Animals in India who were already in danger suffered greatly as a result of Sansar Chand's actions. Similarly, the population of these top predators is already declining. According to the World Wildlife Fund (WWF), poaching is a major reason why India's tiger population has decreased by more than half 100 years ago (WWF n.d.).

Declining Tiger Population The primary motivation for tiger hunting is their bones and skin, which is the cause of the reduction in the tiger population in India. Things start to change as tigers become rare and disappear because they aid in prey management and ecological health. The equilibrium in local habitats is currently upset due to predators (WWF n.d.).

Elephant and Rhino Poaching In the course of Chand's illicit poaching operations, rhinoceroses and elephants suffered greatly. Their numbers are at high risk of going down. Covering such great lands once had rhinos and elephants, but today, poaching and trading in illegal elephant ivory is what has caused them to disappear (WWF n.d.).

8.3.3.5 Reform and Legacy

After learning of Sansar Chand's case, India placed extra attention on reducing wildlife crime. His case showed how action against wildlife trafficking could be taken. The requirement for increased enforcement and better cooperation between national and international groups are needed.

Legal Frameworks Hunting is forbidden by the Wildlife Protection Act of 1972. The circumstances in this situation have made protecting endangered species even more important. It declared that anyone who break poaching rules ought to be detained and dealt with. Following the passage of the Wildlife Conservation Act in

1972, regulations pertaining to the conservation of wildlife became increasingly stringent.

Awareness and Advocacy Public awareness of animal protection, and specifically the illegal rhino traffic, increased significantly as a result of the case. Elephants and tigers are its main subjects. However, in various events, wildlife advocates and environmental activists use this case as an example. The group advocates for stricter laws and more effective means of enforcing them (WWF n.d.).

8.3.3.6 Conclusion

This incident highlights the prevalence of wildlife trafficking as well as the impact organised crime has on environmental protection initiatives. Despite this, his imprisonment has had a major influence on environmental crime. In the fight to protect wildlife and the environment, this decision represents a huge step forward. It highlights the need of comprehending global politics in the battle against wildlife crimes, as both increased enforcement and collaboration are being employed.

8.3.4 Oil Spillage in Mumbai Coast (MSC Chitra Incident, 2010)

The events involving the MSC Chitra are among the most significant oil spill incidents in the region. An incident occurred in August 2010 near Mumbai, India. The event involved the collision of two large vessels: the oil tanker Khalijia III and the cargo ship MSC Chitra. The accident resulted in a large oil spill, which seriously damaged the coastal ecosystem, especially the fishing industry, marine life, and coastal residents.

The event happened in the busy Mumbai Harbour while ships were navigating one of the busiest sea lanes in India. More than 1000 tonnes of heavy fuel oil were spilt into the Arabian Sea by the ship, resulting in a major environmental disaster (Cedre n.d.). The Indian Coast Guard, environmental organisations, and local officials responded to the incident with great urgency in order to contain the spill and lessen its effects. Figure 8.7 depicts the devastating impact of an oil spill near Mumbai, highlighting the damage to marine and coastal environments, and underscoring the need for prompt response and mitigation measures to protect these ecosystems.

8.3.4.1 Cause of the Incident

Navigational mistakes and poor water visibility caused two vessels to collide in the MSC Chitra incident. According to Indian Maritime Administration reports, the tragedy happened when the ships were navigating a crowded sea channel off the coast of Mumbai (Indian Maritime Administration 2010).

Navigational Errors The incident's investigation revealed that the vessels' poor coordination and navigation were the main causes of the collision. The collision is

Fig. 8.7 Oil spill near Mumbai affecting marine and coastal environments. (Courtesy: *Mumbai oil spill*, by Reuters, 2011)

said to have occurred because the MSC Chitra did not maintain a sufficient distance from the Khalijia III.

Environmental Conditions In addition, bad weather, such as extreme winds, high waves, and poor visibility, might have made the situation worse and hindered rescue attempts (Indian Meteorological Department 2010). The Khalijia III's fuel tank suffered a major break as a result of the collision, resulting in a crude oil spill. A wide variety of marine life and coastal ecosystems were impacted by the spill, which spread quickly due to the Arabian Sea's strong currents (Cedre n.d.).

8.3.4.2 Extent of the Oil Spill

Over 1000 tonnes of crude oil were spilt into the Arabian Sea as a result of the incident, and estimates suggest that this oil scattered over more than 30 nautical miles of coastline (Indian Coast Guard 2010). The marine ecosystem suffered greatly as a result of the oil leak that reached Mumbai's coastline and nearby coastal areas. Mangroves, beaches, and coastal ecosystems were all contaminated by the spill, which caused long-lasting environmental damage.

Marine Life Impact Numerous marine animals, such as fish, turtles, and seabirds, were affected by the oil spill; they suffocated, became poisoned, and suffered long-term damage to their reproductive systems (Marine Conservation Group 2011).

Mangroves and Coastal Ecosystems Mangrove trees, coral reefs, and fishing grounds along the Mumbai coastline were all severely damaged. As natural barriers against coastal erosion and as spawning grounds for many marine species, mangroves suffered severe harm from oil (Environmental Protection Agency [EPA] 2010). Investigations are still ongoing to determine the long-term effects on these ecosystems since oil exposure may prevent mangrove tree growth and regeneration (World Wildlife Fund 2012).

8.3.4.3 Impact on Local Communities and Economy

Locals, particularly those involved in fishing, tourism, and coastal agriculture, were greatly impacted by the oil leak.

Fishing Industry One of the industry's most severely affected was fishing. Fishing grounds were closed and fish catches were reduced as a result of oil contamination, endangering the livelihoods of thousands of fishermen (National Fisheries Department 2010). Due to the health hazards related with the contamination, fishermen reported that the oil leak made the waters unsafe for fishing and fish consumption.

Tourism and Beaches The leak had a negative impact on Mumbai's coastal areas, which include well-known beaches like Versova and Juhu. Beaches were damaged by the oil disaster, making them dangerous for both visitors and locals, which reduced tourism (Mumbai Tourism Board 2010). Cleaning the site was costly and took many people a long time.

Economic Losses The Ministry of Economic Affairs (2011) projected that the event would result in substantial financial losses for numerous businesses, as well as the tourism and fishing sectors, amounting to millions of dollars, including impacts on fishing and other coastal enterprises.

8.3.4.4 Response and Clean-Up Efforts

In response to the oil disaster, the Indian Coast Guard (ICG) and local authorities took the initiative. In order to contain and remove the oil from the water's surface, booms and skimmers were used in the remediation operations. Chemical dispersants were used to break down the oil and lessen its spread.

Use of Booms and Skimmers The instruments were utilised to control and monitor the oil and inhibit the propagation of the sickness along the shoreline. Barriers limited the oil supply to the impacted area. Specialised nets were employed to retrieve it from the water's surface (Indian Coast Guard 2010).

Manual Clean-Up Local volunteers, including fishers, were solicited for assistance and organise initiatives to facilitate beach clean-up efforts. The substantial leak rendered clean-up challenging. Neglecting to remove the majority of the oil subsequently resulted in complications for marine organisms. Shortly after the event, the issue typically persists in the ocean (Clean Oceans Initiative 2011).

Long-Term Monitoring and Rehabilitation Plans were established to evaluate the circumstances surrounding the leak. What occurs in marine and coastal habitats following the expedited clean-up process? What ensues as a consequence? The assessment of maritime biodiversity is ongoing, primarily aimed at the restoration of mangroves and analogous ecosystems as the ecosystem is vulnerable to numerous concerns (World Wildlife Fund 2012).

8.3.4.5 Legal and Regulatory Impact

Because of the MSC Chitra disaster in the maritime business, there was a clear need for greater better rules and improved ways of following them. It received a great number of suggestions.

Responses to the leak aimed at stopping similar problems in the future:

Stronger Maritime Safety Regulations: Enhancing preparedness among the operating vessels. When the commercial district is congested near Mumbai, the government and international organisations such as the International Maritime Organisation (IMO) conduct analyses and safety regulations (International Maritime Organisation 2011).

Environmental Laws and Accountability: This is to guarantee that those accountable for the spills faced prosecutions and appropriate legal actions. The owners of MSC Chitra were accountable for the damage, while those operating Khalijia III were tasked with doing the repairs. The corporations involved in the drilling were required to allocate funds for environmental remediation. In 2011, the Ministry of Environment and Forests assessed the findings.

8.3.4.6 Conclusion

The incident involving the MSC Chitra is the most significant oil spill in India's history. The tragedy inflicted considerable harm on coastal regions and their ecosystems, highlighting the necessity for improved maritime regulations and robust strategies for managing oil spills at sea. The catastrophe brought to light the dangers faced by coastal communities and businesses that depend on the maritime environment.

Future disaster relief will require increased preparedness, sophisticated spill response systems, and greater collaboration between government agencies, shipping companies, and communities.

8.3.5 Bhopal Gas Tragedy Revisited

A devastating industrial tragedy in history, the Bhopal Gas Tragedy took place on the evening of December 2–3, 1984. The Union Carbide India Limited (UCIL) pesticide plant in Bhopal, Madhya Pradesh, leaked a toxic gas that released the highly dangerous chemical methyl isocyanate (MIC) into the surrounding area. In addition to causing widespread poisoning and thousands of deaths and injuries, the gas

Fig. 8.8 Remnants and memorial of the 1984 Bhopal Gas Tragedy, as seen 30 years later. (Courtesy: *Bhopal Gas Tragedy: 30 years later, we revisit the site of disaster*, by The Indian Express, 2014)

quickly travelled into nearby residential areas. Environmental, public health, and the city's socioeconomic structures were all severely impacted by the disaster.

An estimated 3000 people died from gas exposure in the first few days, and official figures later increased to 15,000 deaths in the following years (Government of India 2004). Over 500,000 people were affected, and many of them suffered from long-term health issues such respiratory conditions, cognitive impairment, and birth defects (Eckerman 2005). The immediate effects of the gas leak were catastrophic because, without adequate warning or planning, a whole population was exposed to deadly pollutants. Figure 8.8 captures the remnants and memorial of the 1984 Bhopal Gas Tragedy, 30 years after the disaster, serving as a haunting reminder of the devastating industrial accident and its lasting impact on the lives of those affected.

8.3.5.1 Cause of the Incident

A combination of human mistake, mechanical failure, and inadequate safety procedures at the UCIL plant led to the gas leak. The main source of the spill was a violent gas escape into the atmosphere brought on by the explosion of a methyl isocyanate storage tank.

Technical Failures The safety systems of the factory were not properly maintained, and critical safety equipment, such as refrigeration units designed to keep the MIC tank from reaching dangerous levels, was either deactivated or not operating (Eckerman 2005). The factory has a history of safety violations, and several examinations revealed shortcomings in its management and maintenance practices.

Inadequate Safety Measures According to expert and environmental agency investigations, UCIL failed to implement necessary safety and precautionary measures to lessen the risks associated with handling and storing MIC. The tragedy was caused in large part by inadequate training for plant workers, poor safety procedures, and a lack of community safety plans (Gupta 2002).

Inadequate Emergency Response An inadequate emergency response system made the catastrophe worse. When the leak occurred, no prompt notification was sent to the nearby villages. Due to inadequate public communication strategies, many of the people were unaware of the danger until it was too late.

8.3.5.2 Environmental and Health Impact

The environmental consequences of the Bhopal Gas Tragedy were significant and extensive. The immediate repercussions of the gas spill were catastrophic although the enduring ramifications of exposure to MIC persist throughout Bhopal.

Health Effects Numerous immediate and long-term health problems, such as severe respiratory distress, blindness, lung damage, and neurological illnesses, were brought on by the chemical exposure. In the years following the incident, the affected population saw higher rates of cancer, birth defects, and miscarriages (Sriramachari 2004). The survivors continue to experience health issues, and many continue to seek payment for medical treatment and rehabilitation.

Soil and Water Contamination The disaster caused significant environmental damage in addition to direct human casualties. Long-lasting ecological impact was caused by the chemicals that leaked from the factory, which harmed nearby groundwater and land. The facility's toxic waste remains in the soil, making the nearby towns' water supplies unsafe to drink (Greenpeace 1999).

Ecosystem Disruption The biodiversity of the area was disturbed when MIC spread into the nearby environment, affecting the local animals. Pollution of the air and water has a long-lasting effect on microbes, plants, and animals, making ecosystem restoration more difficult (Dhara and Dhara 2002).

8.3.5.3 Legal and Social Consequences

Legal Proceedings The legal fallout from the disaster has been marked by insufficient compensation for the victims, political interference, and delays. The Indian government and Union Carbide Corporation (UCC) came to an agreement in 1989 for $470 million in compensation for the deaths; nonetheless, this decision has been widely lambasted as being woefully inadequate (Government of India 2004). Despite the constant demands of victims and activists for more restitution and accountability, many aspects of the case are still unresolved, and court battles have continued for decades.

Corporate Accountability The participation of Dow Chemical and Union Carbide has generated criticism. Despite UCC's subsequent acquisition by Dow Chemical, the legal system has mostly failed to hold the firm entirely accountable for the harm that resulted. Justice has yet to be served for the victims of the gas leak as a number of lawsuits and court cases remain pending (Amnesty International 2004).
Social and Economic Impact The disaster had a significant impact on Bhopal's societal structure. As many families lost their primary breadwinners, the city's economy collapsed, and survivors were forced into poverty due to the long-term health effects. Survivors' and their communities' mental health suffered as a result of the embarrassment surrounding the disaster since many felt abandoned by the government and the companies responsible (Jasanoff 2007).

8.3.5.4 Response and Rehabilitation Efforts

Relief Operations The Indian government's and other NGOs' initial relief efforts focused on providing the affected people with emergency medical care and short-term housing. However, these efforts fell short, and many survivors were left to suffer with chronic illnesses without access to appropriate medical care (CSE 2009).

Health and Medical Aid Over the years, many organisations have provided medical care to the survivors; yet, the health issue still exists. For people affected by gas exposure, the government has set up a number of health institutions; nevertheless, these facilities are inadequate and underfunded for the large number of affected individuals (ICMR 2010).

Environmental Restoration Prolonged delays due to bureaucratic impediments and insufficient funding have hindered efforts to remediate the contaminated property. Various environmental organisations are striving for a comprehensive remediation of the area affected by chemical incidents that have left hazardous substances (Greenpeace 1999).

8.3.5.5 Lessons Learned and Global Impact

International Standards: This prompted global conversations regarding environmental regulations, industry protocols, and governmental responsibilities for industrial safety. New regulations were enacted to enhance safety at industrial sites, and protocols for responding to dangerous chemical incidents were established (Shrivastava 1992).

Corporate Responsibility Numerous individuals throughout the environmental and activist movements have utilised this case study to illustrate the peril of prioritising profits over the welfare of others.

The Degree of Corporate Accountability for Environmental Incidents (Amnesty International 2004) The tragedy serves as a reminder that governments must prioritise safety enforcement and businesses must adhere to these protocols.

Global Environmental Movement The global endeavour to uphold environmental justice, aimed at safeguarding disadvantaged populations from environmental and industrial hazards, was severely disrupted by the Bhopal incident. Consequently, global recommendations for human rights and environmental health have been established (Jasanoff 2007).

8.3.5.6 Conclusion

Bhopal Gas Tragedy is remembered as one of the greatest industrial tragedies ever. Even after decades of legal and social efforts, their families have not obtained fairness in the court's problems and issues with community health and pollution remain unsolved. The case shows how being irresponsible in industry, not having proper safety rules, and major companies not wanting to accept the consequences of what they do. It shows that we must have greater accountability from businesses and better ways to handle disasters rules for industries and support for systems should be introduced.

8.4 Conclusion

The case studies covered in this chapter demonstrate the variety of information and approaches used to investigate environmental crimes in India. Industrial pollution, illicit mining, and animal trafficking and the disposal of toxic waste are two examples of activities that pose serious environmental risks.

Sustainable Development and Public Health All cases have numerous significant problems that hinder the effectiveness of environmental stewardship, including inadequate enforcement, errors in procedures, delays in receiving justice, and government meddling. Also, these investigations show how important it is becoming to use the legal system for advocacy: getting people involved, working together across multiple departments, and holding people responsible for their actions.

In their activities, concerned authorities, non-governmental organisations, and legal groups make it clear that, even under a strict system, you can assist people grow if the conditions are right. Our top priority must be conservation.

The government must swiftly improve India's handling of environmental violation charges, new legislation, augmented investigative authority, enhanced collaboration across organisations, and expanded outreach to impacted individuals. The data obtained from these case studies can guide policymakers.

Experts in this field and the regulatory authorities aim to strengthen the framework ofenvironmental justice inside the country.

References

AlJazeera (2018, May 23) Indian police accused of killing at least 11 protesters. Retrieved from https://www.aljazeera.com/news/2018/5/23/indian-police-accused-of-killing-at-least-11-protesters

Amnesty International (2004) Clouds of injustice: Bhopal disaster 20 years on. Amnesty International Publications, London

Bahuguna A (2014, June 22) Sansar Chand: the kingpin of the wildlife trade in India. Down To Earth. https://www.downtoearth.org.in/blog/sansar-chand-the-kingpin-of-the-wildlife-trade-in-india-44919

Bar & Bench (2018) Vedanta-Sterlite-Copper-Plant-Thoothukudi-NGT.pdf. Retrieved from https://images.assettype.com/barandbench/import/2018/12/Vedanta-Sterlite-Copper-Plant-Thoothukudi-NGT.pdf

Cedre (n.d.) MSC Chitra oil spill: incident overview. Centre of documentation, research and experimentation on accidental water pollution. Retrieved from https://www.cedre.fr/en/

Clean Oceans Initiative (2011) Oil spill clean-up efforts in India: The MSC Chitra response. Retrieved from http://www.cleanoceans.org

CSE (Centre for Science and Environment) (2009) Bhopal gas tragedy: the saga continues. CSE Report.

Dhara VR, Dhara R (2002) The Union carbide disaster in Bhopal: a review of health effects. Arch Environ Health 57(5):391–404

Eckerman I (2005) The Bhopal saga: causes and consequences of the world's largest industrial disaster. Universities Press, Hyderabad

Ej Atlas (n.d.) Sterlite copper smelter unit, Tamil Nadu, India. Retrieved from https://ejatlas.org/conflict/sterlite-copper-smelter-unit-india

Environmental Protection Agency (EPA) (2010) Impact of oil spills on coastal ecosystems: a case study of the MSC Chitra incident. Retrieved from https://www.epa.gov/oil-spills

Foil Vedanta (2020) Why the Vedanta-Sterlite judgement matters. Retrieved from https://www.foilvedanta.org/articles/why-the-vedanta-sterlite-judgement-matters/

Government of India (2004) Report of the Bhopal Gas Leak Disaster (Processing of Claims) Act, 1985 and its implementation. Ministry of chemicals and fertilizers, Department of chemicals and petrochemicals. Retrieved from https://www.chemicals.gov.in

Greenpeace (1999) The Bhopal legacy: toxic waste and contaminated groundwater. Greenpeace, Bhopal

Gupta A (2002) Corporate social responsibility and safety: the Bhopal gas tragedy. Indian Institute of Management, Ahmedabad

ICMR (Indian Council of Medical Research) (2010) Health effects of exposure to toxic gas at Bhopal: Research reports 1985–1994. Ministry of health and family welfare

India Kanoon (2010, October 20) Sansar Chand vs State of Rajasthan. https://indiankanoon.org/doc/1658270/

Indian Coast Guard (2010) MSC Chitra oil spill: Response and clean-up operations. Government of India. Retrieved from http://www.indiancoastguard.gov.in

Indian Express (2019, June 13) Hamirpur illegal mining case: CBI raids Prajapati's residences. Retrieved from https://indianexpress.com/article/india/hamirpur-illegal-mining-case-cbi-raids-prajapatis-residences-5777784/

Indian Maritime Administration (2010) Report on the MSC Chitra oil spill incident. Ministry of Shipping, Government of India. Retrieved from http://www.shipmin.gov.in

Indian Meteorological Department (2010) Weather conditions during the MSC Chitra incident. Retrieved from http://www.imd.gov.in

International Maritime Organization (IMO) (2011) International maritime safety regulations: A post-incident review. Retrieved from https://www.imo.org

Jasanoff S (2007) Bhopal: law, knowledge and ambiguity. Osiris 22(1):287–302

Marine Conservation Group (2011) Marine life and oil spills: the long-term effects of the MSC Chitra disaster. Retrieved from http://www.marineconservation.org

Ministry of Economic Affairs (2011) Economic impact of the MSC Chitra incident. Government of India. Retrieved from http://www.economicaffairs.gov.in

Mumbai Tourism Board (2010) Impact of oil spills on Mumbai's tourism industry. Retrieved from http://www.mumbaitourism.gov.in

National Crime Records Bureau (2022) Increase in offences under 4 of 7 environment-related acts in 2022. Down to Earth. Retrieved from https://www.downtoearth.org.in/environment/increase-in-offences-under-4-of-7-environment-related-acts-in-2022-ncrb-93222

National Fisheries Department (2010) Impact of the MSC Chitra oil spill on the fishing industry. Government of India. Retrieved from http://www.fisheries.gov.in

National Green Tribunal (NGT) (2020) Enforcement & Monitoring Guidelines for Sand Mining. Retrieved from https://greentribunal.gov.in/sites/default/files/news_updates/1746085046.pdf

OpIndia(2024,December)SterliteCopperPlantclosure:Regretoverjoblossesfuelsprotests.Retrieved from https://www.opindia.com/2024/12/sterlite-copper-plant-closure-regret-over-job-losses-fuels-protests/

Reuters (2024, February 29) India top court rejects Vedanta's plea to reopen Sterlite Copper's smelting unit – lawyer. Retrieved from https://www.reuters.com/world/india/india-top-court-rejects-vedantas-plea-reopen-sterlite-coppers-smelting-unit-2024-02-29/

SANDRP (2020, May 27) UP riverbed mining overview: NGT, CBI, Govts cannot stop the menace. Retrieved from https://sandrp.in/2020/05/27/up-riverbed-mining-overview-ngt-cbi-govts-cannot-stop-the-menace/

SANDRP (2023, January 20) 2022: Riverbed mining destroying Indian river eco-system & freshwater species. Retrieved from https://sandrp.in/2023/01/20/2022-riverbed-mining-destroying-indian-river-eco-system-freshwater-species/

Sharma M (2019a) Environmental crime: understanding the impact and regulatory responses in India. Environ Stud J 45(2):65–81

Sharma R (2019b) Best practices for environmental crime scene investigation. Indian J Forensic Stud 11(1):45–60

Shrivastava P (1992) Bhopal: anatomy of a crisis. Paul Chapman Publishing, London

Singh A (2021) Air pollution and its enforcement in India: a study of regulatory failures. J Environ Prot 22(3):112–120

Sriramachari S (2004) The Bhopal gas tragedy: an environmental disaster. Curr Sci 86(7):905–920

The New Indian Express (2018, November 28) NGT panel says closure of Sterlite plant unsustainable. Retrieved from https://www.newindianexpress.com/states/tamil-nadu/2018/Nov/28/closure-of-sterlite-plant-at-tuticorin-not-sustainable-against-natural-justice-committee-to-ngt-1904533.html

The News Minute (2018, March 27) Sterlite – here's the proof: How the copper plant impacts health of Thoothukudi people. Retrieved from https://www.thenewsminute.com/delve/sterlite-here-s-proof-how-copper-plant-impacts-health-thoothukudi-people-78772

Times of India (2018, March 27) Sterlite polluting Thamirabarani, shut it down: Stalin. Retrieved from https://timesofindia.indiatimes.com/city/chennai/sterlite-polluting-thamirabarani-shut-it-down-stalin/articleshow/63475848.cms

Tiwari S, Prasad D (2020) Environmental crime and its control in India: a study of legal framework and enforcement mechanisms. J Environ Law Policy 3(2):78–93

Verma S (2018a) Anthropogenic activities and the degradation of natural resources in India. Environ Issues Rev 27(1):31–42

Verma P (2018b) Preserving environmental evidence: legal and procedural guidelines. Environ Law Rev 5(3):120–135

Wikipedia (2024) Uttar Pradesh sand mining scandal. Retrieved from https://en.wikipedia.org/wiki/Uttar_Pradesh_sand_mining_scandal

Wildlife Conservation Trust (2024) The menace of sand mining in the Gangetic plains: an explainer. Retrieved from https://www.wildlifeconservationtrust.org/the-menace-of-sand-mining-in-the-gangetic-plains-an-explainer/

Wildlife Protection Act (1972) The Wild Life (Protection) Act, 1972. https://www.indiacode.nic.in/bitstream/123456789/1726/1/a1972-53.pdf
Wildlife Trust of India (2024, January 15) Nabbed! The don of wildlife crime. https://www.wti.org.in/news/nabbed-the-don-of-wildlife-crime/
World Health Organization (WHO) (2022) Air quality and health in India: a report on pollution levels. Retrieved from https://www.who.int/air-quality-india-report
World Wildlife Fund (2012) The long-term effects of oil spills on mangroves and marine biodiversity. Retrieved from https://www.worldwildlife.org
World Wildlife Fund (n.d.) Tiger. https://www.worldwildlife.org/species/tiger

Ecosystem Restoration and Remediation in the Context of Forensic Science and Environmental Analysis

9

Annwesha Mazumdar, Rajdip Khaniya, Dipak Kumar Mahida, Vishal M. Makwana, Ankita Patel, and Mahipal Singh Sankhla

Abstract

Ecosystem restoration and remediation are important for recovering from environmental harm and maintaining ecological equilibrium. For forensic sciences and environmental studies, these methodologies are essential for both estimating and remediating pollution, habitat destruction, and human impact. Forensic environmental investigation is a methodical analysis using scientific procedures, alongside the current legal requirements, for the purpose of identification of the origin of contamination according to causality principle and the subjects responsible for this contamination, in order to be used in the legal process relating to the environmental offenses. Sophisticated analytical methods such as remote sensing, GIS technology, and high-resolution mass spectrometry contribute to track pollutants commitment and ecosystem damage. The incorporation of forensic science into ecosystem restoration provides a means to undertake cost-effective, targeted remediation, based on evidence, thereby achieving a more efficient recovery of natural ecosystems. New technologies, including bioremediation, phytoremediation, and nanotechnology-based detoxification methods, are being developed for the restoration of polluted environment in an eco-friendly manner. AI and machine learning are also enhancing forensics by predicting patterns of

A. Mazumdar
Regional Forensic Science Laboratory, Durgapur, West Bengal, India

R. Khaniya · D. K. Mahida (✉) · A. Patel
Department of Biochemistry and Forensic Science, Gujarat University, Ahmedabad, Gujarat, India

V. M. Makwana
Department of Biosciences, Veer Narmad South Gujarat University, Surat, Gujarat, India

M. S. Sankhla
School of Basic and Applied Sciences, K.R. Mangalam University, Gurugram, Haryana, India

M. S. Sankhla et al. (eds.), *Introducing the Synergy Between Forensic Science and Environmental Analysis*, https://doi.org/10.1007/978-981-95-0199-1_9

contamination and seeking best matches for restoration efforts. Furthermore, forensic analysis on soil and water helps to understand the long-term environmental influences on ecology and assist a policy and enforcement approach. In this chapter, the convergence of forensic science with environmental analysis is considered, with a focus on the approaches employed to identify environmental damage, legal aspects of ecosystem degradation, and novel remediation technologies. Integrating forensic investigations with restoration science, the chapter is a study of the critical role played by interdisciplinary efforts to save environmental health and secure ecological justice.

Keywords

Ecosystem · Restoration · Forensic science · Environmental harm · Ecological justice

9.1 Introduction

9.1.1 Overview of Ecosystem Degradation

Decomposition is the most serious environmental problem that is evolving as a result of natural actions and human activity, such as deforestation, pollution, climate change, and misuse of land. This degradation results in loss of biodiversity, depletion of the balance in nature as well as in reduction of the ecosystem potential regarding the provision of life support services, such as air or water cleansing, soil fertility maintaining, or climate regulation. The effects of degradation stretch beyond the environment to human health, food security, and the economy (Makwana et al. 2023).

The examination of ecosystem degradation assists in the determination of the cause and extent of environmental injury, which supports legal and regulatory actions against the responsible party in forensic science. Environmental assessment is important in appraising the intensity of degradation, localizing fragile sites, and planning strategies of conserving the resources. Remote sensing technique, biological indicators, and chemical analysis methods served as the main tools to evaluate ecosystem health and identification of the warning signs of degradation.

Restoration and remediation attempt to reverse damaging effects through conservation policies, replanting efforts, minimization of pollution, and habitat rehabilitation. Quantification of the degradation process is essential for the creation of such policies for ecological resilience and long-term sustainability. Amid the backdrop of the increasing global environmental pressures, the restoration of degraded ecosystems is essential to protect biodiversity and welfare of future generations.

9.1.2 Cause of Ecosystem Degradation

Ecosystem degradation is a multifaceted problem resulting from the accumulation of anthropogenic and natural drivers. With increasing human activities, industrial pollution, deforestation, and environmentally unsustainable agricultural practices have led to significant degradation of the ecosystems around the world. In many areas, there are heavy metal, POP (persistent organic pollutants), and mycotoxin pollutants which pollute the water, earth, and air, destroying the ecological processes that are necessary for diversity. Cutting of trees and deforestation—such as agricultural, logging, etc.—all result in habitat loss, soil loss, and reduced carbon sequestration potential, contributing to a cycle which places higher levels of pressure on the climate (Mahida et al. 2024; Ondiko et al. 2022). Ecosystem is also degraded due to inherent reasons. Climate change, including temperature and precipitation changes, has profound impacts on species ranges and ecosystem functioning. Such changes can overload natural systems and exacerbate current vulnerabilities as ecosystems become more vulnerable to further degradation (Bellard et al. 2012). Furthermore, natural disasters such as hurricanes, wildfires, and volcanic eruptions can bring rapid and profound impacts to the ecosystems that may take decades to recover from. The combination of human-induced and natural impacts generates a web of challenges for the conservation and restoration of ecosystems, and well-informed and adaptive management approaches are required to prevent ongoing degradation and enhance ecosystem resilience.

9.1.3 Effects of Ecosystem Degradation

I would add that damage to ecosystem—with its broad and complex implications—calls attention to environmental systems but also systems with social dimensions. It leads to loss of biodiversity as pollutants, habitat loss, and climate change break down ecological networks, diminish species diversity, and reduce ecosystem resilience. When species are lost or extinguished, complete food chains and ecological relationships are compromised and the instability is enhanced. The loss of this biodiversity also hampers the resilience of ecosystems to rebound from such environmental stress, making them more susceptible in the future to disturbances (Makwana et al. 2024a, b).

A second major outcome is the loss of ecosystem services, which are necessary for human survival and quality of life. The deterioration of natural habitats depletes clean water and air, degrades fertile soil, and undermines climate-regulating functions. These variations have a direct impact on food production, economic productivity, and public health. For instance, degradation of the soil may result in low crop harvest and, consequently, in food insecurity, whereas pollution of the air and water can result in respiratory and waterborne diseases.

The outcomes of ecosystem health avoidable or untoward diseases and conditions result as toxicants are released into the surroundings and sequestered by living organisms. These toxicants build up as they are passed up the food web due to biomagnification and bioaccumulation, being harmful to wildlife and human health. Forensic toxicology is important for detecting such toxicants and tracing their origins and implications for ecosystem and human health.

When the ecosystems are destroyed, on the other hand, it poses enormous public health risk. Pollution of water bodies and soils, along with exposure to noxious substances, can lead to the spread of disease and negatively impact both air and water quality. Bad air, for instance, causes respiratory diseases, and food and water chock full of chemical pollutants may cause a multitude of chronic diseases. It is important for the habitat restoration and human health, which raised the need of remediation and conservation of attractive areas and the coordination of local large-scale measures (Adla et al. 2022).

9.1.4 Importance of Restoration and Remediation

Remediation and restoration represent crucial practices for reversing ecological degradation and thus restoring systems structure, function, and biodiversity. One of their greatest values is that they increase biodiversity by reintroducing native species and restoring habitat, making ecosystems more resilient (Young 2000). These efforts also enhance ecosystem services by restoring essential processes like carbon sequestration, water filtration, and soil fertility that are essential to a functioning balance of nature (de Groot et al. 2002). And also returning damaged environments could be a key way to fight climate change as freshly planted forests and wetlands are carbon sinks, reducing the stockpile of greenhouse gases (Griscom et al. 2017). Aside from its environmental benefits, these are also serving for the human welfare with bearing in mind reducing the exposure to the environmental pollutants and enhancing air, water, and soil quality (Münzel et al. 2023). By addressing both human health concerns ecological, remediation and restoration ensure long-term environmental sustainability and stability (Khatsuriya et al. 2023, 2024).

9.1.5 Role of Forensic Science in Environmental Analysis

Environmental assessment is one of the major realms of application of forensic science and can be used especially in the deterioration of ecological systems. Forensic science can determine when biosecurity is breached and attributes this to a relationship to using advanced analytical capabilities to determine who is responsible, how long it has been, and the magnitude of natural resource or environment degradation (Carter et al. 2007). It is also helpful for impact assessment that focuses on ecological damage, and it allows the estimation of potential risk to environmental and human health (Myers et al. 2013). Moreover, forensic science also plays an important role in the law and regulations by providing evidence for environmental jurisprudence and for policymaking. Apart from these applications, it encourages the innovation in remediation method and sticks in phytoremediation, bioremediation, and microbial intervention to recycle the developed ecosystems (Megharaj et al.

2011). Environmental management is significantly enhanced by attending to conservation and rehabilitation of natural systems through introducing forensic science as part of the scientific approach.

9.2 Ecosystem Degradation: Causes and Consequences

9.2.1 Anthropogenic Activities and Pollution

Human-induced activities disrupt natural ecological systems and most times results in pollution and degradation of the environment. Pollution is predominantly derived from industrial processes, with compounds such as chemicals, heavy metals, and particles being released during the manufacture and generation of energy. Moreover, the emission of toxic industrial effluents in aquatic bodies pollutes water and represents a serious threat to aquatic life. Pollution caused by use of pesticides and fertilizers in agriculture is another reason for concern due to the water eutrophication and soil acidification they cause. In addition to this, livestock farming's methane emissions also have a big influence on climate change as they contribute to greenhouse gas concentration. Urbanization exacerbates environmental stress also by ineffective waste management, where microplastics and chemicals pollute terrestrial and aquatic environments. What's more is the constantly growing volume of car emissions adds to air pollution, which in turn results in the formation of smog and breathing diseases (Prakash and Verma 2022; Chavda et al. 2024).

Various human activities cause pollution of different forms, which also adversely affect ecosystems and human health. Air pollution comes from smokestacks, car exhaust, and agricultural fields, which pump out gases and fine particles and in turn cause climate change and lung disease. Pollution of water when it occurs from agricultural runoff, oil spills, and raw sewage pollutes water bodies, destroys aquatic life, and degrades quality of water. Soil contamination is also serious as hydrocarbons, metals, and persistent organic pollutants tend to accumulate in soil and can influence the growth of plants, can reduce fertility of soil, and can even allow the introduction of toxic substances in the food chain. Solutions to these pollution sources involve sustainable practices, laws and regulations, and technological developments to prevent pollution and restore the health of the environment.

9.2.1.1 Consequences

The consequences of anthropogenic pollution are vast since biodiversity, human health, and socioeconomic as well as economic stability are all impacted. The loss of biodiversity is one of the major detrimental effects as the pollutions disturb the natural ecosystems and change the habitats of the living species, and some of the species become extinct, and the process has an adverse impact on the ecological functions. Environmental exposure to contaminants is also a significant health risk, with exposure to pollutants associated with respiratory disease, genetic damage, and chronic health problems. Contaminated air, water, and soil can increase the risk of disease, especially among the more vulnerable people, adding another level in the pressure on public health systems. As for pollution, apart from its environmental and health repercussions, it also imposes a severe economic burden through costs associated with healthcare, environmental restoration, and loss of productivity. These are costs that are harmful to the regional economies and the national fare that serve to confirm necessity of finding new production methods and a more preventive commitment to the laws. To deal with these problems, a comprehensive strategy is required with a strong emphasis on environmental preservation and pollution control along with the recognition of the importance of eco-friendly technology in the restoration and maintenance of natural habitat (Landrigan et al. 2018).

9.2.2 Effect on Biodiversity and Ecosystem Services

The impact of humans on the environment is considerable and widespread resulting in serious ecological and social implications. A drastic impact is the loss of species diversity as the destruction of habitats and environmental pollution are eliminating a number of species, resulting in a destabilized ecosystem (Pimm and Raven 2000). Moreover, there are interferences in critical ecosystem services such as pollination, natural drainage, and climate regulation from human activities. The reduction of such services undermines ecosystem resilience, as well as agricultural productivity, water quality, and the overall environmental condition. Human-caused changes to the environment not only have their ecological consequences but also they have social effects on a larger scale. Economies that rely on the land or sea to sustain the population through, for example, fishing, farming, or forestry can suffer when ecosystems decline. This depletion of resources can contribute to food insecurity, forced displacement, and increased susceptibility to environmental disasters (Environmental Degradation and Human Well-Being: Report of the Millennium Ecosystem Assessment 2005). Addressing these challenges requires proactive conservation efforts, sustainable resource management, and policies that balance human development with environmental preservation to ensure long-term ecological and social stability.

9.3 Case Studies of Major Environmental Disasters

Man-induced environmental disasters have resulted in great ecological and social damages. Among all one of the worst have been those of Bhopal Gas Tragedy (1984) where a deadly gas leak caused death and widespread health issues and environment contamination to up to 500,000 people (Broughton 2005). In the Deepwater Horizon (2010), thousands of barrels of crude oil were released into the Gulf of Mexico causing massive damage in the form of marine pollution, loss of aquatic ecosystems, and livelihood losses for fishing coastal communities (States and Guard 2011).

The Chernobyl disaster (1986) was a release of nuclear radioactive materials that caused long-term damage to the local ecology, genetic mutations to wildlife, and cancer in humans (Møller and Mousseau 2006). In the realm of persistent environmental destruction, the ongoing destruction of the Amazon Rainforest serves as proverbial canary in the coal mine as large-scale deforestation and agriculture expansion have resulted in lost biodiversity, disrupted climate regulation, and the release of trapped carbon, all feeding into the vicious cycle of global climate change (Malhi et al. 2008). These environmental calamities underscore the compelling necessity for more stringent environmental regulations, responsible industrial practices, and greater international cooperation to avoid future environmental disaster and to protect the environment and human health.

9.4 Environmental Forensics in Ecological Restoration

9.4.1 Principles and Objectives of Environmental Forensics

Environmental forensic science is one tool of ecological restoration that involves observing empiricism to analyze performance and heal the Earth. This is an interdisciplinary field with knowledge in the areas of chemistry, geology, forensics, and environmental science tasked with the goal of localizing the pollution source and planning a feasible cleanup. A fundamental premise of environmental forensics is to identify the source of contamination whether it results from agricultural runoff, industrial discharges, or natural mobilizations so that effective and efficient remediation measures can be enacted (Morrison and Murphy 2005).

Chemical fingerprinting of pollutants is a widely employed modality for pollutant screening that utilizes modern analytical techniques, for example, GC-MS (gas chromatography-mass spectrometry) and stable isotope analysis, in tracing the pollutants and their distribution in ecosystems and quantifying their level of ecological significance (Rossi 2025). Monitoring of the environment in restoration projects is also critical for long-term success, including periodic evaluation of air, water, and soil quality to track progress of remediation and to identify new soils at risk for contamination (Mir et al. 2023).

Additionally, using techniques of historical pollution reconstruction, such as sediment core analysis and dendrochronology, we could uncover past pollution events and their leftover impact on the health of ecosystems and aid in the planning of a more comprehensive restoration approach (Zhang 2019). In addition to its science-based applications, environmental forensics also assists in the legal and regulatory processes by providing environmental forensics evidence that helps to develop policy, bring legal action against polluters, and enforce remediation obligations (Agardy 2005). These principles promote environmental forensics with an aim to assisting in the recovery actions, ensuring liability on the part of the polluters, and protecting the long-term environmental health.

9.4.1.1 Objectives of Environmental Forensics in Ecological Restoration

Forensic science is critical to the process of restoration of disturbed habitats to determine the damage, responsibility, and resource reallocations and remediations that are required. A specific application is to quantify environmental damage whereby forensics measure pollution and loss of biodiversity and ecosystem services in order to derive well-informed restoration strategies (Lipton and LeJeune 2018). Further, forensic approaches are valuable for the assessment of blame for pollution, such as natural vs. human sources, so that responsible parties can be held accountable and remediated (Boehm and Murphy 2015).

Not only does environmental forensics help pinpoint sources, but it also facilitates litigation and legislation, generating essential information for lawsuits, regulatory adherence, and the creation of socially sound laws and regulations (Mudge 2008). Another important contribution involves increasing knowledge of remediation processes so that forensic studies provide input for selecting more efficient methods of remediation, such as bioremediation, phytoremediation, and chemical stabilization, favoring ecosystem restoration (Bica et al. 2009). Moreover, forensic science contributes to preventing future environmental harm and therefore facilitates a more proactive implementation of, for example, pollution abatement, sustainable land use planning, and the improvement of laws and regulations to safeguard natural ecosystems (Mudge 2008).

Environmental forensics can play an important role in having confidence in both the legal and technical aspects of restoration when it is based in sound scientific principles on the landscape. This integrative discipline ranges from electron science to policy to enforcement and is critical for sustainability of the environment for longtime scales.

9.4.1.2 Analytical Tools and Techniques

Environmental forensics employs instrumental methods to detect, trace, and quantify contaminants and has a significant impact on tracing the sources of pollution, the assessment of environmental damages, and remediation. Chromatography is one of the key techniques which permits the separation and identification of complex mixtures of contaminants. VOCs are measured using gas chromatography (GC)-based techniques for a majority of the analysis performed on water, air, and soil samples to provide a reliable contaminant profile (Frysinger et al. 2002). On the other hand, HPLC is suitable for analyzing polar and nonvolatile organic pollutants, such as pharmaceuticals and pesticides, in environmental matrices; however, heavy metals are typically analyzed using techniques like AAS or ICP-MS (Ahmed 2024). Mass spectrometry (MS) is a further basic technique that has been established as indispensable for the identification and quantification of environmental pollutants. One of the most applied devices for this purpose based on high sensitivity is gas chromatography-mass spectrometry (GC-MS) for the identification of variety of hydrocarbons, pesticides, and other organic pollutants (Santos and Galceran 2003). Similarly, emerging pollutants, such as EDCs and pharmaceuticals, have been monitored using liquid chromatography-mass spectrometry (LC-MS) (Petrovic and Barceló 2006). ICP-MS (inductively coupled plasma mass spectrometry) is also widely used for trace metal analysis of environmental samples and can give accurate determination of toxic metals (Beauchemin 1989). Along with common chromatographic and spectrometric analyses, the isotope study is a key tool aimed at pollution sources and transformation processes' detections. Stable isotope analysis could distinguish between anthropogenic and natural pollution by comparing isotopic composition of elements such as nitrogen, carbon, and sulfur (Pichlmayer et al. 1998). Radiogenic isotope analysis: This method is used to monitor heavy metal contaminate and contamination time frame (Billström 2008; Mahida et al. 2022, 2023). Furthermore, the compound-specific isotope analysis (CSIA) can be used to infer degradation pathways and verify the efficiency of bioremediation options (Casiraghi et al. 2022). The use of such cutting-edge scientific techniques allows environmental forensics to provide reliable, detailed data for environmental site decision-making.

9.4.1.3 Applications in Identifying Pollutants and their Sources

It is only through such methods that it is possible to appropriately identify contaminants, trace them to their source, and support remediation and legal action. Various analytical techniques allow scientists and decision-making bodies to track pollution to particular sources, estimate its environmental effects, and design remedial action plans. In industrial and environmental forensics, heavy metals are commonly traced using ICP-MS, while GC-MS is applied for co-existing organic contaminants, facilitating source identification of complex pollutant mixtures. These approaches help to detect some heavy metals, petrochemical components, and POPs (persistent organic pollutants), which serve to identify pollution sources, and contribute to the execution of remediation strategies (Wahab et al. 2020). Equally important is that of finding the sources of agricultural runoff contributing to water pollution from pesticides, herbicides, and fertilizers.

9.4.1.4 New Insights: Finding Out the Source of Water Pollution

New method makes it possible to differentiate between harmful nitrates from fossil fuels and those from synthetic or organic fertilizers. Techniques including liquid chromatographymass spectrometry (LCMS) and stable isotope analysis are now being used to differentiate between pollutants originating from synthetic and organic agriculture, and from fossil fuels, in order to prevent the occurrence of contaminants which are harmful to the aquatic environment and soil health (Adu-Gyamfi et al. 2024). In urban areas, identifying the sources of pollutants is required to distinguish pollutants from vehicle exhausts, wastewater, various forms of human mobilization, etc. Spectroscopic methods allow to detect the presence of polycyclic aromatic hydrocarbons (PAH) and volatile organic compounds (VOC) contributing to water and air quality decrease in dense areas (Ho and Lee 2002). For oil spill fingerprinting in the sea, the origin of petroleum contamination is attributed to crude oil sources or refined petroleum product origins. They say molecular marker analysis and measurements of the ratio of carbon isotopes in crude oil provide an accurate way to determine which contaminants are in the oil and direct cleanup and establish liability for the environmental damage (Wang et al. 2006) Likewise, Forensic Microbial

Source Tracking uses a DNA-based microbial profiling technique to identify fecal contamination in aquatic systems. The approach identifies human, livestock, and wildlife sources of contamination, which can be used to direct mitigative measures toward improving water quality and public health (dela Peña et al. 2021). In addition, studies of metal contamination are based on isotope ratio analysis to distinguish whether contaminants stem from mining, industrial, and natural weathering. Through recognition of these specific isotopic "fingerprints," environmental forensics is able to offer vital information used in pollution policymaking and action since it detects not just the type but the source of the pollutants (Deng et al. 2016). Employing these state-of-the-art forensic measures guarantees that pollution sources are determined with all due certainty, the guilty parties are punished, and remediation efforts are efficiently initiated, working to guarantee meaningful protection of the environment and the legacy of sustainable living.

9.5 Remediation Strategies and Techniques

Environmental restoration is essential to restoring polluted or contaminated areas to human health and the environment (Schwitzguebel et al. 2009). This process is very important in that contaminants are taken out or are stabilized from key environmental media like soil, groundwater, sediment, and surface water (Caliman et al. 2011). Application of successful strategies that can be categorized as either physical, chemical, or biological, for example, soil vapor extraction, bioremediation, and advanced nanotechnology, can significantly reduce the impacts of pollutants. These remedial projects are vitally necessary in order to restore society to a cleaner, safer ecology for our people and children.

9.5.1 Physical and Chemical Remediation Methods

9.5.1.1 Physical Remediation Methods

Soil Vapor Extraction (SVE) Soil vapor extraction applies a vacuum to the soil to extract volatile and semi-volatile contaminants (Cao et al. 2021). Wells are drilled into the contaminated area, where a vacuum pump sucks in the fumes. The vented vapors are then processed in any of a number of conventional, for example, with activated carbon or thermal oxidation. This is also an efficacious approach for remediation of volatile organic compounds (VOCs) in soil and groundwater, for example, gasoline, benzene, and trichloroethylene (TCE) (Parmar and Rao 2008).

Thermal Desorption In thermal desorption, contaminated soil is excavated and heated to a temperature high enough to vaporize the contaminants (Stegemeier and Vinegar 2001). The temperature can range from low (90–320 °C) to high (320–560 °C) depending on the contaminants. The vapors are recovered and subjected to a treatment such as condensation, adsorption, or thermal decomposition. This procedure is applied to soils that are contaminated with hydrocarbons, pesticides, and other organic contaminants. It is effective for a wide range of contaminants, including polychlorinated biphenyls (PCBs) and polycyclic aromatic hydrocarbons (PAHs) (Zhao et al. 2019; Mahida et al. 2025).

Pump and Treat This approach requires the extraction of polluted groundwater to the surface and subsequent processing to remove pollutants (Keely 2019). Water treatment, for instance, using air stripping, activated carbon adsorption, ion exchange, or chemical precipitation. The pure water is subsequently reinjected in the ground or discharged. Pump and treat is a fairly typical use where groundwater is contaminated by VOCs, heavy metals, and any other dissolved contaminants.

Encapsulation Encasement refers the containment of contaminated material, often in a casing, to prevent the contaminated particle from contacting the surrounding environment and avoid spreading of contamination (Meunier et al. 2010). This can be achieved using techniques like capping (covering the contaminated soil) and with the placement of vertical barriers (building walls within the contaminated zone). This type of method is particularly useful at sites where it may not be feasible to remove all the contaminants (e.g., landfills or very large contaminated soil site).

9.5.1.2 Chemical Remediation Methods

In Situ Chemical Oxidation (ISCO) It is in situ chemical oxidation injection of an oxidizing agent (such as hydrogen peroxide, potassium permanganate, or ozone) to the contaminated zone (Seol et al. 2003). These chemicals then chemically react with the pollutants, transforming them into something less nasty. The reaction can be facilitated by the addition of a catalyst or the pH attacked. ISCO is applied for the treatment of organic contaminants (e.g., petroleum hydrocarbons, chlorinated solvents, pesticides). It is applied for groundwater and soil remediation.

In Situ Chemical Reduction (ISCR) Cometah (2018) also highlighted that contaminants in the soil can be reduced into less toxic or less mobile forms through the process of in situ chemical reduction which uses reducing agents (e.g., zero-valent iron (ZVI) or sodium dithionite). The reducing agents are delivered into the contaminated region, and the reducing agents contact the contaminants. ISCR is suitable for chlorinated solvents, heavy metals, and other refractory compounds. This includes the progressive hydraulic ore exhaustion by extracting valuables from or through pore spaces and interstices of a solid substrate and using the substrate after increasing the bulk porosity directly for dirt water treatment, sewage treatment, water treat, and water purification.

Soil Washing Soil washing involves excavating contaminated soil and using water, sometimes combined with chemical additives to wash away contaminants (Liu et al. 2021). It separates contaminants from soil particles, with the clean soil and a contaminated wash solution. The wash solution is next processed to decontaminate the same. Soil washing is appropriate for heavy metal, hydrophobic organics, and radionuclide contaminated soils. It is frequently used for cleaning of sites or for redevelopment of brownfields.

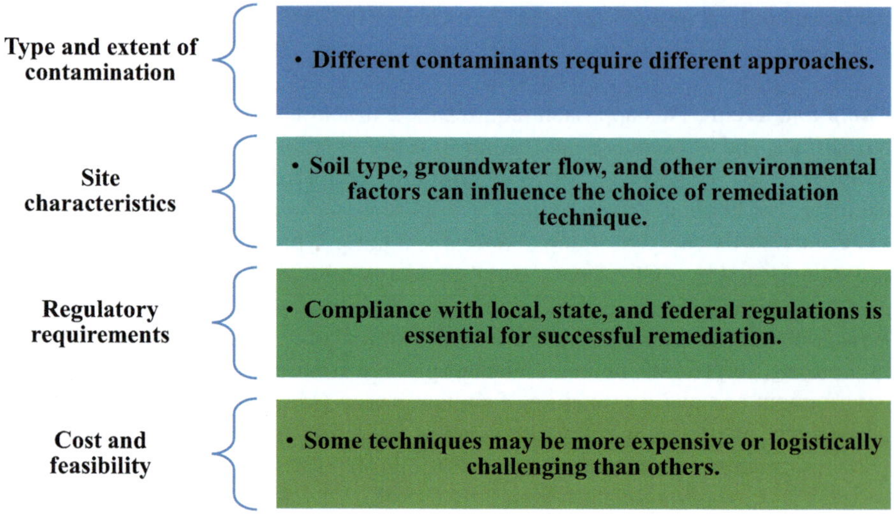

Fig. 9.1 Factors affecting the selection of the correct remediation method

Stabilization/Solidification Stabilization/solidification involves mixing contaminants with binding agents (such as cement, lime, or fly ash) to reduce their mobility and toxicity (Dillard 2019). The process can stabilize contaminants by changing their chemical form or physically encapsulating them within a solid matrix. This method is used for treating contaminated soils, sludges, and sediments with heavy metals, radionuclides, and organic pollutants. It is commonly used in landfills and industrial waste treatment. Figure 9.1 outlines the key factors influencing the selection of an appropriate remediation method, highlighting the complexities involved in choosing the most effective approach for environmental cleanup and restoration.

9.6 Bioremediation: Microbial and Phytoremediation Approaches

9.6.1 Microbial Bioremediation

Microbial bioremediation utilizes microorganisms, bacteria, and fungi, in particular, to degrade pollutants and to convert them into less hazardous forms (Joshi et al. 2024). There are two primary techniques:

Bioaugmentation: This involves introducing specific strains of microorganisms that are efficient at degrading the target contaminants (Cycoń et al. 2017). These microbes are naturally occurring or have been genetically manipulated to increase their biodegradation potential.

Bio stimulation: This process augments the activity of the native microorganisms by “feeding” them with nutrients, air, or other substances that will stimulate their growth and metabolic action (Adams et al. 2015). This helps the native microbes to degrade contaminants more effectively.

9.6.2 Phytoremediation

Phytoremediation uses plants to clean up polluted environments (Suresh and Ravishankar 2004). It's a sustainable method that can be used for various types of pollutants. Table 9.1 shows advantages and challenges of eco-friendly remediation approaches. There are several techniques within phytoremediation:

Phytoextraction: Plants absorb contaminants through their roots and accumulate them in their biomass (Bhargava et al. 2012). The plants are then collected and disposed of safely.

Phytodegradation: Plants degrade contaminants through metabolic processes within their tissues (Singh and Jain 2003).

Phytostabilization: Plants immobilize contaminants in the soil, reducing their mobility and preventing their spread (Bolan et al. 2011).

Rhizofiltration: The roots of plants are used to filter contaminants from water, especially heavy metals and radionuclides (Kristanti et al. 2021).

Phytohydraulics: Plants are used to control the movement of water and contaminants within the soil (Khan et al. 2023).

9.6.3 Nanotechnology in Environmental Remediation

Nanotechnology suggests innovative resolutions for environmental remediation by using materials at the nanoscale to clean up polluted sites (Bhawana and Fulekar 2012).

9.6.3.1 Nanomaterials Used in Remediation

Nanoscale Zero-Valent Iron (nZVI) High reactivity and large surface area-to-volume ratio (Deng et al. 2022). Used for reduction and damaging contaminants like chlorinated solvents and heavy metals.

Table 9.1 Advantages and challenges of eco-friendly remediation approaches

Advantages	Challenges
Eco-friendly—Natural process that minimizes environmental impact	Time-consuming—Can take months or a year to yield quality results
Sustainable—It utilizes natural process and organisms promoting environmental health.	Site-specific—Effectiveness depends upon the site factors such as soil type and contaminant type.
Cost-effective	Regulatory consideration—Environmental guidelines and regulations are essential.

Table 9.2 Advantages and challenges of using nanomaterials in environmental remediation

Advantages	Challenges
High efficiency—Nanomaterials offer high reactivity and efficiency in contaminant removal.	Challenges in scaling up laboratory techniques for field applications
Versatility—Can be tailored for specific contaminants and environmental conditions	Potential analysis of environmental and health impacts of nanomaterials
Cost-effective	Regulatory consideration—Need environmental guidelines and regulations for use of nanotechnology in remediation

Carbon-Based Nanomaterials Carbon nanotubes and graphene. Effective in adsorbing pollutants from water and soil due to their high surface area (Madima et al. 2020).

Metal Oxide Nanoparticles Titanium dioxide (TiO_2) and zinc oxide (ZnO). Used in photocatalytic deprivation of organic pollutants and heavy metals (Gusain et al. 2019; Gautam et al. 2023).

Polymeric Nanomaterials Versatile and can be tailored for specific contaminants. Used in adsorption and stabilization of contaminants (Hussain et al. 2024).

9.6.3.2 Techniques and Applications

Adsorption: Contaminants are captured on the surface of nanomaterials. Effective for removing heavy metals, dyes, and organic pollutants from water (Okoro et al. 2022).

Photocatalysis: Nanomaterials like TiO_2 are activated by light to degrade organic pollutants (Mahlambi et al. 2015). Used for treating wastewater and air pollution.

Chemical Reduction: Nanoparticles like nZVI chemically reduce contaminants to less harmful forms (Stefaniuk et al. 2016). Effective for treating chlorinated solvents and heavy metals.

Filtration: Nanomaterials are used in filters to remove contaminants from water or air. Used in water purification systems and air filters. Table 9.2 shows advantages and challenges of using nanomaterials in environmental remediation.

9.7 Case Studies of Successful Remediation Projects

The Hudson River PCB Cleanup (USA) The Hudson River was polluted with polychlorinated biphenyls (PCBs) from industrial actions (McCarty et al. 2003). The US Environmental Protection Agency (EPA) employed a dredging and removal strategy to dig and dispose of polluted sediment. Major decrease in PCB levels, refining water quality and ecosystem health.

Oil Spill Bioremediation in the Gulf of Mexico (US) The Deepwater Horizon oil spill discharged millions of barrels of crude oil into the Gulf of Mexico, producing widespread environmental damage (Atlas and Hazen 2011). A combination of physical removal, chemical dispersants, and bioremediation techniques, including the use of oil-degrading bacteria. Significant reduction in oil contamination, aiding in the recovery of marine and coastal ecosystems.

Sustainable Remediation Techniques in Europe (Concawe Report) Various contaminated sites across Europe required sustainable remediation techniques. Concawe commissioned a study to gather and publish 10 European case studies demonstrating sustainable remediation techniques (Daines et al. 2011). Identification of key success factors and practical approaches for deploying sustainable remediation on operational sites.

Minamata Bay Mercury Cleanup (Japan) Minamata Bay was contaminated with mercury from industrial discharge, causing severe health issues in the local population. The Japanese government carried out dredging and capping of contaminated sediments, along with soil washing techniques (Akito et al. 2014). The decrease in mercury levels has played a significant role in reinstating the bay's ecosystem and improving public health.

9.8 Ecosystem Restoration Frameworks

Ecosystem restoration frameworks provide an organized process for improving an ecosystem (Wagner et al. 2015). They include assessment and planning, habitat restoration, species reintroduction, continued monitoring, adaptive management, and community participation. Such frameworks also address the implementation of actions that restore such functions as reforestation, wetland restoration, and invasive species control and aim to support the recovery of ecosystem processes, biodiversity, and resilience (Gann et al. 2019). Such models lead to sustainability through the cooperation with local actors and stakeholders, bringing long-term environmental and social advantages. It is essential for creating healthier and sustainable ecosystems. Figure 9.2 shows the various objectives of ecosystem restoration frameworks.

9.8.1 Restoring Degraded Landscapes

Rehabilitation measures for degraded land are significant to ecosystem remediation. The purpose of this mechanism is to restore the ecosystem functionality, the biodiversity, and resilience of degraded ecosystems.

Soil Rehabilitation Adding compost, manure, or mulch to improve soil fertility and structure (Magdoff and Weil 2004). This leads to better nutrient availability,

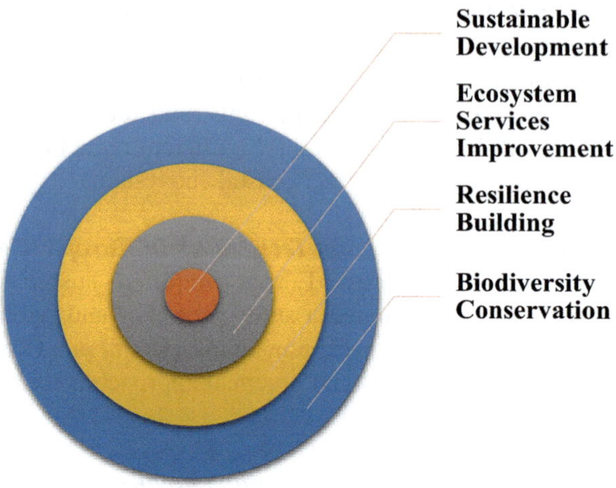

Fig. 9.2 Objectives of ecosystem restoration frameworks

water retention, and microbial activity. Implimentation of techniques such as contour plowing, terracing, planting cover crops, etc., to control soil erosion. These practices are able to stabilize soils and minimize nutrient leaching. Soil test before planting to check for nutrient deficiencies and pH. Recommendations: Based on the findings, recommendations are made for soil conditioning with lime or gypsum to neutralize imbalance and allow the soil a fighting chance of defending itself (Osman 2012).

Reforestation Selecting and planting native tree and plant species that are well adapted to the local climate and soil conditions (Withrow and Johnson 2006). It's good for biodiversity and for the stability of the ecosystem. Afforestation: Planting trees in areas that were previously not forested to make new woodland. This contributes to carbon sequestration, cleaner air, and habitat for wildlife. Improvement of a more diverse and sustainable land-use system that entails the inclusion of trees and shrubs into agricultural fields (Lovell 2010). This can recover soil health, enhance crop yields, and provide additional income sources for farmers.

Wetland Restoration Restoring natural water flow patterns by removing barriers such as dams and levees and reestablishing floodplains (Reich and Lake 2015). This is beneficial to wetland hydrology and aquatic organisms. Control pollution inputs from agricultural runoff, industrial discharges, and urban stormwater to enhance water quality in wetlands (O'geen et al. 2010). This can be achieved through best management practices (BMPs) and the creation of buffer zones. Planting native wetland species and controlling the invasive plants in order to restore a natural vegetation profile with the associated biological diversity. This is good for wildlife habitat and helps make ecosystems resilient.

Invasive Species Management Implementing targeted eradication programs to remove invasive species through mechanical, chemical, or biological control methods. This helps in restoring native species and ecosystem balance (Culliney 2005). Establishment of surveillance programs to detect new taxa of invasive species in the early stages of their invasion and to prevent their spread. Early detection facilitates quick response and successful treatment. Educating communities about the effects of invasive species and involving them in the removal process. Public outreach can be a useful tool for preventing the introduction and establishment of invasive species.

Habitat Connectivity Developing corridors that connect isolated habitats so that species can move between them to the access resources and increase the gene flow which in turn makes them more resistant to environmental changes (Christie and Knowles 2015). Integration of green infrastructure including green roofs, urban parks, and green belts within urban planning to maintain habitat connectivity and biodiversity in urban areas. Creating riparian buffers along rivers and streams for habitat, water quality, and stream bank stability (Cole et al. 2020). These buffers create continuous habitat corridors and enhance ecosystem services.

Community Engagement On-site mobilization of community members, government departments, NGOs, and stakeholders in the restoration effort. Cooperative work helps to keep the restoration work functioning throughout the years. Educational programs and outreach to help people understand the importance of the restoration of ecosystem and to tell about the restoration and to the communities to get their participation. Encouraging sustainable land-use practices, including sustainable agriculture, forestry, and fisheries to help limit the pressure on natural ecosystems and recovery programs. Ecological rehabilitation as a way for living in a sustainable world restoration of degraded landscapes calls for a holistic, integrated approach to a complex set of ecological, social, and economic issues. By the pursuit of these methods, we can promote healthier, more resilient ecosystems, encourage biodiversity, and leave an enduring environment for our children's children. Figure 9.3 shows various methods which can be opted for restoring land degradation.

9.8.2 Reintroduction of Native Species

The reintroduction of native species is a critical component of ecosystem restoration (Gibbs et al. 2008). The purpose is to return to a natural balance by reintroducing species that have been extirpated (not just endangered) from their original home ranges by human intervention and other contributory factors.

Assessment of Habitat Suitability Knowing what the reintroduced species needs in order to survive and thrive in a given area, for example, food, water, shelter, and breeding spots (Berger-Tal and Saltz 2014). This ensures that the chosen habitat can sustain the species in terms of survival and reproduction. Evaluating the existing habitat condition such as the vegetation canopy, water quality, and presence of other

Fig. 9.3 Restoring degraded landscapes

species. First, they are more likely to succeed if reintroduced in good-quality habitats. Determining potential threats such as predators, human impact, or environmental dangers that may affect the reintroduced animal (Fernández et al. 2006). Mitigating these threats is crucial for long-term success.

Genetic Diversity and Source Populations A similar question concerns the existence of suitable gene pools that will facilitate the adaptability of the reintroduced individuals to new environmental conditions as well as the ability to avoid inbreeding depression (Williams and Hoffman 2009). This builds a strong and healthy population. Choosing source populations from areas with a similar environment as the recipient area (Osborne and Seddon 2012). In both instances, the opportunity for successful long-term adaptation to environmental change may be increased.

9.8.2.1 Phased Release and Acclimatization

Soft Release Gradually conditioning animals to the new environment by soft release (De Milliano et al. 2016). This might include penning or holding areas where animals can acclimatize to natural conditions before they are released.

Hard Release Direct release of individuals into the wild without an acclimatization period (De Milliano et al. 2016). This technique is employed when the fitness of a species is high in the area or when a soft release is not possible.

Monitoring and Adaptive Management Collecting prerelease information on habitat quality, species occurrence, and ecological relationships as a baseline for monitoring (Moehrenschlager and Lloyd 2016). Monitoring the survival, reproduction, and movements of released individuals. The monitoring will detect problems and inform adaptive management. Adapting management in response to monitoring. This may involve supplementary feeding, environmental modifications, or further releases to support reintroduction success.

Community Involvement and Education Engaging the local stakeholder (local communities, GOIs, NGOs, and others) in the reintroduction. Working together for a common goal creates more backing and in turn provides sustainability for the future of the project. Educating the public about the need to reintroduce species and why it's great for the ecosystem. Education encourages community support and involvement.

9.9 Case Studies

Wolves in Yellowstone National Park: The reintroduction of wolves in Yellowstone has had trickle down effects for the ecosystem and for the health and balance of the species (Sims et al. 2020).

Iberian Lynx in Spain: The number of the critically endangered animal has, to a large extent, been increased through conservation measures, including reintroduction programs, according to the Spanish government (Vargas et al. 2008).

Beavers in the United Kingdom: Reintroducing beavers has restored wetland habitats and improved biodiversity, showcasing the benefits of keystone species in ecosystem restoration (Stringer and Gaywood 2016).

9.10 Monitoring and Adaptive Management

9.10.1 Monitoring

Monitoring involves the systematic collection of data to evaluate the progress and effectiveness of restoration efforts. Key components include:

Baseline Data Collection: Conducting comprehensive surveys to gather data on species presence, population sizes, habitat conditions, and ecological processes before restoration begins (Morrison 2013). Identifying undisturbed reference sites to serve as benchmarks for comparison with restored areas.

Indicators of Success: Monitoring species diversity, abundance, and population dynamics (Thomas 2005). Key species may include indicator species, keystone species, and endangered species. Assessing vegetation cover, soil health (Cardoso et al. 2013), water quality, and habitat structure to determine the health of the ecosystem. Measuring ecological processes such as nutrient cycling, water infiltration, and carbon sequestration.

Data Collection and Management: Performing periodic field surveys to collect information on species and habitat status. Monitoring broadscale changes of vegetation and use of land with satellite imagery craw and other modern remotesensing techniques (Manfreda et al. 2018). Mobilizing local communities and volunteers in data production to extend coverage and build ownership at the community level (Dickinson et al. 2012). Maintaining and controlling reports/securities in central databases accessible to public and for analysis. Applying statistical analyses to data to detect trends, patterns, or relationships. Using GIS to map spatial data and to study change through time.

9.10.2 Adaptive Management

Adaptive management is a dynamic and iterative process that involves making informed adjustments to management strategies based on monitoring results and new scientific knowledge (Williams and Brown 2014). Adaptive management involves two phases of learning. Technical learning is an iterative process of decision-making, monitoring, and assessment. Social and institutional learning requires revisiting set-up aspects throughout the deliberative phase. The deliberative phase of adaptive management, key decision-making elements are refined, such as stakeholder involvement, objective determination, management options, projections of consequences, and monitoring protocols (Linkov et al. 2006). The framework's iterative phase integrates features from the deliberative phase for successive decision-making and learning (Williams and Brown 2014). Decision-making involves identifying actions based on existing information and expected consequences of management activities on the ground. Follow-up monitoring helps assess resource status, advise future decisions, and evaluate outcomes. Assessment utilizes monitoring data to assess management effectiveness, determine resource status, and better comprehend resource changes. Figure 9.4 shows deliberative phase and iterative phase for adaptive management.

Setting Clear Objectives: Defining precise and measurable goals for the restoration program with ecological, social, and economic aspects (Gann et al. 2019). Developing certain measures which will be used to measure how much progress has been made in meeting the goal.

Implementation of Management Actions: Carrying out measures like habitat restoration, species reintroduction, invasive species removal, and community engagement to meet the goals (Wagner et al. 2015). Allocating resources like funding and equipment resources to accomplish management actions.

Continuous Monitoring and Evaluation: Monitor the effects of management on a regularly basis to evaluate management success and detect the appearance of any problems. Comparison of squad testing results to establish progress and determine if goals are being achieved.

Feedback and Learning: Information gathered through monitoring and evaluation can be used for adaptively managing the strategies. This could entail changing behaviors, reallocating resources, or changing goals. Adaptively applying management actions and learning from them over time and incorporating new scientific understanding to enhance restoration efforts.

Stakeholder Involvement: Engagement of local communities, government ministries, nongovernment organizations, and others in the process of adaptive management. Teaming up makes changes in management reflect more varied thinking. Keeping all decision-making processes transparent and informing the stakeholders on your progress.

9.11 Legal and Ethical Aspects of Ecosystem Restoration

Restoration of ecosystems is one such endeavor toward combating climate change and the loss of biodiversity. But it's not without its issues, legal, ethical. These dimensions must be understood if we are to have restoration processes that are effective and fair. The laws and policies surrounding ecological restoration have changed significantly in recent decades. Global accords such as the Convention on Biological Diversity (CBD) and the UN Decade on Ecosystem Restoration (2021–2030) contain ambitious goals for resetting destroyed ecosystems (Leadley et al. 2022). But for these agreements to take effect, they need strong national laws and policies to support them. Other legal obstacles to restoration include a legal culture that is in opposition to restoration objectives, the requirement for restoration mechanisms in specific, and the difficulties of addressing multiple legal regimes (Foster and Bell-James 2024).

However, facilitative policies and high law can help promote restoration. These legal issues ("tenure" and "jurisdiction") constrain the ability to restore landscapes (Foster and Bell-James 2024). Coordination mechanisms are necessary to address these challenges and to guarantee that restoration actions are effectively carried out. Ecosystem restoration is a form of environmental moral repair. This view is one which focuses on the ethical imperative to rehabilitate human-damaged ecosystems. Based on the principle of reparative justice, which is designed to remedy a wrong done to the environment, ethical recovery and restoration need to involve local

Fig. 9.4 Two-phase learning in adaptive management: (i) deliberative phase and (ii) iterative phase

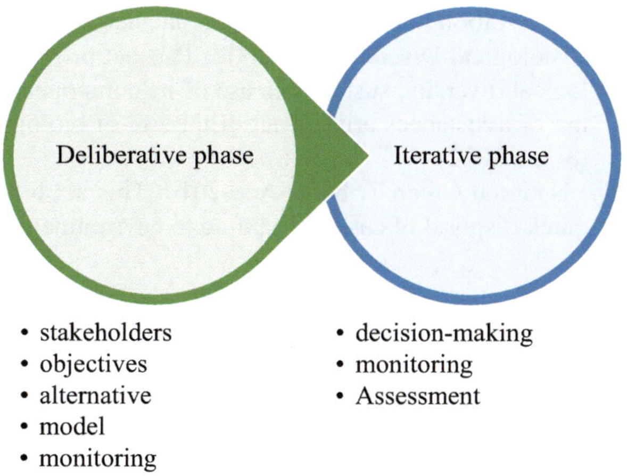

communities and other stakeholders. This guarantees that the restoration process is integrated and that it addresses the needs and rights of the people who are affected most directly by the projects. Rehabilitation works must create sustainable livelihood opportunities for local communities. Doing so involves offering alternative livelihood and income-earning opportunities that are conducive to conservation aims, and which help to relieve some of the pressure on the natural resource base, while also advancing longer-term sustainability. Regulatory environmental laws and regulations are essential to conserving natural resources for the America people, generations, and for our sustainable development (Vilks 2022). These laws govern the interaction between human activities and the environment, aiming to mitigate the impact of activities such as industrial operations, agriculture, waste disposal, and urban development. Here are some key aspects of environmental laws and regulations:

9.11.1 Environmental Laws and Regulations

The Environment (Protection) Act, 1986: This act offers the framework for the safety and development of the environment and the hindrance of risks to human beings, other living creatures, plants, and property (Rajasugunasekar and Devi 2022).

The Water (Prevention and Control of Pollution) Act, 1974: This act targets to prevent and regulate water pollution and preserve the wholesomeness of water in the country (Harwalkar 1974).

The Air (Prevention and Control of Pollution) Act, 1981: This act seeks to prevent, regulate, and decrease air pollution, ensuring the quality of air in the country (Reitze 2001).

The Wildlife (Protection) Act, 1972: This act delivers for the safety of wild animals, birds, and plants and matters connected therewith or ancillary or incidental thereto (Tripathi and Roy 2021).

The Forest (Conservation) Act, 1980: This act aims to preserve forests and regulate deforestation to sustain ecological balance (Kumar 2015).

The Biological Diversity Act, 2002: This act provides for the maintenance of biological diversity, sustainable use of its components, and fair and equitable sharing of assistances arising out of the use of biological resources and knowledge (Singh 2022).

The National Green Tribunal Act, 2010: This act forms a specialized body for the quick disposal of cases pertaining to environmental issues (Gill 2019).

9.11.2 International Environmental Agreements

The Paris Agreement: An international treaty aimed at limiting global warming to well below 2 °C above pre-industrial levels and pursuing efforts to limit it to 1.5 °C (Rogelj et al. 2016).

The Convention on Biological Diversity (CBD): An international treaty aimed at conserving biological diversity, promoting sustainable use of its components, and ensuring fair and equitable sharing of benefits arising from genetic resources (Smagadi 2006).

The Ramsar Convention: An international treaty for the conservation and sustainable use of wetlands (Gardner and Davidson 2011).

9.11.3 Responsibilities of Polluters: The "Polluter Pays" Principle

The "Polluter Pays" principle is a fundamental concept in environmental law and policy. It posits that those who produce pollution should bear the costs of managing it to prevent damage to human health or the environment (Khan 2015). This principle is designed to incentivize polluters to reduce their environmental impact and invest in sustainable practices. Components of the "Polluter Pays" principle are responsibility, costs, and accountability (Barthakur 2021).

Responsibility: Polluters are held responsible for the environmental damage they cause. This includes both direct and indirect pollution.

Costs: Polluters must cover the costs associated with pollution control, cleanup, and environmental restoration. This can include:

- Costs for reducing emissions and discharges.
- Costs for environmental cleanup and remediation.
- Compensation for damage to natural resources and ecosystems.

Accountability: The principle ensures that polluters are accountable for their actions. This can involve legal and regulatory measures to enforce compliance.

9.11.4 Implementation Mechanisms

Governments may impose taxes or levies on activities that generate pollution (Bovenberg and Goulder 2002). The money can then be used to help finance environmental cleanup and restoration initiatives. In the Emissions Trading System, the total amount of pollution from factories and power plants is capped, allowing a market for companies to buy and sell emissions permits. Firms can sell excess permits if they reduce their emissions. Those causing pollution could be taken to court. This might involve fees, penalties, and mandates to contribute to cleanup and restoration efforts (Poulter 1998).

9.11.4.1 Advantage of the "Polluter Pays" Principle

Environmental Protection: By making polluters pay for the harm they cause, the principle incentivizes companies to accept cleaner technologies and practices (Heine et al. 2020).

Economic Efficiency: It safeguards that the costs of pollution are internalized by the polluters rather than being borne by society as a whole (O'Connor 1997). This can lead to more well-organized allocation of resources.

Innovation: The financial problem of pollution control can push innovation in cleaner technologies and processes.

Fairness: It aligns with the belief of fairness by ensuring that those responsible for pollution are held accountable for its consequences.

9.11.4.2 Challenges and Considerations

Effectively implementing the "Polluter Pays" principle requires robust regulatory frameworks and enforcement mechanisms (Schwartz 2010). There may be concerns about the economic impact on industries and businesses, particularly small- and medium-sized enterprises. Ensuring that the principle is applied consistently across different countries and regions can be challenging, given varying levels of development and regulatory capacity.

9.11.4.3 Ethical Considerations in Restoration and Remediation

Restoration and remediation are addressed in order to rehabilitate a destroyed environment to a natural or natural-like (i.e., ecologically functioning) state, suitable for the ecological health of the area and the welfare of human beings (Burger 2008). But these initiatives are faced by a series of ethical dilemmas that must be carefully handled just to make their results equitable, inclusive, and sustainable. Acknowledging that nature has an individual value, independent of human usefulness, the rehabilitation of the destroyed should seek to respect and conserve the intrinsic value of the species and ecosystems (Gann et al. 2019). That restoration projects are first and foremost about conserving biodiversity and the diversity of life, including those that may not seem to have human economic or aesthetic value, is intended to help guarantee that restoration works to save the most biologically diverse, the web of life, and all life alike, including though whose worth is not purely economic or aesthetic. Recovery and remediation must be based on the needs and rights of future generations. This includes not sacrificing the capacity of future generations to satisfy their own needs to the actions of the present day. These efforts also need to be sustainable, in that they support ecological integrity and resiliency in the long term. Restoration initiatives aimed at addressing environmental inequities often focus on communities disproportionately affected by pollution and degradation, typically comprising low-income and minority populations. Equitable apportionment of the benefits and liabilities of restoration and remediation. This involves paying for damaged livelihoods of people who are affected by environmental degradation and letting them participate in the decision-making process as well—involving the local community and stakeholders in the planning, implementation, and monitoring of restoration initiatives (Evans et al. 2018). Thus, they know that their wisdom, values, and priorities are being honored and built upon. Recognizing the cultural and historical connection of communities to natural resources, any attempts to repair them should honor and protect these links. Bringing full transparency to decisions; spreading all information regarding the goals, instruments, and outcomes of restoration measures; and bringing environmental criminals to account, that means holding polluters responsible for cleanup, recognizing that ecosystems are dynamic and intricate. Restoration Considerations: The reclamation of degraded sites must be responsive to monitoring outcomes and new scientific knowledge (Gann et al. 2019).

9.12 Challenges and Future Directions

9.12.1 Scientific and Technological Barriers

There are several scientific and technical challenges to ecosystem restoration that can constrain its effective development and advance. One bottleneck is the imperfect performance of the prediction models. Many models of climate and ecology do not have the ability to predict specific local consequences of ecosystems at the size of viral impact (Bramer et al. 2018). Enhancements in these models with large amounts of biological and climatic information are necessary for better prediction between seasons and to make decisions (Weaver et al. 2013).

A second challenge is that there are a number of missing values in the data. There is much uncertainty—especially for ecosystems in remote or understudied regions—about where species occur, the quality of their habitats, and their responses to environmental changes (Thuiller et al. 2008). Long-term ecological research and comprehensive monitoring programs are required to address these gaps.

Another major challenge is technological access. Although remote sensing, genetic analysis, and climate modelling methodologies are highly valuable for ecosystem restoration, they are not available or too costly in many areas (Camarretta 2020). Equal opportunity access to these technologies is essential for global conservation and restoration.

What's more, the integration of traditional knowledge with modern science could enhance restoration. In addition, restoration efforts need to consider the fact that natural communities are both complicated and diverse. Ecosystems are dynamic and interrelated, with many factors impacting their health and stability (Ives and Carpenter 2007). Construction and implementation of solutions returning to the state of such complexity are mediated by interdisciplinary work and turning to new methods. It highlights the requirement for dissemination and application of scientific knowledge. By mismatching science and on-the-ground restoration, scientists and restoration practitioners are out of sync. That includes translating complicated scientific discoveries into straightforward guidelines to be implemented, as well as training and support for practitioners.

9.12.2 Socioeconomic and Policy Challenges

Restoring ecosystems faces overwhelming socioeconomic and policy challenges that restrain effectiveness and progress. One of the challenges that need to be addressed is how to conciliate conservation efforts with economic development, particularly in developing countries where human life depends considerably on natural resources (Garnett et al. 2007). Comprehensive policies that promote sustainable development and alternative livelihoods should be upheld to avoid the breakdown of local ecosystems under the livelihood pressures. Rehabilitation can be hindered by lack of financial resources. In order to realize effective ecological restoration on a potentially large scale, investment is often substantial and typically inadequate (Clark et al. 2018). This response may include such mechanisms as payment for ecosystems services, conservation finance, or public-private partnerships to mobilize and support the necessary financial and other resources.

Enforcement and execution of policies adds to the problem. Even in places with good policies on the books, they are not implemented or are undermined by corruption or lack of political will. The networks also need to strengthen governance and transparency and engage the community in policy design and implementation to resolve these problems. Local communities have the opportunity as monitors and implementers of conservation activities to ensure long-term success of restoration efforts (Evans et al. 2018). In addition, integrating climate change into existing policy is incredibly important. Restoration plans need to be flexible and adaptive to respond to dynamic conditions. Achieving this will need further research, observation, and more collaboration between scientists, government officials, and stakeholders.

9.12.3 Integrating Forensic Science in Sustainable Restoration

Applying forensic science in support of sustainable rehabilitation offers a valuable mechanism to conduct investigations of environmental crimes in an open and evidence-based way, focused on providing substantive scientific evidence. The forensic tools, for example, DNA fingerprints and chemical or trace sections fingerprinting, are particularly useful in identification of degradation and restoration of damaged (Dawson and Mayes 2015). They enable precise damage attribution to the specific impacts, be it pollution or illegal activities, and support focused restoration. For instance, in forensic analysis, poisons in soil and water can be tracked back to sources, like a factory, which can inform cleaning and remediation efforts. The same forensic science is also employed to combat wildlife crime—such as poaching and the illegal trade. Genetic fingerprinting and isotope analysis techniques are used to determine the species and origin of trafficked species, assisting law enforcement to arrest and prosecute criminals. Forensic science contributes to the protection of endangered species and habitats. In addition, forensics technologies help in monitoring and assessing the restoration efforts (Chango et al. 2024). Assessment of environmental samples pre- and post-restoration treatments provides science-based feedback that helps ensure restoration efforts are both effective and sustainable. Forensic science fosters trust and collaboration between stakeholders during conservation projects, which, in turn, increases transparency and accountability in conservation projects.

9.13 Case Studies in Ecosystem Restoration and Forensic Analysis

9.13.1 Oil Spill Cleanup and Restoration

Exxon Valdez Oil Spill (1989): The Exxon Valdez oil spill occurred when an oil tanker ran aground in Prince William Sound, Alaska, and dumped nearly 11 million gallons of crude oil (Margulies 2002). Cleanup techniques such as booms and skimmers were employed, as were chemical dispersants. Bioremediation methods were employed to accelerate the natural decomposition of oil by microbes (Abatenh et al. 2017). The long-term restoration projects involved habitat restoration, assessment of affected wildlife populations, and ongoing research to determine the environmental effects of the spill.

Deepwater Horizon Oil Spill (2010): The Deepwater Horizon oil spill was one of the biggest marine oil spills in history, releasing approximately 210 million gallons of oil into the Gulf of Mexico (Omene 2019). This response involves the use of dispersants, in situ burning, and mechanical recovery. Bioremediation served a critical role in the degradation of oil. Rehabilitation programs to restore the marine and coastal ecosystem and to compensate for lost resources of affected populations also were initiated.

9.13.2 Heavy Metal Contamination and Soil Remediation

Bangalore Dump Site, India: The Bingipura dump in Bangalore contained heavy metals such as copper, zinc, iron, chromium, cadmium, nickel, and lead (Sumalatha et al. 2022). Immobilizing agents such as lime, cement, sodium hydroxide, sawdust, and dried leaves were employed to immobilize the heavy metals. It was found that lime in combination with sawdust gave the optimum immobilization efficiency and leaching of heavy metal (Sumalatha et al. 2022).

Cyprus Heavy Metal Contamination: This case study investigated heavy metal contamination of Cyprus. The soil and water have overpass the critical threshold for arsenic, cadmium, chromium, and nickel (Akün 2020). The contaminated soil and water were remediated with phytoremediation approaches: phytoaccumulation and phytostabilization. The research proved effectiveness of the use of selected plant species to cleanup or stabilize heavy metals and to reduce the risk of metal pollution on human health (Akün 2020).

9.13.3 Wetland Restoration Projects

A. Sembakkam Lake, Chennai, India: The Sembakkam Lake Restoration Project sought to rehabilitate a degraded urban wetland in Chennai (Jadeja et al. 2022). The project comprised of cleaning up the lake, eliminating invasive plant species, expanding water storage capacity, and evolving recreational areas. The renewal activities resulted in improved water quality, more biodiversity, and better flood mitigation.

B. Fishlake Wetland Restoration, United Kingdom.

In the United Kingdom, the Fishlake Wetland restoration project sought to restore a 60-ha wetland near the River Don (Richards et al. 2020). Such work involved dropping flood banks, sculpturing fish habitat, and planting reeds that will double as in-channel fish homes. Ecosystem: The new wetlands have created a varied, species-rich habitat and reduced flooding risk, as well as provide a range of recreational facilities for the wider community.

The examples in ecological restoration and forensic investigation indicate the broad and imaginative toolkit we need to have in our arsenal to solve the environmental dilemmas that challenge us. From clearing up after catastrophic oil spills to removing heavy metal contamination or rehabilitating critical wetlands, all point to the need for tailored solutions and concerted action. Efforts to restore oil spills such as after Exxon Valdez and Deepwater Horizon provide examples of the importance of both fast response and long-term recovery to mitigate damage to the environment. Heavy metal cleanup in places like Bangalore waste and Cyprus serves as an example of how advanced practices and community involvement lead to sustainable solutions. The wetland restoration at places like Sembakkam Lake and Fishlake is proving the significance of habitat restoration and building resilient ecosystems. Incorporation of forensic analysis into ecosystem restoration enhances our ability to conserve and to rehabilitate natural settings. International cooperation would need to be strengthened and local communities incorporated in conservation efforts to ensure long-term success.

9.14 Conclusion

The scientific and technological barriers are key to the success of ecosystem restoration. We can address all of these problems by investing in research, improving predictive models, making the technology available, and encouraging cooperation. It is important to address social and policy issues in order to make ecological restoration work. Through promoting inclusive, sustainable development, assuring adequate finance, strengthening governance, and incorporating climate resilience, we can create a better and more sustainable system for restoring and sustaining the world's ecosystems. The use of forensic analysis into ecosystem restoration increases the accuracy, efficiency, and accountability of conservation activities, ensuring that restoration operations are scientifically sound and long-term sustainable.

References

Abatenh E, Gizaw B, Tsegaye Z, Wassie M (2017) The role of microorganisms in bioremediation – a review. Open J Environ Biol 2(1):038–046

Adams GO, Fufeyin PT, Okoro SE, Ehinomen I (2015) Bioremediation, biostimulation and bioaugmention: a review. Int J Environ Bioremed Biodegrad 3(1):28–39

Adla K, Dejan K, Neira D, Dragana Š (2022) Chapter 9 – Degradation of ecosystems and loss of ecosystem services. In: Prata JC, Ribeiro AI, Rocha-Santos T (eds) One health. Academic, pp 281–327. https://doi.org/10.1016/B978-0-12-822794-7.00008-3

Adu-Gyamfi J, Skrzypek G, Imfeld G, Heng L (2024) Tracing the sources and fate of contaminants in agroecosystems. Springer. https://doi.org/10.1007/978-3-031-47265-7

Agardy FJ (2005) CHAPTER 2 – Forensic (legal) perspectives. In: Agardy FJ, Nemerow NL (eds) Environmental solutions. Academic, pp 15–33. https://doi.org/10.1016/B978-012088441-4/50003-4

Ahmed R (2024) High-performance liquid chromatography (HPLC): principles, applications, versatility, efficiency, innovation and comparative analysis in modern analytical chemistry and in pharmaceutical sciences. Clin Invest (Lond) 14(9):524–535

Akito M, Shinichiro Y, Akihiro H, Michiaki K, Ikuko S, Akihide T, Hirokatsu A (2014) Reevaluation of Minamata Bay, 25 years after the dredging of mercury-polluted sediments. Mar Pollut Bull 89(1–2):112–120

Akün ME (2020) Heavy metal contamination and remediation of water and soil with case studies from Cyprus. In: Heavy metal toxicity in public health, vol 3. IntechOpen, London

Atlas RM, Hazen TC (2011) Oil biodegradation and bioremediation: a tale of the two worst spills in US history. Environ Sci Technol 45(16):6709–6715

Barthakur A (2021) Polluter pays principle as the key element to environmental law. Int J Sci Res Publ 11(3):274–277

Beauchemin D (1989) The ICP-MS approach to environmental studies. Microchim Acta 99(3):273–281. https://doi.org/10.1007/BF01244682

Bellard C, Bertelsmeier C, Leadley P, Thuiller W, Courchamp F (2012) Impacts of climate change on the future of biodiversity. Ecol Lett 15(4):365–377. https://doi.org/10.1111/j.1461-0248.2011.01736.x

Berger-Tal O, Saltz D (2014) Using the movement patterns of reintroduced animals to improve reintroduction success. Curr Zool 60(4):515–526

Bhargava A, Carmona FF, Bhargava M, Srivastava S (2012) Approaches for enhanced phytoextraction of heavy metals. J Environ Manag 105:103–120

Bhawana P, Fulekar M (2012) Nanotechnology: remediation technologies to clean up the environmental pollutants. Res J Chem Sci ISSN 2231:606X

Bica I, Alboiu IN, Iancu I, Dimache A, Stefanescu M, Voicu A, Lazaroaie M, Cirstea D, Onutu I, Petrisor IG (2009) Innovative forensic assessment and remediation strategies for polluted aquifers. Environ Forensic 10(3):214–228. https://doi.org/10.1080/15275920903140478

Bilström K (2008) Radiogenic isotopes and their applications within a range of scientific fields. In: Subías I, Bauluz B (eds) Instrumental techniques applied to mineralogy and geochemistry (pp. 111–131). Madrid: Spanish Society of Mineralogy

Boehm PD, Murphy BL (2015) Chapter 1 – Applications of environmental forensics. In: Murphy BL, Morrison RD (eds) Introduction to environmental forensics, 3rd edn. Academic, pp 3–20. https://doi.org/10.1016/B978-0-12-404696-2.00001-1

Bolan NS, Park JH, Robinson B, Naidu R, Huh KY (2011) Phytostabilization: a green approach to contaminant containment. Adv Agron 112:145–204

Bovenberg AL, Goulder LH (2002) Environmental taxation and regulation. In: Handbook of public economics, vol 3. Elsevier, pp 1471–1545

Bramer I, Anderson BJ, Bennie J, Bladon AJ, De Frenne P, Hemming D, Hill RA, Kearney MR, Körner C, Korstjens AH (2018) Advances in monitoring and modelling climate at ecologically relevant scales. In: Advances in ecological research, vol 58. Elsevier, pp 101–161

Broughton E (2005) The Bhopal disaster and its aftermath: a review. Environ Health 4(1):6. https://doi.org/10.1186/1476-069X-4-6

Burger J (2008) Environmental management: integrating ecological evaluation, remediation, restoration, natural resource damage assessment and long-term stewardship on contaminated lands. Sci Total Environ 400(1–3):6–19

Caliman FA, Robu BM, Smaranda C, Pavel VL, Gavrilescu M (2011) Soil and groundwater cleanup: benefits and limits of emerging technologies. Clean Techn Environ Policy 13:241–268

Camarretta N (2020) Bridging the divide between remote sensing and forest ecological restoration: applications for effectiveness monitoring (Doctoral dissertation, University of Tasmania).
Cao W, Zhang L, Miao Y, Qiu L (2021) Research progress in the enhancement technology of soil vapor extraction of volatile petroleum hydrocarbon pollutants. Environ Sci: Processes Impacts 23(11):1650–1662
Cardoso EJBN, Vasconcellos RLF, Bini D, Miyauchi MYH, dos Santos CA, Alves PRL, de Paula AM, Nakatani AS, de Pereira JM, Nogueira MA (2013) Soil health: looking for suitable indicators. What should be considered to assess the effects of use and management on soil health? Sci Agric 70:274–289
Carter DO, Yellowlees D, Tibbett M (2007) Cadaver decomposition in terrestrial ecosystems. Naturwissenschaften 94(1):12–24. https://doi.org/10.1007/s00114-006-0159-1
Casiraghi G, Pedretti D, Beretta GP, Masetti M, Varisco S (2022) Assessing a large-scale sequential in situ chloroethene bioremediation system using compound-specific isotope analysis (CSIA) and geochemical modeling. Pollutants 2(4):462–485. https://doi.org/10.3390/pollutants2040031
Chango X, Flor-Unda O, Gil-Jiménez P, Gómez-Moreno H (2024) Technology in forensic sciences: innovation and precision. Technologies 12(8):120
Chavda B, Makwana VM, Gor T, Patel A, Sankhla MS, Mahida DK (2024) Nanoparticle-based bioremediation approach for plastics and microplastics. In: Bioremediation of environmental toxicants. CRC Press, pp 112–126
Christie MR, Knowles LL (2015) Habitat corridors facilitate genetic resilience irrespective of species dispersal abilities or population sizes. Evol Appl 8(5):454–463
Clark R, Reed J, Sunderland T (2018) Bridging funding gaps for climate and sustainable development: pitfalls, progress and potential of private finance. Land Use Policy 71:335–346
Cole LJ, Stockan J, Helliwell R (2020) Managing riparian buffer strips to optimise ecosystem services: a review. Agric Ecosyst Environ 296:106891
Culliney TW (2005) Benefits of classical biological control for managing invasive plants. Crit Rev Plant Sci 24(2):131–150
Cycoń M, Mrozik A, Piotrowska-Seget Z (2017) Bioaugmentation as a strategy for the remediation of pesticide-polluted soil: a review. Chemosphere 172:52–71
Daines K, Dow R, Lethbridge G, Smith J, Fort R, de Ibarra M et al (2011) Prepared for the CONCAWE Water Soil Waste Management Group by its Special Task Force WQ/STF-33 and Arcadis
Dawson LA, Mayes RW (2015) Criminal and environmental soil forensics: soil as physical evidence in forensic investigations. In: Introduction to environmental forensics. Elsevier, pp 457–486
de Groot RS, Wilson MA, Boumans RMJ (2002) A typology for the classification, description and valuation of ecosystem functions, goods and services. Ecol Econ 41(3):393–408. https://doi.org/10.1016/S0921-8009(02)00089-7
De Milliano J, Di Stefano J, Courtney P, Temple-Smith P, Coulson G (2016) Soft-release versus hard-release for reintroduction of an endangered species: an experimental comparison using eastern barred bandicoots (Perameles gunnii). Wildl Res 43(1):1–12
dela Peña LBRO, Labrador KL, MAG N, Bolo NR, Rivera WL (2021) Microbial source tracking of fecal contamination in Laguna Lake, Philippines using the library-dependent method, rep-PCR. J Water Health 19(5):762–774. https://doi.org/10.2166/wh.2021.119
Deng W, Li X, An Z, Yang L (2016) The occurrence and sources of heavy metal contamination in peri-urban and smelting contaminated sites in Baoji. China Environ Monit Assess 188(4):251. https://doi.org/10.1007/s10661-016-5246-y
Deng J, Yoon S, Pasturel M, Bae S, Hanna K (2022) Interactions between nanoscale zerovalent iron (NZVI) and silver nanoparticles alter the NZVI reactivity in aqueous environments. Chem Eng J 450:138406
Dickinson JL, Shirk J, Bonter D, Bonney R, Crain RL, Martin J, Phillips T, Purcell K (2012) The current state of citizen science as a tool for ecological research and public engagement. Front Ecol Environ 10(6):291–297

Dillard A (2019) Solidification/stabilization. Fundam Site Remed 199

Environmental Degradation and Human Well-Being: Report of the Millennium Ecosystem Assessment. (2005). Population and development review, 31(2), 389–398. https://doi.org/10.1111/j.1728-4457.2005.00073.x

Evans K, Guariguata MR, Brancalion PH (2018) Participatory monitoring to connect local and global priorities for forest restoration. Conserv Biol 32(3):525–534

Fernández N, Kramer-Schadt S, Thulke H-H (2006) Viability and risk assessment in species restoration: planning reintroductions for the wild boar, a potential disease reservoir. Ecol Soc 11(1):6

Foster R, Bell-James J (2024) Legal barriers and enablers to upscaling ecological restoration. Restor Ecol 32(7):e14203

Frysinger GS, Gaines RB, Reddy CM (2002) GC × GC—a new analytical tool for environmental forensics. Environ Forensic 3(1):27–34. https://doi.org/10.1080/713848318

Gann GD, McDonald T, Walder B, Aronson J, Nelson CR, Jonson J, Hallett JG, Eisenberg C, Guariguata MR, Liu J (2019) International principles and standards for the practice of ecological restoration. Restor Ecol 27(S1):S1–S46

Gardner RC, Davidson NC (2011) The Ramsar convention. In: Wetlands: integrating multidisciplinary concepts. Springer, London, pp 189–203

Garnett ST, Sayer J, Du Toit J (2007) Improving the effectiveness of interventions to balance conservation and development: a conceptual framework. Ecol Soc 12(1):2

Gautam K, Mahida DK, Patel A (2023) Green synthesized nanoparticles for development of latent fingerprints. In: Friction ridge analysis: applications of nanoparticles for latent fingerprint development. Springer, Singapore, pp 129–141

Gibbs JP, Marquez C, Sterling EJ (2008) The role of endangered species reintroduction in ecosystem restoration: tortoise–cactus interactions on Española Island. Galápagos Restoration Ecology 16(1):88–93

Gill GN (2019) The national green tribunal: evolving adjudicatory dimensions. Envtl Pol'y & L 49:153

Griscom BW, Adams J, Ellis PW, Houghton RA, Lomax G, Miteva DA, Schlesinger WH, Shoch D, Siikamäki JV, Smith P, Woodbury P, Zganjar C, Blackman A, Campari J, Conant RT, Delgado C, Elias P, Gopalakrishna T, Hamsik MR, Fargione J (2017) Natural climate solutions. Proc Natl Acad Sci 114(44):11645–11650. https://doi.org/10.1073/pnas.1710465114

Gusain R, Gupta K, Joshi P, Khatri OP (2019) Adsorptive removal and photocatalytic degradation of organic pollutants using metal oxides and their composites: a comprehensive review. Adv Colloid Interf Sci 272:102009

Harwalkar D (1974) Water (Prevention and Control of Pollution) Act, 1974

Heine D, Faure MG, Dominioni G (2020) The polluter-pays principle in climate change law: an economic appraisal. Climate Law 10(1):94–115

Ho KF, Lee SC (2002) Identification of atmospheric volatile organic compounds (VOCs), polycyclic aromatic hydrocarbons (PAHs) and carbonyl compounds in Hong Kong. Sci Total Environ 289(1):145–158. https://doi.org/10.1016/S0048-9697(01)01031-2

Hussain E, Ahtesham A, Shahadat M, Ibrahim MNM, Ismail S (2024) Recent advances of clay/polymer-based nanomaterials for the treatment of environmental contaminants in wastewater: a review. J Environ Chem Eng 12(3):112401

Ives AR, Carpenter SR (2007) Stability and diversity of ecosystems. Science 317(5834):58–62

Jadeja NB, Banerji T, Kapley A, Kumar R (2022) Water pollution in India–current scenario. Water Security 16:100119

Joshi SS, Jadhav MPK, Chourasia NK, Jadhav MPJ, Pathade KN, KR, K. (2024) Bioremediation strategies for soil and water pollution harnessing the power of microorganisms. Migr Lett 21(S8):1–20

Keely JF (2019) Performance evaluations of pump-and-treat remediations 1. In: EPA environmental engineering sourcebook. CRC Press, pp 31–57

Khan MR (2015) Polluter-pays-principle: the cardinal instrument for addressing climate change. Laws 4(3):638–653

Khan S, Masoodi TH, Pala NA, Murtaza S, Mugloo JA, Sofi PA, Zaman MU, Kumar R, Kumar A (2023) Phytoremediation prospects for restoration of contamination in the natural ecosystems. Water 15(8):1498

Khatsuriya PA, Makwana VM, Dodia PP (2023) Avaniya Bandhara: a potential biodiversity hotspot for avifauna in Bhavnagar, Gujarat, India. Munis Entomol Zool 18(2):1388–1406

Khatsuriya PA, Makwana VM, Dodia PP (2024) Biodiversity resilience amidst climate shifts: a global perspective. In: Global challenges for the environment and climate change. IGI Global, pp 75–99

Kristanti RA, Ngu WJ, Yuniarto A, Hadibarata T (2021) Rhizofiltration for removal of inorganic and organic pollutants in groundwater: a review. Biointerface Res Appl Chem 11:12326–12347

Kumar PD (2015) Defending the green realm: the Forest Conservation Act 1980 of India in theory and practice. Institute of Social and Economic Change

Landrigan PJ, Fuller R, NJR A, Adeyi O, Arnold R, Basu NN, Baldé AB, Bertollini R, Bose-O'Reilly S, Boufford JI, Breysse PN, Chiles T, Mahidol C, Coll-Seck AM, Cropper ML, Fobil J, Fuster V, Greenstone M, Haines A et al (2018) The Lancet Commission on pollution and health. Lancet 391(10119):462–512. https://doi.org/10.1016/S0140-6736(17)32345-0

Leadley P, Archer E, Bendandi B, Cavender-Bares J, Davalos L, DeClerck F, Gann GD, Gonzales EK, Krug CB, Metzger JP (2022) Setting ambitious international restoration objectives for terrestrial ecosystems for 2030 and beyond. PLOS Sustainability Transform 1(12):e0000039

Linkov I, Satterstrom FK, Kiker G, Batchelor C, Bridges T, Ferguson E (2006) From comparative risk assessment to multi-criteria decision analysis and adaptive management: recent developments and applications. Environ Int 32(8):1072–1093

Lipton J, LeJeune K (2018) Step 2: determining and quantifying environmental damage. In: Lipton J, Özdemiroğlu E, Chapman D, Peers J (eds) Equivalency methods for environmental liability: assessing damage and compensation under the European Environmental Liability Directive. Springer, Dordrecht, pp 57–88. https://doi.org/10.1007/978-90-481-9812-2_4

Liu J, Zhao L, Liu Q, Li J, Qiao Z, Sun P, Yang Y (2021) A critical review on soil washing during soil remediation for heavy metals and organic pollutants. Int J Environ Sci Technol 19(2–3):1–24

Lovell ST (2010) Multifunctional urban agriculture for sustainable land use planning in the United States. Sustainability 2(8):2499–2522

Madima N, Mishra S, Inamuddin I, Mishra A (2020) Carbon-based nanomaterials for remediation of organic and inorganic pollutants from wastewater. A Rev Environ Chem Lett 18:1169–1191

Magdoff F, Weil RR (2004) Soil organic matter management strategies. In: Soil organic matter in sustainable agriculture. CRC Press, Boca Raton, pp 45–65

Mahida DK, Pithiya G, Makwana VM, Patel A (2022) Determination of heavy metals accumulation in urban roadside flora from Ahmedabad, India, and their implication for pollution control. Int J Med Toxicol Legal Med 25(3 and 4):213–219

Mahida DK, Makwana VM, Sankhla MS, Patel A, Dodia P (2023) Accumulation of heavy metals in roadside plants and their role in phytoremediation. In: Anthropogenic environmental hazards: compensation and mitigation. Springer, Cham, pp 119–141

Mahida DK, Patel A, Rawal R (2024) The determination and speciation of arsenic in rice. In: Arsenic in rice. Apple Academic Press, pp 227–248

Mahida DK, Patel A, Sankhla MS, Chavda B, Makwana VM, Kumar K, Rawal R (2025) Assessing the ecotoxicological impact of copper-based Nano-pesticides: a comprehensive review. Biointerface Res Appl Chem 15(2):30

Mahlambi MM, Ngila CJ, Mamba BB (2015) Recent developments in environmental photocatalytic degradation of organic pollutants: the case of titanium dioxide nanoparticles—a review. J Nanomater 2015(1):790173

Makwana VM, Mahida DK, Dodia PP (2023) Unraveling the decline: exploring global patterns in shorebird populations. In: Perspectives on global biodiversity scenarios and environmental services in the 21st century. IGI Global, pp 192–223

Makwana VM, Khatsuriya PA, Matli AG, Dodia PP (2024a) Composition and conservation status of avifauna in urban non-protected Important Bird Area (IBA) site of Western India. J Tropical Biodivers Biotechnol 9(2):87311

Makwana VM, Patel FP, Dodia PP (2024b) Urbanization's ecological toll: exploring its role in biodiversity decline. In: Impact of societal development and infrastructure on biodiversity decline. IGI Global, pp 245–270

Malhi Y, Roberts JT, Betts RA, Killeen TJ, Li W, Nobre CA (2008) Climate change, deforestation, and the fate of the Amazon. Science 319(5860):169–172. https://doi.org/10.1126/science.1146961

Manfreda S, McCabe MF, Miller PE, Lucas R, Pajuelo Madrigal V, Mallinis G, Ben Dor E, Helman D, Estes L, Ciraolo G (2018) On the use of unmanned aerial systems for environmental monitoring. Remote Sens 10(4):641

Margulies P (2002) The Exxon Valdez oil spill. The Rosen Publishing Group, Inc, New York

McCarty JP, Hoffman D, Rattner B, Burton G, Cairns J (2003) The Hudson River—PCB case study. In: Handbook of ecotoxicology. CRC Press, Boca Raton, pp 813–831

Megharaj M, Ramakrishnan B, Venkateswarlu K, Sethunathan N, Naidu R (2011) Bioremediation approaches for organic pollutants: a critical perspective. Environ Int 37(8):1362–1375. https://doi.org/10.1016/j.envint.2011.06.003

Meunier CF, Dandoy P, Su B-L (2010) Encapsulation of cells within silica matrixes: towards a new advance in the conception of living hybrid materials. J Colloid Interface Sci 342(2):211–224

Mir M, Maleki S, Rahdari V (2023) Ecosystem restoration and degradation monitoring using ecological indices. Int J Environ Sci Technol 20(2):1713–1724. https://doi.org/10.1007/s13762-022-04694-8

Moehrenschlager A, Lloyd N (2016) Release considerations and techniques to improve conservation translocation success. In: Reintroduction of fish and wildlife populations. University of California Press, Oakland, pp 245–280

Møller AP, Mousseau TA (2006) Biological consequences of Chernobyl: 20 years on. Trends Ecol Evol 21(4):200–207. https://doi.org/10.1016/j.tree.2006.01.008

Morrison ML (2013) Wildlife restoration: techniques for habitat analysis and animal monitoring, vol 1. Island Press

Morrison R, Murphy B (2005) Environmental forensics: contaminant specific guide

Mudge SM (2008) Environmental forensics and the importance of source identification. In: Hester RE, Harrison RM (eds) Environmental forensics, vol 26. The Royal Society of Chemistry, p 0. https://doi.org/10.1039/9781847558343-00001

Münzel T, Hahad O, Daiber A, Landrigan PJ (2023) Soil and water pollution and human health: what should cardiologists worry about? Cardiovasc Res 119(2):440–449. https://doi.org/10.1093/cvr/cvac082

Myers SS, Gaffikin L, Golden CD, Ostfeld RS, Redford KH, Ricketts TH, Turner WR, Osofsky SA (2013) Human health impacts of ecosystem alteration. Proc Natl Acad Sci 110(47):18753–18760. https://doi.org/10.1073/pnas.1218656110

O'Connor M (1997) The internalisation of environmental costs: implementing the polluter pays principle in the European Union. Int J Environ Pollut 7(4):450–482

O'geen A, Budd R, Gan J, Maynard J, Parikh S, Dahlgren R (2010) Mitigating nonpoint source pollution in agriculture with constructed and restored wetlands. Adv Agron 108:1–76

Okoro HK, Pandey S, Ogunkunle CO, Ngila CJ, Zvinowanda C, Jimoh I, Lawal IA, Orosun MM, Adeniyi AG (2022) Nanomaterial-based biosorbents: adsorbent for efficient removal of selected organic pollutants from industrial wastewater. Emerging Contam 8:46–58

Omene L (2019) Assessing the socioeconomic vulnerability to oil spill from the deep-water horizon oil spill: a case study on coastal counties in the Gulf of Mexico

Ondiko JH, Karanja AM, Ombogo O (2022) A review of the anthropogenic effects of climate change on the physical and social environment. OAlib 09(02):1–14. https://doi.org/10.4236/oalib.1107751

Osborne PE, Seddon PJ (2012) Selecting suitable habitats for reintroductions: variation, change and the role of species distribution modelling. In: Reintroduction biology: integrating science and management, pp 73–104

Osman KT (2012) Plant nutrients and soil fertility management. In Soils: Principles, properties and management (pp. 129–159). Dordrecht: Springer Netherlands

Parmar GR, Rao N (2008) Emerging control technologies for volatile organic compounds. Crit Rev Environ Sci Technol 39(1):41–78

Petrovic M, Barceló D (2006) Liquid chromatography–mass spectrometry in the analysis of emerging environmental contaminants. Anal Bioanal Chem 385(3):422–424. https://doi.org/10.1007/s00216-006-0450-1

Pichlmayer F, Schöner W, Seibert P, Stichler W, Wagenbach D (1998) Stable isotope analysis for characterization of pollutants at high elevation alpine sites. Atmos Environ 32(23):4075–4085. https://doi.org/10.1016/S1352-2310(97)00405-6

Pimm SL, Raven P (2000) Extinction by numbers. Nature 403(6772):843–845. https://doi.org/10.1038/35002708

Poulter SR (1998) Cleanup and restoration: who should pay. J Land Res Envtl L 18:77

Prakash S, Verma AK (2022) Anthropogenic activities and biodiversity threats. Int J Biol Innov 04(01):94–103. https://doi.org/10.46505/ijbi.2022.4110

Rajasugunasekar D, Devi RS (2022) Legal framework in India to protect the environment. Ecological environment: a new perspective, 468

Reich P, Lake PS (2015) Extreme hydrological events and the ecological restoration of flowing waters. Freshw Biol 60(12):2639–2652

Reitze AW (2001) Air pollution control law: compliance and enforcement. Environmental Law Institute

Richards DR, Moggridge HL, Warren PH, Maltby L (2020) Impacts of hydrological restoration on lowland river floodplain plant communities. Wetl Ecol Manag 28:403–417

Rogelj J, Den Elzen M, Höhne N, Fransen T, Fekete H, Winkler H, Schaeffer R, Sha F, Riahi K, Meinshausen M (2016) Paris agreement climate proposals need a boost to keep warming well below 2 C. Nature 534(7609):631–639

Rossi I (2025) Chemical fingerprinting: advances in analytical techniques for environmental monitoring. Revista Latinoamericana de Química 52(1):10–12. https://doi.org/10.52783/rlq.17

Santos FJ, Galceran MT (2003) Modern developments in gas chromatography–mass spectrometry-based environmental analysis. J Chromatogr A 1000(1):125–151. https://doi.org/10.1016/S0021-9673(03)00305-4

Schwartz P (2010) The polluter-pays principle. In: Research handbook on international environmental law. Edward Elgar Publishing, Northampton

Schwitzguebel J-P, Kumpiene J, Comino E, Vanek T (2009) From green to clean: a promising and sustainable approach towards environmental remediation and human health for the 21st century. Agrochimica 53(4):1–29

Seol Y, Zhang H, Schwartz FW (2003) A review of in situ chemical oxidation and heterogeneity. Environ Eng Geosci 9(1):37–49

Sims C, Aadland D, Finnoff D, Hochard J (2020) What are the benefits of delisting endangered species and who receives them?: lessons from the gray wolf recovery in Greater Yellowstone. Ecol Econ 174:106656

Singh B (2022) Biological Diversity Act 2002 and its Importance for Conservation of Biological Resources. Cent Asian J Theor Appl Sci 3(6):565–571

Singh O, Jain R (2003) Phytoremediation of toxic aromatic pollutants from soil. Appl Microbiol Biotechnol 63:128–135

Smagadi A (2006) Analysis of the objective of the convention on biological diversity: their interrelation and implementation guidance for access and benefit sharing. Colum J Envtl L 31:243

Stefaniuk M, Oleszczuk P, Ok YS (2016) Review on nano zerovalent iron (nZVI): from synthesis to environmental applications. Chem Eng J 287:618–632

Stegemeier GL, Vinegar HJ (2001) Thermal conduction heating for in-situ thermal desorption of soils. In: Hazardous and radioactive waste treatment technologies handbook, CRC Press, Boca Raton, Florida. pp 1–37

Stringer AP, Gaywood MJ (2016) The impacts of beavers Castor spp. on biodiversity and the ecological basis for their reintroduction to Scotland, UK. Mammal Rev 46(4):270–283

Sumalatha J, Sivapullaiah P, Prabhakara R (2022) Immobilization remediation of a heavy metals contaminated soil: a case study of dump site at Bangalore, India. J Inst Eng (India) Series A 103(1):105–114
Suresh B, Ravishankar G (2004) Phytoremediation—a novel and promising approach for environmental clean-up. Crit Rev Biotechnol 24(2–3):97–124
Thomas J (2005) Monitoring change in the abundance and distribution of insects using butterflies and other indicator groups. Philos Trans R Soc B Biol Sci 360(1454):339–357
Thuiller W, Albert C, Araújo MB, Berry PM, Cabeza M, Guisan A, Hickler T, Midgley GF, Paterson J, Schurr FM (2008) Predicting global change impacts on plant species' distributions: future challenges. Persp Plant Ecol Evol Syst 9(3–4):137–152
Tripathi A, Roy A (2021) Analysis of legislative framework concerning exotic species in India. Int J Law Manage Human 4:5180
United States. Coast Guard, & National Response Team (US) (2011) On Scene Coordinator Report: Deepwater Horizon Oil Spill. US Department of Homeland Security, US Coast Guard
Vargas A, Sánchez I, Martínez F, Rivas A, Godoy JA, Roldan E, Simón M, Serra R, Pérez M, Ensenat C (2008) The Iberian lynx Lynx pardinus conservation breeding program. Int Zoo Yearbook 42(1):190–198
Vilks A (2022) Legal framework for environmental protection in the context of sustainable development. Eur J Sustain Dev 11(4):169–169
Wagner AM, Larson DL, DalSoglio JA, Harris JA, Labus P, Rosi-Marshall EJ, Skrabis KE (2015) A framework for establishing restoration goals for contaminated ecosystems. Integr Environ Assess Manag 12(2):264–272
Wahab SUKA, Samah MAA, Sabuti AA, Yunus K, Chowdhury AJK, John A, Aris MSM, Mohamad MAN, Hamid AAA (2020) Environmental forensic study: tracing of pollution sources using environmetric technique in Balok and Tunggak Rivers near Gebeng industrial area, Kuantan, Pahang, Malaysia. Desalin Water Treat 191:118–125. https://doi.org/10.5004/dwt.2020.25271
Wang Z, Stout SA, Fingas M (2006) Forensic fingerprinting of biomarkers for oil spill characterization and source identification. Environ Forensic 7(2):105–146. https://doi.org/10.1080/15275920600667104
Weaver CP, Lempert RJ, Brown C, Hall JA, Revell D, Sarewitz D (2013) Improving the contribution of climate model information to decision making: the value and demands of robust decision frameworks. Wiley Interdiscip Rev Clim Chang 4(1):39–60
Williams BK, Brown ED (2014) Adaptive management: from more talk to real action. Environ Manag 53:465–479
Williams SE, Hoffman EA (2009) Minimizing genetic adaptation in captive breeding programs: a review. Biol Conserv 142(11):2388–2400
Withrow-Robinson B, Johnson R, State University. Extension Service O (2006) Selecting native plant materials for restoration projects : ensuring local adaptation and maintaining genetic diversity. : Corvallis, Or. : Extension Service, Oregon State University.
Young TP (2000) Restoration ecology and conservation biology. Biol Conserv 92(1):73–83. https://doi.org/10.1016/S0006-3207(99)00057-9
Zhang X (2019) The history of pollution elements in Zhengzhou, China recorded by tree rings. Dendrochronologia 54:71–77. https://doi.org/10.1016/j.dendro.2019.02.004
Zhao C, Dong Y, Feng Y, Li Y, Dong Y (2019) Thermal desorption for remediation of contaminated soil: a review. Chemosphere 221:841–855

Analytical Strategies for Source Identification and Environmental Justice

10

Girraj Sharma, Nilesh Gupta, Akash Kumar, Baljeet Yadav, and Janmejay Singh Tomar

Abstract

Accurate pollution source identification is essential for effective environmental regulation and sustainable resource management. This process involves determining the origin, nature, and quantity of pollutants & emerging contaminants, enabling targeted interventions, and reducing reliance on generalized assumptions. Pollution sources are majorly categorized as point or nonpoint for water pollution, mobile or stationary for air pollution, each category represent unique monitoring and certain challenges. These can be overcome by using advanced methodologies & techniques, including environmental sampling, remote sensing integrated with GIS, chemical fingerprinting, and statistical modelling, enhance source attribution by linking pollutants & emerging contaminants to their origins with high spatial and temporal resolution. However, the exact identification of sources remains difficult because of analytical constraints and imperfect data quality and overlapping processes that occur. The detection technologies need improvement and standardization in monitoring frameworks to overcome these

G. Sharma (✉)
Jindal Institute of Behavioural Sciences, O.P. Jindal Global University, Sonipat, Haryana, India

N. Gupta
Faculty of Pharmacy, Medi-Caps University, Rau, Indore, Madhya Pradesh, India

A. Kumar
Dr. A.P.J. Abdul Kalam Institute of Forensic Science and Criminology, Bundelkhand University, Jhansi, Uttar Pradesh, India

B. Yadav
School of Basic and Applied Sciences, KR Mangalam University, Gurugram, Haryana, India

J. S. Tomar
Department of Pharmacy, Institute of Professional Studies, Gwalior, Madhya Pradesh, India

M. S. Sankhla et al. (eds.), *Introducing the Synergy Between Forensic Science and Environmental Analysis*, https://doi.org/10.1007/978-981-95-0199-1_10

challenges. This chapter will also explain the ability to identify pollution sources effectively enables regulator-ship based on science which leads to declassification protocols and enables protective strategies that protect both nature and public well-being. Moreover, it is also well integrated with the United Nations Sustainable Development Goals; Clean Water & Sanitation (SDG 6), Sustainable Cities and Communities (SDG 11), Responsible Consumption & Production (SDG 12), Life below Water (SDG 14), Life on Land (SDG 15), Peace, Justice & Strong Institutions (SDG 16).

Keywords

Surveillance · Remote sensing · GIS · Chemical fingerprinting · Point and non-point sources · Gas chromatography · Mass spectrometry · GCMS · Fourier transform infrared spectroscopy · SDG-6 · SDG-11 · SDG012 · SDG-14 · SDG-15 · SDG-16

10.1 Introduction

The pollution source identification is one of the most important and complex process for the determination of the root cause of origin and type and quantity of pollutant & emerging contaminants discharged into the environment. It also play pivatol role in distinguishing between natural and anthropogenic sources of pollution (Liu et al. 2022). This process is essential because it allow pollution control programs to target their actions more properly and implementing effective measures through actual data rather than guessworks & assumptions (Chou et al. 2019; Mushtaq et al. 2023). In addition, this process can also identify specific pollution sources from industries, vehicles, or agriculture area, and governments, environmental organizations or regulatory bodies can frame and implement accurate and precise affordable solutions and which can also fix accountability of polluters and minimize public health risks to mitigate this problem (Wang et al. 2024). On the one hand, effective source identification can help to restore ecosystem, develop sustainable urban planning, and create pollution prevention systems which can lead to provide toward pure air, clean water, and healthy soil for environment and communities to sustain their life (Elmqvist et al. 2015). However, if pollution source identification remained unidentified, then the management process becomes ineffective, less impactful, costly, and more expensive (Khanam et al. 2023). Therefore, environmental policymakers need pollution identification as a foundation to find responsible parties and develop specific prevention plans and execute data-driven environmental policies.

The main goal of identifying pollution sources is to fix accountability of polluter, develop specific prevention plans, and execute data-driven environmental policies (Awewomom et al. 2024). These goals can help to locate exact point of origin of contamination, and further authorities can implement precise control measures according to origin of source rather than broad, ineffective measures (Kuppan et al.

2024). Thus, this process may help with the determination and quantification of pollution from different areas, and the responsible authorities must first identify the greatest pollution threats and then implement urgent measures while establishing permanent prevention procedures (Wang et al. 2024). Additionally, this accurate approach of source identification also smooths regulatory bodies to take action against polluters and fix responsibility for remediation and provides critical data for public health protection (Liu et al. 2024). By using this organized methods, organizations achieve better allocation of resources for environmental management operations that also help achieve sustainable development targets and ecological restoration initiatives (Liu et al. 2024; Wang et al. 2024). The above part of the chapter is clearly explaining how various goals of United Nations Sustainability Development Goal like SDG 6 (Clean Water and Sanitation), SDG 11 (Sustainable Cities and Communities), SDG 13 (Climate Action), SDG 15 (Life on Land), and SDG 16 (Peace, Justice and Strong Institutions) are important one.

10.2 Types of Pollution Sources

The sources of pollution are different but it mainly stems from natural and anthropogenic activities (Akhtar et al. 2021). It may also occur from various source such as industrial operations, vehicle discharges, agricultural drainage and inadequate waste handling operations that represent the key human sources producing adverse substances that contaminate environmental compartments air, water and soil (Dehkordi et al. 2024). In continuation industrial facilities release sulfur dioxide, nitrogen oxides, particulate matter and vehicle exhaust gases substantially produce air environmental pollution (Nakhjiri and Kakroodi 2024). Using chemical fertilizers and pesticides by farmers at their fields results in the contamination of water and soil (Vasco et al. 2021). Additionally, it also occur through the degradation of natural ecosystems by urban development activities and deforastation which intensifies environmental contamination levels (Adla et al. 2022). Various pollutants emerging from these origins generate serious environmental dangers which endanger human health and are briefly discussed below (SDG-6, 11, 13, 15 & 16):

10.2.1 Point Sources (SDG-6 , 11, 13 & 15)

Point sources of pollution refer to identifiable, stationary locations that discharge contaminants directly into the environment through a confined release point, making them easier to monitor and regulate compared to diffuse sources (Wilhelm 2009). Common examples include industrial wastewater discharge pipes, sewage treatment plant outfalls, smokestacks from power plants, and underground injection wells. These sources are typically regulated through permitting systems (such as the US EPA's National Pollutant Discharge Elimination System) that set limits on allowable emissions (Ferrier et al. 2005). While point sources are more straightforward to control due to their fixed nature, they can still pose significant environmental and health risks if improperly managed, particularly when releasing toxic chemicals, heavy metals, or nutrient-rich effluents that disrupt ecosystems (Saravanan et al. 2021). The evaluation and management of point source pollution needs ongoing monitoring along with state-of-the-art environmental treatment techniques to achieve proper enforcement of environmental standards for preserving air and water and soil quality (Novotny 2003).

10.2.2 Nonpoint Sources (SDG-6 , 11, 13 & 15)

The pollutant known as nonpoint source (NPS) pollution when it spread from unknown locations or lacking defined discharge points, which increases the difficulty level of tracking and managing this source of contamination ("Nonpoint Source Water Pollution" 2003). These pollutants are typically transferred from one location to the other by rainfall or snowmelt which may carry natural or anthropogenic contaminants across land until they reach water bodies or groundwater ("Nonpoint Source Water Pollution" 2003). Common examples include *agricultural runoff* (containing pesticides, fertilizers, and animal waste), *urban stormwater* (carrying oil, heavy metals, and debris from roads and rooftops), and *sediment erosion* from construction sites or deforested areas (Wang et al. 2022). Unlike point sources, NPS pollution accumulates gradually from numerous small contributors, requiring comprehensive watershed management strategies, such as buffer strips, retention ponds, and public education campaigns, to effectively mitigate its environmental impact (Nuruzzaman et al. 2025). The *Chesapeake Bay pollution* in the United States exemplifies NPS challenges, where excess nutrients from farms and urban areas have caused algal blooms and oxygen-depleted "dead zones," necessitating large-scale regional cooperation for restoration (Nsenga Kumwimba et al. 2023).

10.2.3 Mobile vs. Stationary Sources (SDG-6 , 11, 13 & 15)

Mobile and stationary pollution sources represent two broad categories of pollution origins, differing primarily in their movement and emission patterns (Li et al. 2023). Stationary sources are fixed-location emitters like power plants, factories, and refineries, which release pollutants (e.g., SO_2, particulate matter) through smokestacks or wastewater systems (Reitze 2005). These are typically regulated through permits and emission caps (e.g., Clean Air Act requirements). In contrast, *mobile sources* include vehicles (cars, trucks), aircraft, and ships that emit pollutants (CO, NO_x, hydrocarbons) while moving, making them harder to control due to their dispersed nature (Pohanish 2017). While stationary sources often dominate industrial emissions, mobile sources significantly contribute to urban air pollution; for example, vehicular exhaust accounts for ~50% of NO_x emissions in US cities (Filonchyk and Peterson 2024). Effective management requires distinct strategies, stationary sources benefit from scrubbers and cleaner technologies, while mobile sources need fuel standards (e.g., Euro 6 norms) and electric vehicle transitions (Winkler et al. 2018). The *Los Angeles smog crisis* underscores this interplay, where both traffic (mobile) and industrial emissions (stationary) necessitated combined regulatory approaches (Wang 2023).

10.3 Methods for Identifying Pollution Sources (SDG-6, 11, 13, 15 & 16)

Tracking of pollution sources requires using gathering of information from various sources like field-based observations, analytical methods, mathematical models, artificial intelligence and machine learning to determine origin points (Xu et al. 2024; Olawade et al. 2024). Identifying the origin of pollution requires chemically fingerprinting technique analysis of pollutants to determine their specific sources as well as isotopic analysis technique to differentiate between natural and human-made origins through isotope ratio comparisons (Négrel et al. 2012). The major tools employed in pollution pattern mapping and possible source link detection are geographic information systems (GIS) together with remote sensing technologies (Tikader et al. 2024). Multiple emitters get their contributions evaluated by source apportionment models through positive matrix factorization (PMF) and chemical mass balance (CMB) which employ monitoring data for their statistical measurements (Rudel and Perovich 1994). Complex environmental settings are assessed by tracking pollution sources using microbial source tracing in conjunction with microbial source tracking methods and land use evaluation (Ballesté et al. 2020). The framework operates through multiple interrelated systems to precisely track pollution sources which provide both specific administrations and remedy operations (Liu et al. 2023). These are explained below.

10.3.1 Environmental Monitoring

The process of source pollution identification through environmental monitoring uses field sampling with remote sensing alongside data analysis methods (Li et al. 2025). Air quality monitoring stations and water sampling and soil testing together monitor pollutant concentrations yet GC-MS and GC-MS tools assist specialists in identifying particular pollutants (Kiani et al. 2024). The combination of satellite imagery and drones represents remote sensing technologies which generate extensive spatial data to locate important emission sources (Sun et al. 2024). GIS systems unite data from diverse sources to construct pollution mapping as well as pinpoint the sources of contamination (Vaye et al. 2022; Srivastava et al. 2012). Additionally, real-time sensors along with tracer studies perform analysis of pollutant movement to allow authorities a clear view of what causes pollution by tracking down industrial discharges or agricultural runoff or urban emissions. Thus, these methods collectively support and enhance targeted pollution control strategies (Meyer et al. 2019) while Fig. 10.1 illustrates the analysis scheme.

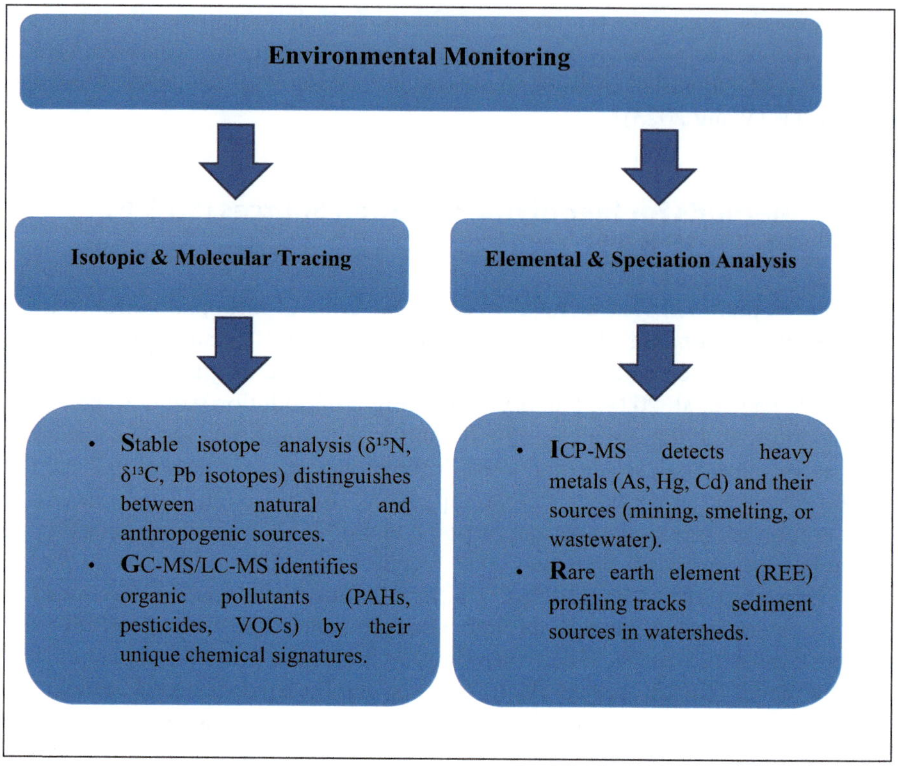

Fig. 10.1 The scheme of analysis

10.3.2 Remote Sensing and GIS

Traditional approach of pollution identification and detection includes field measurements, and laboratory testing has specific limitations and boundary constraints (Artiola and Brusseau 2019). In contrast, modern methods of detection and identification like RS combine with GIS become a powerful technique in environmental monitoring as well as pollution source identification (Korchenko 2019). These geospatial technologies allow the acquisition, integration, analysis, and visualization of spatial data across large areas and time periods, providing a synoptic and dynamic view of environmental conditions (Yu et al. 2019). Remote sensing techniques like drone-based sensors and satellite imagery can efficiently identify and detect pollution hotspots by observing variations in land and thermal sensing of vegetation (Jońca et al. 2022). Additionally, the combination of hyperspectral and multispectral imaging systems enables users to detect pollutants including oil pollutants in water alongside harmful metals in soil samples. GIS links with ground observation including water and air quality tests alongside database entries to build visual pollution distribution models (TavallaieNejad et al. 2025). Moreover, GIS facilitates source tracking of contaminants by analyzing spatial patterns through its ability to

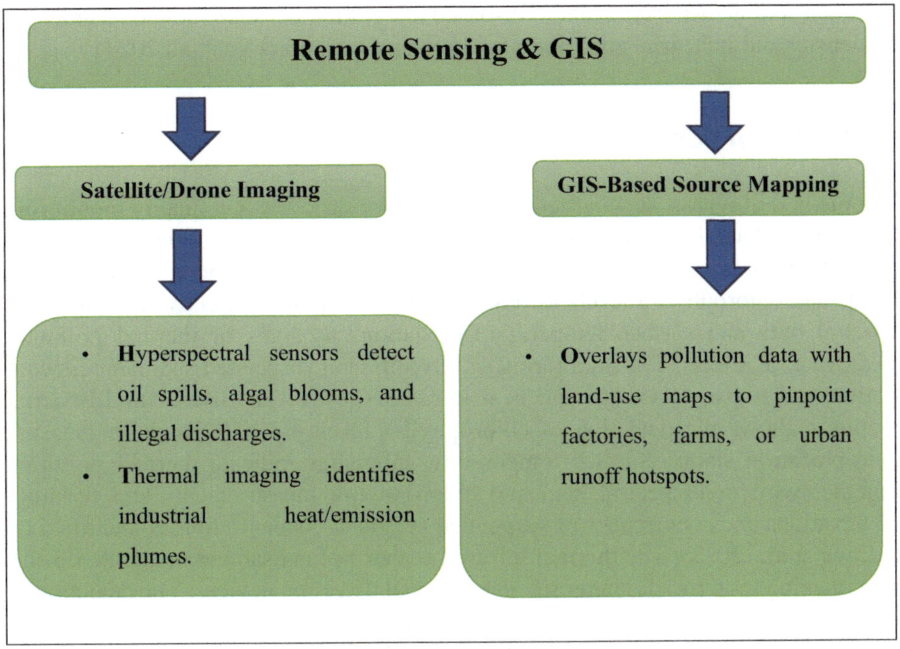

Fig. 10.2 The analysis scheme used for environmental monitoring

generate detailed maps which identify the origins from industrial sites and agricultural areas and urban waste facilities (Kaplan et al. 2024). Additionally, pollution dispersion modelling that derives from GIS-based systems operators can enhance their environmental management practices along with regulatory choices (Samad et al. 2023). These technologies integrate to perform accurate large-scale detection and source discovery which allows for better strategic pollution remediation efforts (Basnyat et al. 2000; Hou et al. 2017). Figure 10.2 shows the analysis scheme used for environmental monitoring.

10.4 The Key Application of RS and GIS for Environmental Pollution Monitoring Is Discussed Below

10.4.1 Air Pollution

Remote sensing can effectively and successfully identify and detect aerosols, trace gases, and air quality indicators by employing satellite with moderate resolution imaging spectroradiometer, tropospheric monitoring instrument, and ozone monitoring instrument (Martin 2008). Satellite instruments equipped with sensors deliver measurements of aerosol optical depth together with information about the trace gas concentrations of NO_2, SO_2, CO, and O_3 (Dhankar et al. 2024). Researchers can create a comprehensive understanding of pollution sources through the analysis of pollutant spatial distribution and GIS-based integration between these data and traffic density and industrial zone and population information (Xu et al. 2024).

10.4.2 Water Pollution

The method of remote sensing proves essential for surface water quality monitoring through the detection of multiple water characteristics including turbidity and suspended sediment concentration measured using visible and near-infrared (NIR) bands and chlorophyll-a levels as indicators of eutrophication derived from multispectral data and surface temperature deviations as signs of thermal pollution (Adjovu et al. 2023). The monitoring of turbidity and suspended sediment concentration makes use of visible and near-infrared spectra. The use of multispectral methods allows the detection of chlorophyll-a levels by NIR bands to measure eutrophication status. Surface temperature anomalies indicate thermal pollution together with multispectral data and chlorophyll-a measurements and turbidity observations and occurrences of suspended sediment concentration as identified by Adjovu et al. (2023). The thermal infrared sensor in Landsat and moderate resolution imaging spectroradiometer serve as thermal detection devices. The detection of heat discharge from industrial cooling relies on the combination of spectroradiometer instruments systems (Almeida et al. 2021). Risk assessment through the utilization of GIS allows the mapping of thermal anomalies between different areas. GIS technology enables spatial investigations between industrial facilities and sewage outfalls and urban runoff areas understanding of water quality and pollution sources (Naimaee et al. 2024).

10.4.3 Soil and Land Pollution

Soil contamination identification occurs through multispectral and hyperspectral sensors that monitor vegetation changes together with spectral reflections and soil water levels (Tshanga et al. 2024). Waste disposal sites as well as mining areas and oil spill locations possess unique spectral patterns that distinguish them from other land sections. GIS interprets spatial patterns together with land use information alongside topographical data and hydrological flow data to find sources as well as dispersion routes (Singh and Fiorentino 1996; DeVantier and Feldman 1993).

10.4.4 Marine and Coastal Pollution

Remote sensing equipment employs optical and synthetic aperture radar sensors (SAR) to identify oil spills as well as algal blooms and sediment plumes present in coastal regions. The surveillance capacity of SAR operates efficiently when there is cloud cover and during nighttime operations. The analysis of drift patterns and source tracking receives support from GIS tools that build models with wind direction data and ocean current (Hafeez et al. 2019).

10.4.5 Industrial and Urban Pollution

Enforcement of industrial emissions together with illegal dumping operations and unauthorized landfills becomes more efficient by using high-resolution satellite data and UAV drone information (Fraternali et al. 2024). Environmental justice evaluations become possible through GIS technology that analyzes the geographical relationships between pollution sources and vulnerable populations and sensitive environmental zones (Wilson et al. 2010).

10.5 Chemical Fingerprinting

The underlying method of chemical fingerprinting depends on pollutants creating distinct chemical marks from production techniques and combustion practices and raw material origins and geological source characteristics of their origin (Baragaño et al. 2022). Chemical fingerprinting has developed into a powerful precise scientifically robust methodology among the various tools that researchers have created for this purpose (Zhong et al. 2024). Environmental contaminant analysis through chemical fingerprinting evaluates precise chemical makeup patterns to identify the origin source (Stout and Uhler 2003). The technique depends on sophisticated analytical methods to detect and differentiate specific chemical compound patterns which emerge from each pollution source. Chemical fingerprinting has developed indispensable value for forensic environmental science and other fields like oil spill forensics and groundwater contamination studies and urban air quality assessment during recent decades (Kamber 2009).

The procedure exhibits determinants but the process remains partially unreliable. The chemical signatures in forensic investigations may change through time because of factors such as weathering along with degradation and source contamination (Su et al. 2023). Chemical fingerprinting depends on analytical chemistry and statistical modelling as well as environmental forensics to boost accuracy when interpreting complex datasets (Kolobaric et al. 2023). The identification process in chemical fingerprinting requires different complex analytical methods based on what type of contaminant and environmental matrix scientists examine (Kadadou et al. 2024).

10.5.1 Gas Chromatography and Gas Chromatography-Mass Spectrometry (GC-MS)

GC together with GC-MS stands as the primary analytical tool used to examine volatile as well as semi-volatile organic compounds (Arsene et al. 2011). The methods show exceptional capability for the detection of hydrocarbons together with petroleum products and polycyclic aromatic hydrocarbons (PAHs) found in instances of oil spills and vehicular emissions along with industrial discharges. The complex mixture undergoes separation through GC to provide individual compounds which MS examines using mass-to-charge ratios (Ifegwu and Anyakora 2015). To track the origin of spilled oil, GC-MS technology performs detailed evaluations of hydrocarbon composition patterns and biological markers (Reddy and Quinn 1999).

10.5.2 Isotope Ratio Mass Spectrometry (IRMS)

Stable isotope analysis functions as an effective instrument in chemical fingerprinting systems. Stable isotopes of carbon ($^{13}C/^{12}C$), nitrogen ($^{15}N/^{14}N$), sulfur ($^{34}S/^{32}S$), and oxygen ($^{18}O/^{16}O$) possess naturally occurring ratios alongside human-induced changes which enable researchers to distinguish between different pollution sources (Sankoh et al. 2021). The use of nitrogen isotope analysis determines which type of nitrate pollutant comes from synthetic fertilizers or from animal waste or sewage sources (Curt et al. 2004). The biogeochemical cycling of ecosystems as well as heavy metal source tracing becomes possible through the application of isotope ratio techniques (Wang et al. 2024).

10.5.3 Inductively Coupled Plasma Mass Spectrometry (ICP-MS)

ICP-MS functions as the primary device to detect and measure trace metals together with rare earth elements in different analytical settings. The instrument operates exceptionally well at detecting small amounts of contaminants contained in water and soil and sediment samples (Beauchemin 2006). The device serves as a key instrument for laboratory examinations of mining pollution levels and urban runoff investigations and industrial wastewater assessments. ICP-MS produces elemental profiles which researchers can use to identify the leading source of pollution between different samples (Ettler et al. 2004).

10.5.4 Spectroscopic Techniques

The analysis of pollutants through molecular structures depends increasingly on two techniques known as Fourier transform infrared spectroscopy (FTIR) alongside Raman spectroscopy (Mattsson et al. 2024). The advanced detection tools are especially beneficial for new contaminants including microplastics together with complex organic compounds (Mattsson et al. 2024). The fingerprint signatures produced by each material enable comparison against reference databases to facilitate identification procedures (Käppler et al. 2016; Kinigopoulou et al. 2022).

The chemical fingerprints methods undergo comparisons with source samples of known origin such as industrial effluents petroleum products and agricultural chemicals to determine their origins (Kadadou et al. 2024). PAHs from vehicle exhaust exhibit unique concentration patterns than those emitted during coal combustion systems, and lead isotopes reveal the exact manufacturing or mining places responsible for metal pollution (Grmasha et al. 2023). The precise connection between pollution sources and their origins becomes possible through chemical fingerprinting analysis alongside PCA statistical methods (Shi et al. 2024) which enables environmental investigators to support regulatory action and remediation work.

10.5.5 Statistical and Modelling Approaches

The process of locating pollution sources depends on statistical and modelling methodologies used to evaluate intricate environmental datasets for pattern recognition and relational analysis (A K and Mathew 2024). The pollution source distinction between natural and anthropogenic pollution sources becomes possible through multivariate statistical methods including principal component analysis (PCA) and factor analysis (FA) along with cluster analysis which group similar contaminants and determine main contamination sources (Busico et al. 2018). The measurement process in Positive Matrix Factorization (PMF) and Chemical Mass Balance (CMB) receptor models identifies different sources through a comparison between pollutant data and specific source profiles (Begum et al. 2007). Dispersion models AERMOD and CALPUFF use meteorological data together with topographical information to track pollution movement through air or water, thus determining emission points (Ridzuan et al. 2020). Large datasets can be analyzed effectively through machine learning algorithms which identify nonstandard patterns to advance source identification procedures. The integration of these analytical methods with chemical measurements from the field leads to an effective system for proper pollution source identification which guides development of effective mitigation approaches (Bayraktar and Mutlu 2024).

10.6 Case Studies of Pollution Source Identification

Various case analyses show that state-of-the-art tracking methods succeed in locating industrial pollution sources. As part of their investigations, researchers in Pearl River Delta China unified chemical fingerprinting methods (GC-MS and ICP-MS) with receptor modelling (PMF) to detect heavy metals in sediments originating from electronics production sites and electroplating plants (Cai et al. 2017). Hyperspectral remote sensing drone data revealed illegal wastewater releases in the German Ruhr Valley by analyzing spectral patterns that originated from textile industry chemicals and later received confirmation through PCA analysis (Karlsson et al. 2016). The Central Pollution Control Board of India accomplished unauthorized particulate matter emission detection at Delhi steel mills through real-time sensor networks coupled with GIS heat mapping technology (Tikader et al. 2024). This evidence demonstrates that implementing analytical chemistry with statistical modelling joined by spatial technology enables authorities to fight industry concealment which breaches environmental standards.

Many global urban centers have teamed up with modern investigative methods to detect and control local emission sources in their built environments. The city of Los Angeles used CALPUFF dispersion modelling together with VOC fingerprinting to separate ship emissions from port activities from refinery sources which allowed them to create specific fuel regulations (Molina et al. 2010). Beijing managed to locate ammonia plumes from agricultural activities through the satellite remote sensing system (TROPOMI) in combination with GIS-based hotspots that brought about optimized fertilizer application schedules (Pan et al. 2018). During peak congestion, Mexico City used data-driven techniques to detect unlawful industrial burning activities through artificial intelligence algorithms (Molina et al. 2010).

Different methods for detecting agricultural pollution origins demonstrate success using modern monitoring and predictive models. The Chesapeake Bay received its nutrient overdrafts through watershed modelling via SWAT and $\delta^{15}N/\delta^{18}O$ fingerprints which tracked the problem to particular poultry facilities and heavy corn fertilization zones (Longo et al. 2021). Funded by the European Union, Danish authorities employed fDOM sensors together with multivariate statistics for the Water Framework Directive application to identify between dairy farm runoff and municipal wastewater in particular catchment regions (Pan et al. 2018).

10.7 Challenges in Source Identification

10.7.1 Complex Pollution Pathways

A complex environmental setting makes detecting pollution sources challenging because different contamination paths intersect with dynamic transport patterns and altering pollutant signs (Percy et al. 2012). The ability to separate pollution sources remains impossible mainly because multiple sources such as agricultural runoff and industrial effluents and urban wastewater combine synergistically (Müller et al. 2020). A worse situation occurs from environmental changes because pollutants undergo degradation and chemical modifications during time spans, which modifies their initial source identity through processes of nutrient transformations and metal specification. The comprehension of contamination becomes intricate through dispersal mechanisms because pollutants can migrate extended distances from their points of origin such as PM2.5 particles flowing from wildfires across large territories and agricultural nitrates spreading across watersheds (Kieta 2023). Present-day pollution meets historical pollutants when past chemicals such as PCBs or heavy metals enter the mix with modern contaminants (Mishra et al. 2023). The technical problems here worsen due to operational barriers which involve inadequate monitoring outpost availability combined with high price points for compound-specific isotope testing and restricted abilities to model complex environmental systems (Meckenstock et al. 2004).

10.7.2 Limitations of Detection Methods

Pollution detection techniques encounter important barriers which obstruct precise source identification (Seesaard et al. 2024). The detection capabilities of standard analytical tools reach limitations because they cannot identify ultra-trace concentrations of contaminants or emerging pollutants which fall under detection thresholds (Li et al. 2025). Standard testing methodologies show poor competence in distinguishing between compounds of similar chemical composition regardless of their source origin specifically when pollutants produce overlapping signatures (Dévier et al. 2011). Companies using high-resolution mass spectrometry or isotopic analysis must overcome significant cost barriers and complex operational needs because their adoption remains limited especially across developing nations (Hernández et al. 2012). Field sensors deployed in the field represent valuable monitoring tools, but their accuracy remains inadequate to distinguish between pollution origins because their readings can be affected by environmental factors (Mutunga et al. 2024). Additionally, most detection systems provide only snapshot data, missing temporal variations in emissions that are crucial for accurate source attribution. Most of the pollution detection instruments give pictures of emissions while failing to track important changes in release patterns that help researchers determine pollution origins. The poor methodology applied to detect nonpoint source pollution becomes challenging because contaminants from different origins produce unpredictable interactions in environmental matrices. The environmental management process faces major impediments because contemporary detection methods and emerging nano-sensors coupled with AI-enhanced spectroscopy do not provide sufficient tracking ability to identify pollution sources (Sayyad and Attarde 2025).

10.7.3 Data Accuracy and Reliability Issues

Reliable environmental assessments remain vulnerable because significant measurement issues stand as barriers to proper identification of pollution sources (Li et al. 2025). Different sampling methods and preservation practices and analytic protocols used among studies cause measurement results to become nonequivalent (Singh et al. 2023). Measurement instruments create additional uncertainty because sensor drifts along with calibration problems and detection threshold fluctuations distort results mainly when measuring low contaminant levels (Singh et al. 2023). The absence of data occurs frequently because of inconsistent monitoring routines and inadequate sensor distribution preventing proper understanding of pollution patterns and making it harder to identify pollution sources (Liu et al. 2016). Standard reference materials for examining emerging contaminants are insufficient for verifying analytical outcomes in different laboratory settings. The accuracy problems become more complex when multiple data collection methods need to be integrated by uniting satellite remotely sensed data with ground-based measurement systems since resolution mismatch between sources causes interpretational issues. These reliability issues can produce wrong pollution source conclusions because of insufficient quality assurance/quality control (QA/QC) protocols and standardized data validation frameworks (Sturtevant et al. 2021).

10.8 Mitigation Strategies Based on Source Identification

10.8.1 Regulatory Measures

Effective pollution mitigation requires strict regulatory standards which are developed from accurate identification of pollution origins. The implementation of targeted emission standards by policymakers becomes possible when monitoring and analytical techniques identify specific contamination sources (Khanam et al. 2023). Regulatory areas that use geographic information systems-refined pollution maps can establish their boundaries from locations detected by remote sensing combined with ground-based data collection (Khanam et al. 2023).

10.8.2 Pollution Prevention Techniques

The most sustainable method to reduce pollution involves stopping pollutants right at their origin point which becomes possible through accurate monitoring and enhanced analytical techniques. Major emitting industries should use closed-loop systems to stop wastewater discharges while applying green chemistry standards for designing more secure production methods (Sharma et al. 2024). Agricultural operations improve fertilizer and pesticide efficiency through soil contaminant analysis and nutrient runoff mapping which stops excess chemical substances from polluting water systems (Ravikumar et al. 2024). The process of chemical substance fingerprinting directs material substitution programs to replace dangerous inputs such as lead additives and PFAS with safer alternatives (Roy et al. 2022). Predictive maintenance systems in manufacturing utilize real-time sensor data for manufacturing equipment leak prevention before leaks happen (Roy et al. 2022). Unclear source identification enables prevention measures along with lowering cleanup expenses and promoting circular economic methods of contamination management (Ayvaz and Alpay 2021).

10.8.3 Public Awareness and Policy Interventions

The exact source identification (Feng et al. 2025) is not only determinant of pollution control but also contingent upon effective use of public engagement and evidence-based policy interventions. However, when governments assign specific monitoring data to pinpoint the identity of specific polluting sources that are industrial emissions, vehicular exhaust, and agricultural runoff, they can then launch very targeted public awareness campaigns driven by the knowledge that communities in particular regions have an elevated risk to these types of pollution along with any associated strategies for mitigation such as air quality alerts tied to industrial emissions and awareness of water conservation measures like pesticides application seasons (Khanam et al. 2023; Odubo and Kosoe 2024).

Low-cost sensors in combination with citizen science programs allow residents to collect hyperlocal pollution data in order to increase accountability of identified polluters while the programs also create a platform for community-led advocacy (Anderson and Bateman 2000). Source apportionment studies can be made use of by policymakers to design context-specific regulation, for instance, stricter fuel standard for cities with high transport related PM2.5, as well as zoning laws prohibiting operations of livestock near sensitive watershed (Haakman et al. 2020). Subsidies for clean technologies in high emitting activity sectors or penalties for emissions noncompliance from real-time tracking of such can be used as an economic driver for behavior change. Such publicly accessible contaminant release inventories, which are systems of transparent pollution disclosure, ensure that industries are kept accountable, as are health impact assessments linked to particular pollution sources (Johnston Edwards and Walker 2020). These strategies for long-term environment governance are inclusive and inclusive policymaking, community mobilization combined with scientific source identification SDG 6 (Clean Water and Sanitation), SDG 11 (Sustainable Cities and Communities), SDG 13 (Climate Action), SDG 15 (Life on Land), and SDG 16 (Peace, Justice and Strong Institutions).

10.9 Future Directions and Research Opportunities

The expedition toward pollution source identification in the face of increasing scale and complexity of environmental challenges will be enabled by technology for technologies, interdisciplinary integration, and global collaboration (Awewomom et al. 2024).

10.9.1 AI-Driven Source Attribution

Machine learning and artificial intelligence will play an increasingly vital role in analyzing large, complex datasets from satellites, sensors, and laboratory results. These tools can enhance predictive modelling, identify previously undetected pollution patterns, and enable near-real-time source tracking, especially in dynamic urban or industrial settings (Blasch et al. 2021).

10.9.2 Integration of Big Data and IoT

The expansion of Internet of Things (IoT) devices for environmental sensing combined with cloud-based platforms and big data analytics will allow for continuous, hyperlocal monitoring of pollution. This real-time data stream can improve responsiveness to pollution events and refine long-term trend analysis (Swaminathan et al. 2022).

10.9.3 Advancements in Sensor Technologies

Future innovations in nano-sensors and biosensors will enhance detection sensitivity and specificity, particularly for emerging contaminants such as pharmaceuticals, microplastics, and PFAS. Portable, low-cost sensors could democratize environmental monitoring and expand coverage in under-resourced regions (Manayil Parambil et al. 2025).

10.9.4 High-Resolution Remote Sensing and 4D Modelling

With the evolution of hyperspectral satellites and drone-mounted sensors, future remote sensing will offer unprecedented spatial, spectral, and temporal resolution. Coupling this data with 4D environmental models (spatial + temporal) will enhance our ability to simulate pollution dispersion and transformation across ecosystems (Bolch et al. 2021).

10.9.5 Holistic Environmental Forensics

Integrating chemical fingerprinting with genomic, isotopic, and metabolomic approaches can create multilayered pollution profiles, enabling more precise discrimination among mixed or degraded sources. These methods will be especially valuable in tracing complex industrial mixtures and legacy pollutants (González-Gaya et al. 2021).

10.9.6 Climate-Linked Pollution Dynamics

Future research must explore how climate change variables such as rising temperatures, extreme weather, and changing hydrological cycles affect pollution transport, transformation, and detection. Understanding these interactions will be critical for adapting source identification strategies to a changing environment (Kundzewicz 2008).

10.9.7 Global Harmonization of Data and Standards

Developing international frameworks for standardized monitoring protocols, pollutant classification systems, and open-access data sharing will improve cross-border pollution tracking and foster collaborative research, especially for transboundary air and water pollution issues (Sarfraz et al. 2024).

10.9.8 Community-Based Monitoring and Citizen Science

Expanding participatory monitoring initiatives using mobile sensors and user-friendly platforms will enhance data richness and public engagement. Future efforts may include integrating crowdsourced data into regulatory frameworks and early warning systems (Kangana et al. 2024).

10.9.9 Policy-Driven Research Prioritization

Environmental regulations must be revised, and future directions should be viewed from the lenses of society such as enhanced environmental justice, reducing exposure disparities, and enhancing the resilience of vulnerable ecosystems. Research will align more closely with policy impact metrics and socioeconomic outcomes (Hendricks and Van Zandt 2021).

10.10 Conclusion

Environmental protection along with sustainable development relies heavily on the establishment of pollution source identification as its fundamental base. The exact location identification of contaminants from different sources such as factories or farms or cities becomes essential because it enables effective protection measures through enforcement regulations and health safeguards. Advanced analytical techniques consists of chemical fingerprinting, isotope tracing, remote sensing, and statistical modelling allow us to resolve complex pollution routes regardless of shifting environmental conditions and source. The combination of multidisciplinary methods now enables researchers to precisely determine pollution sources between air, water, and soil even though they face detection issues and data reliability problems as well as complexities from diffuse and leftover pollutants. Moreover, analytical strategies for source identification are not just scientific tools—they are instruments of justice and sustainability. When applied effectively, they help restore environmental balance, protect public health, and empower communities. These outcomes are closely aligned with the Sustainable Development Goals, especially SDG 6 (Clean Water and Sanitation), SDG 11 (Sustainable Cities and Communities), SDG 13 (Climate Action), SDG 15 (Life on Land), and SDG 16 (Peace, Justice and Strong Institutions), reinforcing the global commitment to a cleaner, fairer, and more resilient world

The process of identifying pollution sources serves to construct scientific policy decisions while facilitating revolutionary preventive innovations and exposing data to communities and thus both protects the environment from damage and establishes economic stability alongside fair environmental treatment. Source-specific strategies create real environmental improvements through advanced technical systems and effective governance models which integrate public involvement according to the analyzed worldwide cases. Moving into the future requires funds for developing next-generation monitoring systems and standardized data methods and enhanced cross-domain relationships to break current obstacles and expand effective solutions. Through pollution source identification, we gain access to developing purposeful and equitable ways for resolving our world's most demanding ecological issues.

References

A K N, Mathew A (2024) Sustainable data-driven insights: statistical analysis and artificial intelligence-driven modelling of aerosol concentrations in Hyderabad district, India. Environ Adv 16:100539. https://doi.org/10.1016/j.envadv.2024.100539

Adjovu GE, Stephen H, James D, Ahmad S (2023) Overview of the application of remote sensing in effective monitoring of water quality parameters. Remote Sens 15(7):1938. https://doi.org/10.3390/rs15071938

Adla K, Dejan K, Neira D, Dragana Š (2022) Chapter 9 - Degradation of ecosystems and loss of ecosystem services. One Health: Integrated Approach to 21st Century Challenges to Health (pp. 281–327). Academic Press.

Akhtar N, Ishak MIS, Bhawani SA, Umar K (2021) Various natural and anthropogenic factors responsible for water quality degradation: a review. Water 13(19):2660. https://doi.org/10.3390/w13192660

Almeida CRD, Teodoro AC, Gonçalves A (2021) Study of the urban Heat Island (UHI) using remote sensing data/techniques: a systematic review. Environments 8(10):105. https://doi.org/10.3390/environments8100105

Anderson LM, Bateman TS (2000) Individual environmental initiative: championing natural environmental issues in U.S. business organizations. Acad Manag J 43(4):548–570. https://doi.org/10.2307/1556355

Arsene C, Vione D, Grinberg N, Olariu RI (2011) GC × GC-MS hyphenated techniques for the analysis of volatile organic compounds in air. J Liq Chromatogr Relat Technol 34(13):1077–1111. https://doi.org/10.1080/10826076.2011.587749

Artiola JF, Brusseau ML (2019) The role of environmental monitoring in pollution science. In: Environmental and pollution science. Elsevier, pp 149–162. https://doi.org/10.1016/B978-0-12-814719-1.00010-0

Awewomom J, Dzeble F, Takyi YD, Ashie WB, Ettey ENYO, Afua PE, Sackey LNA, Opoku F, Akoto O (2024) Addressing global environmental pollution using environmental control techniques: a focus on environmental policy and preventive environmental management. Discov Environ 2(1):8. https://doi.org/10.1007/s44274-024-00033-5

Ayvaz S, Alpay K (2021) Predictive maintenance system for production lines in manufacturing: a machine learning approach using IoT data in real-time. Expert Syst Appl 173:114598. https://doi.org/10.1016/j.eswa.2021.114598

Ballesté E, Demeter K, Masterson B, Timoneda N, Sala-Comorera L, Meijer WG (2020) Implementation and integration of microbial source tracking in a river watershed monitoring plan. Sci Total Env 736:139573. https://doi.org/10.1016/j.scitotenv.2020.139573

Baragaño D, Ratié G, Sierra C, Chrastný V, Komárek M, Gallego JR (2022) Multiple pollution sources unravelled by environmental forensics techniques and multivariate statistics. J Hazard Mater 424:127413. https://doi.org/10.1016/j.jhazmat.2021.127413

Basnyat P, Teeter LD, Lockaby BG, Flynn KM (2000) The use of remote sensing and GIS in watershed level analyses of non-point source pollution problems. For Ecol Manag 128(1–2):65–73. https://doi.org/10.1016/S0378-1127(99)00273-X

Bayraktar OM, Mutlu A (2024) Analyses of industrial air pollution and long-term health risk using different dispersion models and WRF physics parameters. Air Qual Atmos Health 17(10):2277–2305. https://doi.org/10.1007/s11869-024-01573-8

Beauchemin D (2006) Inductively coupled plasma mass spectrometry. Anal Chem 78(12):4111–4136. https://doi.org/10.1021/ac060712t

Begum BA, Biswas SK, Hopke PK (2007) Source apportionment of air particulate matter by chemical mass balance (CMB) and comparison with positive matrix factorization (PMF) model. Aerosol Air Qual Res 7(4):446–468. https://doi.org/10.4209/aaqr.2006.10.0021

Blasch E, Pham T, Chong C-Y, Koch W, Leung H, Braines D, Abdelzaher T (2021) Machine learning/artificial intelligence for sensor data fusion–opportunities and challenges. IEEE Aerosp Electron Syst Mag 36(7):80–93. https://doi.org/10.1109/MAES.2020.3049030

Bolch EA, Hestir EL, Khanna S (2021) Performance and feasibility of drone-mounted imaging spectroscopy for invasive aquatic vegetation detection. Remote Sens 13(4):582. https://doi.org/10.3390/rs13040582

Busico G, Cuoco E, Kazakis N, Colombani N, Mastrocicco M, Tedesco D, Voudouris K (2018) Multivariate statistical analysis to characterize/discriminate between anthropogenic and geogenic trace elements occurrence in the Campania Plain, Southern Italy. Environ Pollut 234:260–269. https://doi.org/10.1016/j.envpol.2017.11.053

Cai M, Lin Y, Chen M, Yang W, Du H, Xu Y, Cheng S, Xu F, Hong J, Chen M, Ke H (2017) Improved source apportionment of PAHs and Pb by integrating Pb stable isotopes and positive matrix factorization application (PAHs): a historical record case study from the northern South China Sea. Sci Total Environ 609:577–586. https://doi.org/10.1016/j.scitotenv.2017.07.190

Chou LM, Huang D, Tan KS, Toh TC, Goh BPL, Tun K (2019) Singapore. In: World seas: an environmental evaluation. Elsevier, pp 539–558. https://doi.org/10.1016/B978-0-08-100853-9.00031-2

Curt MD, Aguado P, Sánchez G, Bigeriego M, Fernández J (2004) Nitrogen isotope ratios of synthetic and organic sources of nitrate water contamination in Spain. Water Air Soil Pollut 151(1–4):135–142. https://doi.org/10.1023/B:WATE.0000009889.36833.c0

Dehkordi MM, Nodeh ZP, Dehkordi KS, Salmanvandi H, Khorjestan RR, Ghaffarzadeh M (2024) Soil, air, and water pollution from mining and industrial activities: Sources of pollution, environmental impacts, and prevention and control methods. Results Eng 23:102729. https://doi.org/10.1016/j.rineng.2024.102729

DeVantier BA, Feldman AD (1993) Review of GIS applications in hydrologic modeling. J Water Resour Plan Manag 119(2):246–261. https://doi.org/10.1061/(ASCE)0733-9496(1993)119:2(246)

Dévier M-H, Mazellier P, Aït-Aïssa S, Budzinski H (2011) New challenges in environmental analytical chemistry: identification of toxic compounds in complex mixtures. C R Chim 14(7–8):766–779. https://doi.org/10.1016/j.crci.2011.04.006

Dhankar S, Mishra AK, Kumar K (2024) Satellite derived air pollution climatology over India and its neighboring regions: spatio-temporal trends and insights. Phys Chem Earth Parts A/B/C 136:103769. https://doi.org/10.1016/j.pce.2024.103769

Elmqvist T, Setälä H, Handel S, Van Der Ploeg S, Aronson J, Blignaut J, Gómez-Baggethun E, Nowak D, Kronenberg J, De Groot R (2015) Benefits of restoring ecosystem services in urban areas. Curr Opin Environ Sustain 14:101–108. https://doi.org/10.1016/j.cosust.2015.05.001

Ettler V, Mihaljevic M, Komarek M (2004) ICP-MS measurements of lead isotopic ratios in soils heavily contaminated by lead smelting: tracing the sources of pollution. Anal Bioanal Chem 378(2):311–317. https://doi.org/10.1007/s00216-003-2229-y

Feng T, Shi Y, Wang X, Wan X, Mi Z (2025) Synergies of air pollution control policies: a review. J Environ Manag 377:124655. https://doi.org/10.1016/j.jenvman.2025.124655

Ferrier RC, D'Arcy BJ, MacDonald J, Aitken M (2005) Diffuse pollution – what is the nature of the problem? Water Environ J 19(4):361–366. https://doi.org/10.1111/j.1747-6593.2005.tb00574.x

Filonchyk M, Peterson MP (2024) Analysis of air pollution from vehicle emissions for the contiguous United States. J Geovis Spat Anal 8(1):16. https://doi.org/10.1007/s41651-024-00180-6

Fraternali P, Morandini L, Herrera González SL (2024) Solid waste detection, monitoring and mapping in remote sensing images: a survey. Waste Manag 189:88–102. https://doi.org/10.1016/j.wasman.2024.08.003

González-Gaya B, Lopez-Herguedas N, Bilbao D, Mijangos L, Iker AM, Etxebarria N, Irazola M, Prieto A, Olivares M, Zuloaga O (2021) Suspect and non-target screening: the last frontier in environmental analysis. Anal Methods 13(16):1876–1904. https://doi.org/10.1039/D1AY00111F

Grmasha RA, Abdulameer MH, Stenger-Kovács C, Al-sareji OJ, Al-Gazali Z, Al-Juboori RA, Meiczinger M, Hashim KS (2023) Polycyclic aromatic hydrocarbons in the surface water and sediment along Euphrates River system: occurrence, sources, ecological and health risk assessment. Mar Pollut Bull 187:114568. https://doi.org/10.1016/j.marpolbul.2022.114568

Haakman R, Beenakker I, Geerlings H (2020) Reducing vehicle-related NOx and PM emissions in metropolitan areas: a comparison between the Randstad and the Rhine-Ruhr area. J Clean Prod 247:119175. https://doi.org/10.1016/j.jclepro.2019.119175

Hafeez S, Sing Wong M, Abbas S, Kwok CYT, Nichol J, Ho Lee K, Tang D, Pun L (2019) Detection and monitoring of marine pollution using remote sensing technologies. In: Bachari Fouzia H (ed) Monitoring of marine pollution. IntechOpen. https://doi.org/10.5772/intechopen.81657

Hendricks MD, Van Zandt S (2021) Unequal protection revisited: planning for environmental justice, hazard vulnerability, and critical infrastructure in communities of color. Environ Justice 14(2):87–97. https://doi.org/10.1089/env.2020.0054

Hernández F, Sancho JV, Ibáñez M, Abad E, Portolés T, Mattioli L (2012) Current use of high-resolution mass spectrometry in the environmental sciences. Anal Bioanal Chem 403(5):1251–1264. https://doi.org/10.1007/s00216-012-5844-7

Hou D, O'Connor D, Nathanail P, Tian L, Ma Y (2017) Integrated GIS and multivariate statistical analysis for regional scale assessment of heavy metal soil contamination: a critical review. Environ Pollut 231:1188–1200. https://doi.org/10.1016/j.envpol.2017.07.021

Ifegwu OC, Anyakora C (2015) Polycyclic aromatic hydrocarbons. In: Advances in clinical chemistry, vol 72. Elsevier, pp 277–304. https://doi.org/10.1016/bs.acc.2015.08.001

Johnston Edwards S, Walker TR (2020) An overview of Canada's National Pollutant Release Inventory program as a pollution control policy tool. J Environ Plan Manag 63(6):1097–1113. https://doi.org/10.1080/09640568.2019.1634525

Jońca J, Pawnuk M, Bezyk Y, Arsen A, Sówka I (2022) Drone-assisted monitoring of atmospheric pollution—A comprehensive review. Sustainability 14(18):11516. https://doi.org/10.3390/su141811516

Kadadou D, Tizani L, Alsafar H, Hasan SW (2024) Analytical methods for determining environmental contaminants of concern in water and wastewater. MethodsX 12:102582. https://doi.org/10.1016/j.mex.2024.102582

Kamber BS (2009) Geochemical fingerprinting: 40 years of analytical development and real world applications. Appl Geochem 24(6):1074–1086. https://doi.org/10.1016/j.apgeochem.2009.02.012

Kangana N, Kankanamge N, De Silva C, Goonetilleke A, Mahamood R, Ranasinghe D (2024) Bridging community engagement and technological innovation for creating smart and resilient cities: a systematic literature review. Smart Cities 7(6):3823–3852. https://doi.org/10.3390/smartcities7060147

Kaplan G, Yalcinkaya F, Altıok E, Pietrelli A, Nastro RA, Lovecchio N, Ieropoulos IA, Tsipa A (2024) The role of remote sensing in the evolution of water pollution detection and monitoring: a comprehensive review. Phys Chem Earth Parts A/B/C 136:103712. https://doi.org/10.1016/j.pce.2024.103712

Käppler A, Fischer D, Oberbeckmann S, Schernewski G, Labrenz M, Eichhorn K-J, Voit B (2016) Analysis of environmental microplastics by vibrational microspectroscopy: FTIR, Raman or both? Anal Bioanal Chem 408(29):8377–8391. https://doi.org/10.1007/s00216-016-9956-3
Karlsson TM, Grahn H, Van Bavel B, Geladi P (2016) Hyperspectral imaging and data analysis for detecting and determining plastic contamination in seawater filtrates. J Near Infrared Spectrosc 24(2):141–149. https://doi.org/10.1255/jnirs.1212
Khanam Z, Sultana FM, Mushtaq F (2023) Environmental pollution control measures and strategies: an overview of recent developments. In: Mushtaq F, Farooq M, Mukherjee AB, Ghosh Nee Lala M (eds) Geospatial analytics for environmental pollution modeling. Springer Nature Switzerland, pp 385–414. https://doi.org/10.1007/978-3-031-45300-7_15
Kiani AF, Farwa U, Malik MR, Akhtar N (2024) Chromatography techniques and their applications in environmental analysis: a study of pollutant detection and monitoring in air, water, and soil. Indus J Biosci Res 2(2):938–950. https://doi.org/10.70749/ijbr.v2i02.322
Kieta K (2023) Determining contamination and sources of sediment in response to recent and historical landscape disturbances [Doctor of Philosophy, University of Northern British Columbia]. https://doi.org/10.24124/2023/59430
Kinigopoulou V, Pashalidis I, Kalderis D, Anastopoulos I (2022) Microplastics as carriers of inorganic and organic contaminants in the environment: a review of recent progress. J Mol Liq 350:118580. https://doi.org/10.1016/j.molliq.2022.118580
Kolobaric A, Cozzolino D, Chapman J (2023) Environmental forensics. In: Sharma V, Sauzier G, Lewis SW (eds) Chemometric methods in forensic science. Royal Society of Chemistry, pp 254–293. https://doi.org/10.1039/BK9781839166099-00254
Korchenko O (2019) GIS and remote sensing as important tools for assessment of environmental pollution. In: 19th SGEM international multidisciplinary scientific GeoConference EXPO proceedings. https://doi.org/10.5593/sgem2019/2.1/S07.039
Kundzewicz ZW (2008) Climate change impacts on the hydrological cycle. Ecohydrol Hydrobiol 8(2–4):195–203. https://doi.org/10.2478/v10104-009-0015-y
Kuppan N, Padman M, Mahadeva M, Srinivasan S, Devarajan R (2024) A comprehensive review of sustainable bioremediation techniques: eco friendly solutions for waste and pollution management. Waste Manag Bull 2(3):154–171. https://doi.org/10.1016/j.wmb.2024.07.005
Li Q-Q, Guo Y-T, Yang J-Y, Liang C-S (2023) Review on main sources and impacts of urban ultrafine particles: traffic emissions, nucleation, and climate modulation. Atmos Environ: X 19:100221. https://doi.org/10.1016/j.aeaoa.2023.100221
Li Y, Chen B, Yang S, Jiao Z, Zhang M, Yang Y, Gao Y (2025) Advances in environmental pollutant detection techniques: enhancing public health monitoring and risk assessment. Environ Int 197:109365. https://doi.org/10.1016/j.envint.2025.109365
Liu J, Li J, Li W, Wu J (2016) Rethinking big data: a review on the data quality and usage issues. ISPRS J Photogramm Remote Sens 115:134–142. https://doi.org/10.1016/j.isprsjprs.2015.11.006
Liu Y, Fei Y, Li Y, Bao X, Zhang P, Institute of Hydrogeology and Environmental Geology, & 268 North Zhonghua Street, Shijiazhuang, Hebei 050061, China (2022) A review of the pollution source identification methods and remediation technologies of groundwater. China Geology. https://doi.org/10.31035/cg2022080
Liu X, Tong X, Wu L, Mohapatra S, Xue H, Liu R (2023) An integrated modelling framework for multiple pollution source identification in surface water. J Environ Manage 347:119126. https://doi.org/10.1016/j.jenvman.2023.119126
Liu X, Sathishkumar K, Zhang H, Saxena KK, Zhang F, Naraginti S, Anbarasu K, Rajendiran R, Rajasekar A, Guo X (2024) Frontiers in environmental cleanup: recent advances in remediation of emerging pollutants from soil and water. J Hazard Mater Adv 16:100461. https://doi.org/10.1016/j.hazadv.2024.100461
Longo SB, Isgren E, Clark B (2021) Nutrient overloading in the Chesapeake Bay. Sociol Dev 7(4):416–440. https://doi.org/10.1525/sod.2020.0032

Manayil Parambil A, Priyadarshini E, Paul S, Bakandritsos A, Sharma VK, Zbořil R (2025) Emerging nanomaterials for the detection of per- and poly-fluorinated substances. J Mater Chem A 13(12):8246–8281. https://doi.org/10.1039/D4TA07045C

Martin RV (2008) Satellite remote sensing of surface air quality. Atmos Environ 42(34):7823–7843. https://doi.org/10.1016/j.atmosenv.2008.07.018

Mattsson K, Jocic S, De Lima JA, Hansson L-A, Gondikas A (2024) Nanoplastics in aquatic environments—sources, sampling techniques, and identification methods. In: Microplastic contamination in aquatic environments. Elsevier, pp 381–397. https://doi.org/10.1016/B978-0-443-15332-7.00003-X

Meckenstock RU, Haderlein SB, Schmidt TC, Zwank L, Elsner M, Berg M (2004) Compound-specific stable isotope analysis of organic contaminants in natural environments: a critical review of the state of the art, prospects, and future challenges. Anal Bioanal Chem 378(2):283–300. https://doi.org/10.1007/s00216-003-2350-y

Meyer AM, Klein C, Fünfrocken E, Kautenburger R, Beck HP (2019) Real-time monitoring of water quality to identify pollution pathways in small and middle scale rivers. Sci Total Environ 651:2323–2333. https://doi.org/10.1016/j.scitotenv.2018.10.069

Mishra RK, Mentha SS, Misra Y, Dwivedi N (2023) Emerging pollutants of severe environmental concern in water and wastewater: a comprehensive review on current developments and future research. Water-Energy Nexus 6:74–95. https://doi.org/10.1016/j.wen.2023.08.002

Molina LT, Madronich S, Gaffney JS, Apel E, De Foy B, Fast J, Ferrare R, Herndon S, Jimenez JL, Lamb B, Osornio-Vargas AR, Russell P, Schauer JJ, Stevens PS, Volkamer R, Zavala M (2010) An overview of the MILAGRO 2006 campaign: Mexico City emissions and their transport and transformation. Atmos Chem Phys 10(18):8697–8760. https://doi.org/10.5194/acp-10-8697-2010

Müller A, Österlund H, Marsalek J, Viklander M (2020) The pollution conveyed by urban runoff: a review of sources. Sci Total Environ 709:136125. https://doi.org/10.1016/j.scitotenv.2019.136125

Mushtaq F, Farooq M, Mukherjee AB, Ghosh Nee Lala M (eds) (2023) Geospatial analytics for environmental pollution modeling: analysis, control and management. Springer Nature Switzerland. https://doi.org/10.1007/978-3-031-45300-7

Mutunga T, Sinanovic S, Harrison CS (2024) Integrating wireless remote sensing and sensors for monitoring pesticide pollution in surface and groundwater. Sensors 24(10):3191. https://doi.org/10.3390/s24103191

Naimaee R, Kiani A, Jarahizadeh S, Haji Seyed Asadollah SB, Melgarejo P, Jodar-Abellan A (2024) Long-term water quality monitoring: using satellite images for temporal and spatial monitoring of thermal pollution in water resources. Sustainability 16(2):646. https://doi.org/10.3390/su16020646

Nakhjiri A, Kakroodi AA (2024) Air pollution in industrial clusters: A comprehensive analysis and prediction using multi-source data. Ecol Inform 80:102504. https://doi.org/10.1016/j.ecoinf.2024.102504

Négrel P, Blessing M, Millot R, Petelet-Giraud E, Innocent C (2012) Isotopic methods give clues about the origins of trace metals and organic pollutants in the environment. TrAC Trends Analyt Chem 38: 143–153. https://doi.org/10.1016/j.trac.2012.03.017

Nonpoint Source Water Pollution (2003) Environmental engineering. Elsevier, pp 233–249. https://doi.org/10.1016/B978-075067294-8/50011-7

Novotny V (2003) Water quality: diffuse pollution and watershed management, 2nd edn. Wiley

Nsenga Kumwimba M, Zhu B, Stefanakis AI, Ajibade FO, Dzakpasu M, Soana E, Wang T, Arif M, Kavidia Muyembe D, Agboola TD (2023) Advances in ecotechnological methods for diffuse nutrient pollution control: wicked issues in agricultural and urban watersheds. Front Environ Sci 11:1199923. https://doi.org/10.3389/fenvs.2023.1199923

Nuruzzaman M, Bahar MM, Naidu R (2025) Diffuse soil pollution from agriculture: impacts and remediation. Sci Total Environ 962:178398. https://doi.org/10.1016/j.scitotenv.2025.178398

Odubo TC, Kosoe EA (2024) Sources of air pollutants: impacts and solutions. In: Izah SC, Ogwu MC, Shahsavani A (eds) Air pollutants in the context of one health, vol 134. Springer Nature Switzerland, pp 75–121. https://doi.org/10.1007/698_2024_1127

Olawade DB, Wada OZ, Ige AO, Egbewole BI, Olojo A, Oladapo BI (2024) Artificial intelligence in environmental monitoring: Advancements, challenges, and future directions. Hyg Environ Health Adv 12:100114. https://doi.org/10.1016/j.heha.2024.100114

Pan Y, Tian S, Zhao Y, Zhang L, Zhu X, Gao J, Huang W, Zhou Y, Song Y, Zhang Q, Wang Y (2018) Identifying ammonia hotspots in China using a National Observation Network. Environ Sci Technol 52(7):3926–3934. https://doi.org/10.1021/acs.est.7b05235

Percy KE, Maynard DG, Legge AH (2012) Applying the forest health approach to monitoring boreal ecosystems in the Athabasca Oil Sands region. In: Developments in environmental science, vol 11. Elsevier, pp 193–217. https://doi.org/10.1016/B978-0-08-097760-7.00009-3

Pohanish RP (2017) V. In: Sittig's handbook of toxic and hazardous chemicals and carcinogens. Elsevier, pp 3232–3272. https://doi.org/10.1016/B978-0-323-38968-6.00021-1

Ravikumar S, Vellingiri G, Sellaperumal P, Pandian K, Sivasankar A, Sangchul H (2024) Real-time nitrogen monitoring and management to augment N use efficiency and ecosystem sustainability–A review. J Hazard Mater Adv 16:100466. https://doi.org/10.1016/j.hazadv.2024.100466

Reddy CM, Quinn JG (1999) GC-MS analysis of total petroleum hydrocarbons and polycyclic aromatic hydrocarbons in seawater samples after the North Cape oil spill. Mar Pollut Bull 38(2):126–135. https://doi.org/10.1016/S0025-326X(98)00106-4

Reitze AW (with Environmental Law Institute) (2005) Stationary source air pollution law. Environmental Law Institute

Ridzuan N, Ujang U, Azri S, Choon TL (2020) Visualising urban air quality using AERMOD, CALPUFF and CFD models: a critical review. In: The international archives of the photogrammetry, remote sensing and spatial information sciences, vol XLIV-4/W3-2020, pp 355–363. https://doi.org/10.5194/isprs-archives-XLIV-4-W3-2020-355-2020

Roy MA, Cousins I, Harriman E, Scheringer M, Tickner JA, Wang Z (2022) Combined application of the essential-use and functional substitution concepts: accelerating safer alternatives. Environ Sci Technol 56(14):9842–9846. https://doi.org/10.1021/acs.est.2c03819

Rudel RA, Perovich LJ (1994) Endocrine disrupting chemicals in indoor and outdoor air. Atmos Environ 43(1):170–181. https://doi.org/10.1016/j.atmosenv.2008.09.025

Samad A, Garuda S, Vogt U, Yang B (2023) Air pollution prediction using machine learning techniques – an approach to replace existing monitoring stations with virtual monitoring stations. Atmos Environ 310:119987. https://doi.org/10.1016/j.atmosenv.2023.119987

Sankoh AA, Derkyi NSA, Frazer-williams RAD, Laar C, Kamara I (2021) A review on the application of isotopic techniques to trace groundwater pollution sources within developing countries. Water 14(1):35. https://doi.org/10.3390/w14010035

Saravanan A, Senthil Kumar P, Jeevanantham S, Karishma S, Tajsabreen B, Yaashikaa PR, Reshma B (2021) Effective water/wastewater treatment methodologies for toxic pollutants removal: processes and applications towards sustainable development. Chemosphere 280:130595. https://doi.org/10.1016/j.chemosphere.2021.130595

Sarfraz Z, Sarfraz A, Sarfraz M (2024) International partnerships and agreements for addressing air pollution. In: Ogwu MC, Izah SC (eds) Sustainable strategies for air pollution mitigation, vol 133. Springer Nature Switzerland, pp 513–548. https://doi.org/10.1007/698_2024_1109

Sayyad J, Attarde K (2025) Synergizing nanotechnology and artificial intelligence for society 5.0 advancement through intelligent systems. In: Mahajan S, Das S, Rocha A, Roy DB, Chawla P (eds) NanoMind: exploring synergies in nanotechnology and machine learning, vol 27. Springer Nature Switzerland, pp 225–249. https://doi.org/10.1007/978-3-031-77296-2_10

Seesaard T, Kamjornkittikoon K, Wongchoosuk C (2024) A comprehensive review on advancements in sensors for air pollution applications. Sci Total Environ 951:175696. https://doi.org/10.1016/j.scitotenv.2024.175696

Sharma K, Nagabhooshanam N, Dasi S, Srinivas B, Kaur K, Rajaram A (2024) Design and implementation of closed-loop water reuse systems in urban and industrial settings for maximizing resource recovery and minimizing waste. Desalin Water Treat 320:100850. https://doi.org/10.1016/j.dwt.2024.100850

Shi B, Meng J, Wang T, Li Q, Zhang Q, Su G (2024) The main strategies for soil pollution apportionment: a review of the numerical methods. J Environ Sci 136:95–109. https://doi.org/10.1016/j.jes.2022.09.027

Singh VP, Fiorentino M (1996) Hydrologic modeling with GIS. In: Singh VP, Fiorentino M (eds) Geographical information Systems in Hydrology, vol 26. Springer Netherlands, pp 1–13. https://doi.org/10.1007/978-94-015-8745-7_1

Singh BJ, Chakraborty A, Sehgal R (2023) A systematic review of industrial wastewater management: evaluating challenges and enablers. J Environ Manag 348:119230. https://doi.org/10.1016/j.jenvman.2023.119230

Srivastava PK, Han D, Gupta M, Mukherjee S (2012) Integrated framework for monitoring groundwater pollution using a geographical information system and multivariate analysis. Hydrol Sci J 57(7):1453–1472. https://doi.org/10.1080/02626667.2012.716156

Stout SA, Uhler AD (2003) Distinguishing "background" hydrocarbons from contamination using chemical fingerprinting. Environ Claims J 15(2):241–259. https://doi.org/10.1080/713609448

Sturtevant C, Metzger S, Nehr S, Foken T (2021) Quality assurance and control. In: Foken T (ed) Springer handbook of atmospheric measurements. Springer International Publishing, pp 49–92. https://doi.org/10.1007/978-3-030-52171-4_3

Su H, Li H, Chen H, Li Z, Zhang S (2023) Source identification and potential health risks of fluoride and nitrate in groundwater of a typical alluvial plain. Sci Total Environ 904:166920. https://doi.org/10.1016/j.scitotenv.2023.166920

Sun Y, Wang D, Li L, Ning R, Yu S, Gao N (2024) Application of remote sensing technology in water quality monitoring: from traditional approaches to artificial intelligence. Water Res 267:122546. https://doi.org/10.1016/j.watres.2024.122546

Swaminathan S, Sankar Guntuku AV, Gupta A, Rengaswamy R (2022) Data science and IoT based mobile monitoring framework for hyper-local PM2.5 assessment in urban setting. Build Environ 225:109597. https://doi.org/10.1016/j.buildenv.2022.109597

TavallaieNejad A, Vila MC, Paneiro G, Baptista JS (2025) A systematic review of machine learning algorithms for soil pollutant detection using satellite imagery. Remote Sens 17(7):1207. https://doi.org/10.3390/rs17071207

Tikader M, Mukhopadhyay D, Dabhadker K (2024) Remote sensing and GIS in air pollution mitigation: a bibliometric review of Chhattisgarh, India. Int J Environ Clim Change 14(4):174–193. https://doi.org/10.9734/ijecc/2024/v14i44106

Tshanga MM, Ncube L, Van Niekerk E (2024) Remote sensing insights into subsurface-surface relationships: land cover analysis and copper deposits exploration. Earth Sci Inf 17(5):3979–4000. https://doi.org/10.1007/s12145-024-01423-2

Vasco C, Torres B, Jácome E, Torres A, Eche D, Velasco C (2021) Use of chemical fertilizers and pesticides in frontier areas: A case study in the Northern Ecuadorian Amazon. Land Use Policy 107:105490. https://doi.org/10.1016/j.landusepol.2021.105490

Vaye O, Ngumbu RS, Xia D (2022) A review of the application of comprehensive two-dimensional gas chromatography MS-based techniques for the analysis of persistent organic pollutants and ultra-trace level of organic pollutants in environmental samples. Rev Anal Chem 41(1):63–73. https://doi.org/10.1515/revac-2022-0034

Wang X (with Ottinger RL) (2023) A tale of two cities: a comparison of air pollution governance in the Los Angeles Area of the USA and the Beijing-Tianjin-Hebei Area of China, 1st edn. Wolters Kluwer Law International

Wang M, Chen L, Wu L, Zhang L, Xie H, Shen Z (2022) Review of nonpoint source pollution models: current status and future direction. Water 14(20):3217. https://doi.org/10.3390/w14203217

Wang F, Xiang L, Sze-Yin Leung K, Elsner M, Zhang Y, Guo Y, Pan B, Sun H, An T, Ying G, Brooks BW, Hou D, Helbling DE, Sun J, Qiu H, Vogel TM, Zhang W, Gao Y, Simpson MJ et al (2024) Emerging contaminants: a one health perspective. Innovation 5(4):100612. https://doi.org/10.1016/j.xinn.2024.100612

Wilhelm FM (2009) Pollution of aquatic ecosystems I. In: Encyclopedia of inland waters. Elsevier, pp 110–119. https://doi.org/10.1016/B978-012370626-3.00222-2

Wilson SM, Richard R, Joseph L, Williams E (2010) Climate change, environmental justice, and vulnerability: an exploratory spatial analysis. Environ Justice 3(1):13–19. https://doi.org/10.1089/env.2009.0035

Winkler SL, Anderson JE, Garza L, Ruona WC, Vogt R, Wallington TJ (2018) Vehicle criteria pollutant (PM, NOx, CO, HCs) emissions: how low should we go? Npj Clim Atmos Sci 1(1):26. https://doi.org/10.1038/s41612-018-0037-5

Xu Y, Li P, Zhang M, Xiao L, Wang B, Zhang X, Wang Y, Shi P (2024) Quantifying seasonal variations in pollution sources with machine learning-enhanced positive matrix factorization. Ecol Indic 166: 112543. https://doi.org/10.1016/j.ecolind.2024.112543

Xu J, Zhang Z, Rao L, Wang Y, Letu H, Shi C, Tana G, Wang W, Zhu S, Liu S, Shi E, Wang Y, Chen L, Dong X, Shi J (2024) Remote sensing of tropospheric ozone from space: progress and challenges. J Remote Sens 4:0178. https://doi.org/10.34133/remotesensing.0178

Yu H, Liu X, Kong B, Li R, Wang G (2019) Landscape ecology development supported by geospatial technologies: a review. Eco Inform 51:185–192. https://doi.org/10.1016/j.ecoinf.2019.03.006

Zhong C, Li L, Wang Y-Z (2024) Applications of chemical fingerprints and machine learning in plant ecology: recent progress and future perspectives. Microchem J 206:111447. https://doi.org/10.1016/j.microc.2024.111447

Microplastic Pollution in India's Marine Ecosystems: Exploring the Need and Utilization of Environmental Forensics

11

Sneha Madhavan and U. Sooraj Krishna

Abstract

Microplastic pollution has emerged as a growing global concern, posing severe risks to all ecosystems, particularly marine environments. The accumulation of microplastics in marine flora and fauna leads to physiological disruptions, threatening biodiversity and the stability of the global food chain. Effective mitigation of this issue requires stringent legal interventions, which can be reinforced through environmental forensics. By utilizing advanced forensic techniques, the sources of microplastic pollution can be accurately traced, enabling the enforcement of targeted legal measures to curb marine contamination. This chapter addresses the issue of microplastic pollution in India's marine ecosystems and explores the efficiency of environmental forensics techniques to combat it. Through the combination of scientific research, technological advancements, and legal enforcement, environmental forensics can play a crucial role in preserving marine ecosystems, reducing contamination levels, and promoting sustainable coastal development. Ultimately, adopting such an approach will enhance marine biodiversity, safeguard human health, and ensure the long-term resilience of aquatic environments.

Keywords

Microplastic pollution · Marine ecosystems · Environmental forensics

S. Madhavan (✉)
Department of Forensic Science, University of Calicut, Kerala Police Academy, Thrissur, Kerala, India

U. Sooraj Krishna
Department of Forensic Science, M. E. S. Kalladi College, Mannarkkad, Palakkad, Kerala, India

M. S. Sankhla et al. (eds.), *Introducing the Synergy Between Forensic Science and Environmental Analysis*, https://doi.org/10.1007/978-981-95-0199-1_11

11.1 Introduction

Plastics are man-made, organic polymers originating from the process of polymerization. These materials are produced in a large quantity due to its high convenience of use and demand and also have created the problem of pollution through its debris (Derraik 2002). The plastic materials are either scattered or improperly disposed in landfills or open, unregulated sites where containment measures are lacking. This form of mismanaged waste has the potential to find its way into the ocean through inland waterways, wastewater discharges, or through wind or tides (Jambeck et al. 2015). Plastic debris is classified based on size into three main categories: macroplastics (greater than 25 mm), mesoplastics (ranging from 5 to 25 mm), and microplastics (less than 5 mm). Microplastics are further classified as primary and secondary microplastics. Primary microplastics are produced for specific purposes in dimensions smaller than 5 mm, including items like microbeads and plastic pellets (used as raw materials in plastic manufacturing). Conversely, secondary microplastics, as their name implies, result from the fragmentation of larger plastic items (Kumar and Varghese 2021a). Microplastics are categorized according to their polymer composition, including common types such as polyethylene (PE), polypropylene (PP), and polyethylene terephthalate (PET) (Amrutha et al. 2023; Tiwari et al. 2019; Devi et al. 2020). These base materials used in plastic production, including small pellets and microplastics, have become increasingly prevalent as marine debris. Along with the larger plastics, these smaller plastic particles have the capacity to travel significant distances in aquatic environments (Barnes et al. 2009). As microplastics closely mimic natural prey, there is increased likelihood of interactions and ingestion by secondary consumers (organisms in a food chain or ecosystem that occupy the second trophic level) in the aquatic or marine environments, both unintentionally and deliberately (Jovanović 2017). This is adversely impacting on their natural processes too. The sources of plastic debris to the marine environments can be terrestrial activities (improper waste managements, factories, etc.) or aquatic activities (ship movements, fishing, etc.). But often the process of determining the sources of plastics or, specifically, microplastics is a daunting task. This problem can be tackled by environmental forensic techniques (Kumar and Varghese 2021a).

11.1.1 The Problem of Microplastic in Indian Coastal Environments

The coastal regions of many Indian cities are facing the devastating effect of microplastic pollution. Girgaon Chowpatty in Mumbai, Tuticorin Beach, situated near the port city of Thoothukudi in Tamil Nadu, and Dhanushkodi Beach, positioned at the tip of Pamban Island in Tamil Nadu, are beaches known for high anthropogenic activities. They are subjected to microplastic contamination, with Mumbai beach exhibiting the highest levels (Tiwari et al. 2019). The beaches in Goa, Karnataka, and several other beaches in Maharashtra also reported the presence of microplastics in coastal environments (Maharana et al. 2020). The negative impact of improper waste management in the coastline of Kanyakumari, a major tourist attraction, has polluted the beaches with macro- and microplastics. Sediment samples from the area contained microplastics, identified as secondary microplastics resulting from the breakdown of larger plastic materials (Sundar et al. 2021).

Several rivers were also detected with the presence of microplastics. A study encompassing the Mandovi, Zuari, Chapora, and Sal rivers in Goa revealed the presence of microplastics (Veerasingam et al. 2016b). Similarly, microplastics were identified in the Sharavathi and Netravati River of Karnataka, which finally meets the Arabian Sea (Amrutha and Warrier 2020; Amrutha et al. 2023). Sediments from Brahmaputra and Indus Rivers contained various types of plastics, including polypropylene, polyethylene, polyamide, polyethylene terephthalate, polyvinyl chloride, and polytetrafluoroethylene (Tsering et al. 2021). The discharge of the Gadilam River at the mouth has found to contribute to plastic debris accumulation on Silver Beach in Tamil Nadu (Vidyasakar et al. 2020). Sediment samples collected from estuarine regions of the River Ganga indicated the presence of both microplastics and mesoplastics (Sarkar et al. 2019). It was observed that a substantial amount of microplastics, estimated between one and three billion pieces, is discharged into the Bay of Bengal daily, with a noticeable increase in concentration as it moves from its source to the open sea (Napper et al. 2021). These findings underscore the importance of managing plastic pollution in inland regions as these microplastics travel all their way to the ocean, ending up in the food chain. The urban flooding events may also carry plastic waste into the ocean through these rivers.

Microplastics were recovered from the surface water of Punnakayal Estuary located in Tuticorin, Tamil Nadu, which indicated the adverse impacts of anthropogenic activities such as fishing and tourism. Untreated sewage from nearby urban areas and industrial discharges flowing through the canal in the region have also contributed to amplify the issue. These microplastics also acted as adsorbents for heavy metals, with varying concentrations of elements like Mn (manganese), Zn (zinc), As (arsenic), Pb (lead), Cu (copper), Cd (cadmium), and Cr (chromium). The diverse range of industries in Tuticorin, including copper smelting, heavy water production, chemical and petrochemical facilities, seafood processing, thermal power plants, harbor operations, and textile production, has contributed to microplastic pollution in the coastal region (Selvam et al. 2021a). Microplastics were found in fish species like Epinephelus merra and Rastrelliger kanagurta (the Indian mackerel) collected from these areas (Kumar et al. 2018).

The majority of industries in Kerala, situated in coastal districts and along riverbanks, use backwaters and major rivers as pathways for discharging dissolved and particulate materials, along with wastewater from industries and sewage treatment plants. This discharge eventually reaches the coast, leading to the presence of microplastic particles such as foam, pellets, and film in coastal waters. Fish in the region have been discovered with microplastics in their digestive tracts (Robin et al. 2020). Microplastics were detected in the surface seawater of the northern coast of Kerala, indicating their widespread distribution in this crucial coastal ecosystem (Anu Pavithran 2021). The Cochin backwaters in Kerala constituting a significant estuarine system in the country has two openings connecting it to the Arabian Sea and has freshwater inflow from six major rivers. These backwaters, along with the adjoining coastal areas, are crucial centers for fishing, tourism, cargo transport, and industrial activities. A study that analyzed surface water, sediments, and fish samples, including benthic (Cynoglossus sp.), demersal (Leiognathus sp., Ambassis ambassis, and Clarias batrachus), and pelagic (Stolephorus sp.) species, found microplastics present in the digestive tracts of approximately 15% of the examined fish collected from different depth zones of this region (Suresh et al. 2020). About 16 fish species caught from the fishing harbor at Cochin were found to contain microplastics such as polyethylene (PE) and polypropylene (PP) in their gut (James et al. 2020).

Microplastic particles were identified in various seafood species from the Sal Estuary in Goa, like oysters, clams, mussels, silver/gray mullet, and catfish (Naik et al. 2021). In the Sundarbans mangrove ecosystems in the Bay of Bengal, various species were found to display the transfer, biomagnification, and bioaccumulation of microplastics across distinct trophic levels (Kumar et al. 2022; Sarker et al. 2022). The commercial fish species of Adayar and Ennore regions of south coast of India also contained microplastics like polyethylene, polypropylene, polyamide, and polyester (Janardhanam et al. 2022). Coastal regions in Tamil Nadu, specifically Tinnakara Island and Chennai Beach, were found to be contaminated with microplastics, attributed to international transport activities in the ocean. The non-degradable nature of plastics was evident from the high quality maintained in the identified plastics from these regions, even after extended periods (Mugilarasan et al. 2017).

The highly productive and biodiverse coastal ecosystems in the northern Indian Ocean, particularly the Gulf of Mannar (GoM) and Palk Bay (PB) regions along the southeastern coast of India, showed an increased presence of microplastics. Surface water, sediments, and economically important fish species in these regions were found to contain microplastics (James et al. 2021). The presence of microplastics in India is notably significant compared to eight other countries around the Indian Ocean, including Maldives, Pakistan, Tanzania, Indonesia, Bangladesh, Sri Lanka, Myanmar, and Australia. Beach sand samples collected from these countries revealed abundant microplastics in fiber, with the highest number obtained from India (Balasubramaniam and Phillott 2016) (Table 11.1).

11.1.2 Microplastics as Adsorbents

Microplastics (MPs) have the potential to facilitate the bioaccumulation and/or biomagnification of harmful contaminants (Koelmans et al. 2015). This means that when organisms consume MPs, these build up toxicity either through degradation to nanoplastics or through toxic substances that have adhered to or absorbed into the microplastic particles. As these organisms move up the food chain, the accumulated contaminants can become more concentrated, posing a greater risk to animals and potentially even to humans who consume seafood.

Table 11.1 Impact of microplastics in Indian coastal regions

Indian coastal regions and rivers	Impact of microplastics
Girgaon Chowpatty, Mumbai	High levels of microplastic contamination due to anthropogenic activities
Tuticorin Beach, Tamil Nadu	Microplastics from industrial discharges (e.g., copper smelting, chemical plants), fishing, and tourism; heavy metal adsorption on microplastics; contamination in fish species like Epinephelus merra and Rastrelliger kanagurta
Dhanushkodi Beach, Tamil Nadu	Microplastic contamination from high anthropogenic activities
Beaches in Goa and Karnataka	Microplastic presence reported in coastal environments due to local activities
Kanyakumari, Tamil Nadu	Pollution from macro- and microplastics due to improper waste management; sediment samples identified secondary microplastics
Silver Beach, Tamil Nadu	Plastic debris accumulation from Gadilam River discharge
Punnakayal Estuary, Tamil Nadu	Microplastics detected due to fishing, tourism, untreated sewage, and industrial discharges; adsorbed heavy metals such as Mn, Zn, As, Pb, Cu, Cd, and Cr
Cochin Backwaters, Kerala	Microplastics detected in surface water, sediments, and fish species; found in the digestive tracts of fishes
Bay of Bengal, Chennai Coast	Fishes found to ingest microplastics; contamination attributed to international transport activities and nondegradable plastics
Sal Estuary, Goa	Microplastics identified in seafood species (e.g., oysters, clams, mullets); average concentration of 48 MPs per liter in water and 3950 MPs per kilogram in sediments
Sundarbans Mangrove Ecosystem	Evidence of microplastic transfer, biomagnification, and bioaccumulation across trophic levels in various species
Tinnakara Island, Tamil Nadu	Contamination attributed to international transport and nondegradable nature of plastics
Gulf of Mannar and Palk Bay, Tamil Nadu	Microplastics found in surface water, sediments, and economically important fish species; pollution affecting highly biodiverse ecosystems
River Ganga Estuarine	Sediments contained microplastics and mesoplastics; significant discharge of microplastics into the Bay of Bengal
Netravati River, Karnataka	Highest abundance of microplastics detected near the Arabian Sea

Microplastics were (MPs) found to adsorb metals such as Br (bromine), Cd (cadmium), Cl (chlorine), Cr (chromium), Hg (mercury), Pb (lead), V (vanadium), Fe (iron), Sb (antimony), Ti (titanium), Zn (zinc), Ba (barium), Bi (bismuth), Cu (copper), and Ni (nickel) (Selvam et al. 2021a, b). These microplastics, along with trace metals, were identified in the muscle and intestines of commonly consumed commercial fish species harvested from Thoothukudi, located on the Gulf of Mannar coast along the Indian Ocean in South India. The predominant anthropogenic activities in this region, including various industries associated with copper production, petrochemicals, fertilizer manufacturing, heavy water production, seafood processing, and titanium dioxide production, were assessed as the likely cause of these metals in the fish species (Selvam et al. 2021a, b).

Apart from metals, marine plastic pellets also become adsorbents by either attracting hydrophobic chemicals, such as polychlorinated biphenyls (PCBs), from seawater due to the low polarity and hydrophobic nature of their surfaces, or by incorporating synthetic additives and chemicals within the plastic resin, potentially posing a threat to marine life (Mato et al. 2001). Extensive liver damage was found in fishes due to the chemicals like polycyclic aromatic hydrocarbons (PAHs), polychlorinated biphenyls (PCBs), and polybrominated diphenyl ethers (PBDEs) accumulated in them when exposed to marine plastic for 2 months (Rochman et al. 2013).

11.2 Effects on Marine Health Due to Microplastics

Microplastic exposure in zooplankton, such as copepods, can lead to adverse effects, including reduced hatching success, diminished survival rates, and altered reproductive outcomes. In particular, filter-feeding copepods have been shown to ingest microplastics, with subsequent impacts on their reproductive capabilities and energetic reserves (Rashid et al. 2021). Endocrine disruption was found in certain fishes due to marine plastic debris which impacted upon their reproductive health. Gene expression changes were also found in some male fishes indicating a potential influence of contaminants in the cod liver oil in the diet (Rochman et al. 2014).

Zooplankton readily ingest significant amounts of microplastics due to their easy availability, raising substantial concerns about potential repercussions when higher trophic levels in the ecosystem further consume these microplastics (Gopinath et al. 2020). Understanding the possible impacts on the biota is crucial, as this ingestion could result in far-reaching ecological consequences. The consumption of microplastics (MPs) can disrupt the physiological functioning of animals through various mechanisms, including the leaching of accumulated organic pollutants from the particles into the stomach lining, as observed in fishes. MPs can also cause physical blockages in the digestive system or occupy space in the stomach that would otherwise be available for food consumption, highlighting the multiple pathways through which MPs can have detrimental effects on animal health and well-being (Bakir et al. 2014). Seabirds have been shown to exhibit a tendency to consume microplastic particles, with microplastics found in the stomachs of many seabirds (Ryan 1987). Microplastics were discovered in the digestive systems of several green turtles in Brazil (Tourinho et al. 2010). Startle response, which is the primary risk avoidance mechanism in hermit crabs, has been observed to lead to a reduction in the duration due to exposure to microplastics, potentially rendering them more susceptible to predation in high-risk situations (Nanninga et al. 2020). When crabs inhale MPs through their gills, a minor and temporary alteration in oxygen consumption and ion regulation was observed (Watts et al. 2016). This emphasizes that microplastic pollution has the potential to impact not only the behavioral patterns but also the physiology of organisms.

11.3 Environmental Forensics in Microplastic Pollution Assessment and Source Identification in India

Environmental forensics focuses on identifying and attributing compounds in environmental samples to specific sources, aiding in legal actions against entities violating environmental regulations (Mudge 2008). Addressing microplastic pollution as an environmental issue requires applying forensic techniques to trace its sources and mitigate its adverse effects on natural and marine ecosystems. However, formal investigation strategies for microplastic contamination are yet to be established in India.

Despite this, independent methods have been employed effectively to identify contamination sources. For example, a study at Nattika Beach, Kerala, successfully identified sources of microplastics using forensic investigation techniques. This investigation linked fibers in samples to activities at a nearby net-mending yard, confirming the source based on specific matching properties, including samples from previous years (Ashwini and Varghese 2019). They also found that polyvinyl chloride (PVC) was not frequently detected due to its tendency to sink rather than float to the shore.

Similarly, a study at Calicut Beach, Kerala, analyzed microplastics across three zones: a recreational area, a drainage outlet, and a fishing harbor. Microplastics in the recreational area were traced to polystyrene (PS) foam plates used by food vendors. In the harbor, polystyrene particles were found to originate from floats in fishing nets, while film-type microplastics were linked to packaging materials and paints from fishing vessels. Fiber microplastics sources were predominantly from discarded fishing nets, though sources in some zones remained unidentified. The presence of white pellets during floods occurred in the study period suggested transport by floodwaters from nearby sources. Polycarbonate urethane (PCU) was linked to foam macroplastics in the area. Polyethylene terephthalate (PET), often used in water bottles, appeared in low quantities. Polyvinyl chloride (PVC) found in sediments was traced to discarded schoolbags, emphasizing the dynamic marine environment's role in dispersing pollutants (Kumar and Varghese 2021a, b) (Table 11.2). These studies underscore the importance of forensic methodologies in identifying microplastic sources, enabling targeted interventions and policy development for marine and environmental protection.

There are other countries also that utilized forensic approaches to tackle the problem of microplastics contamination. The significance of forensic fiber identification techniques in categorizing fibers into groups for further investigation using polarized light microscopy was proposed aiding in the identification of potential sources or contaminants in various environmental samples (Woodall et al. 2015; Leistenschneider et al. 2021).

Table 11.2 Sources of microplastics identified using forensic techniques in the study conducted by Varghese et al.

Zone	Proximity	Identified types of microplastics	Sources
Zone I	Recreational activities	Polystyrene (PS)	Foamed plates used by food vendors
Zone II	Drainage outlet leading to the sea	White pellets, fiber microplastics	White pellets transported by floodwaters (2018, 2019); fibers with unidentified sources
Zone III	Fishing harbor adjacent to Calicut Beach	Polystyrene (PS), film-type microplastics	Polystyrene from fishing net floats; film-type microplastics from packaging materials and paints eroded from fishing vessels; fibers from discarded fishing nets
All Zones (General)		Fiber polycarbonate (PCU), polyethylene terephthalate (PET), polyvinyl chloride (PVC)	PCU from suspected foam macroplastics; PET from packaging water bottles (low abundance due to recycling); PVC from discarded schoolbags

11.4 The Process of Source Identification Using Forensic Techniques

The common frameworks proposed outlines three distinct strategies that provide varying levels of precision in identifying the source of microplastics (Kumar and Varghese 2021a, b; Ashwini and Varghese 2019). These strategies are as follows:

11.4.1 Exact Source Identification

In the first strategy, the exact source is pinpointed when a unique signature is discovered that perfectly matches with the source of the microplastics. This is the most precise method of source identification.

11.4.2 Suspected Source Identification

The second strategy involves identifying larger plastic items in the field that are suspected to be sources of the microplastics found in the collected samples. While not as precise as the first strategy, this approach can still narrow down the potential sources.

11.4.3 Distant vs. Nearer Origin

The third strategy helps in determining whether the microplastics have a distant or nearer origin. This is based on the frequency of occurrence of the microplastics in relation to the beach profile. If the observations from these strategies consistently lead to the same conclusion, it confirms the source of the particular microplastics.

11.5 Factors to Be Considered While Identifying the Source of Microplastics

11.5.1 The Age of Microplastics

The age of microplastics plays a crucial role in source determination. Microplastics with smooth surfaces and no biofilm are likely to have originated from recent anthropogenic activities such as beach littering or industrial discharge. In contrast, aged microplastics characterized by worn-out surfaces and the presence of biofilms indicate long-distance transport through marine environments. Their mobility is influenced by factors such as surface and bottom currents, which facilitate their movement across vast distances (Ashwini and Varghese 2019).

The distribution and concentration of microplastics are also subject to seasonal and environmental dynamics, making these factors critical during sampling and analysis.

Rainy seasons: Increased river runoff introduces land-based microplastics into marine environments.

Dry seasons: The absence of significant freshwater input allows microplastics to accumulate in localized areas.

11.5.2 Winds, Currents, and Floods

Wind directions and oceanic currents influence the transport and deposition of microplastics, leading to accumulation along coastlines or dispersion over larger distances. Events such as floods can also result in heightened concentrations of microplastics, as these events facilitate the transport of particles from remote terrestrial areas into marine environments (Devi et al. 2020; Veerasingam et al. 2016a, b).

11.5.3 Compositional Complexity

The compositional complexity of microplastics presents additional challenges in source identification. This complexity is amplified by the use of additives, which can modify the physical and chemical properties of polymers. Variations in polymer types, the intricate composition of particles, and the broad spectrum of potential modifications further complicate the process of accurately tracing microplastics back to their origins.

To ensure comprehensive analysis and interpretation, sampling strategies must account for these factors to derive adequate inferences. This multifaceted approach is essential for a thorough understanding of microplastic sources and their pathways in the environment.

11.6 Sampling Techniques

Sampling is a vital process in forensic science as this step determines the quality and accuracy of results obtained. The choice of sampling methods is crucial in environmental forensics to ensure the integrity of samples. Most studies follow the National Oceanic and Atmospheric Administration (NOAA) guidelines for selecting sample collection sites and techniques (Opfer et al. 2012; Lippiatt et al. 2013).

Collection methods from water sources can involve either bulk sampling or volume reduction techniques (Hidalgo-Ruz et al. 2012). In bulk sampling, the whole of the collected sample is taken for analysis, whereas in volume reduction techniques, only a certain portion of the collected sample is taken. Bulk sampling includes grab methods like Van Veen Grab and pumping, while volume reduction techniques use trawls or nets. Manta trawls, plankton nets, and other nets are frequently employed to collect seawater samples (Patterson et al. 2021; Sathish et al. 2020; Chauhan et al. 2021). When water is collected using motor pumps, they are pre-cleaned, and sediments from the pumped water are filtered using steel sieves. Underwater sediments and beach sediments are collected using grab samplers, steel scoops, trowels, shovels, and spade (Chauhan et al. 2021; Razeghi et al. 2021; Gwinnett et al. 2021, Karthik et al. 2018).

The selection of an appropriate mesh size for the collection of samples is critical in sampling using nets as it significantly impacts the size range of microplastics analyzed. The Marine Strategy Framework Directive (MSFD) guidelines recommend using a mesh size of 333 μm, with commonly employed sizes ranging from 300 to 390 μm, which is considered practical for reliable and meaningful samples in open net sampling situations (Marrone and Pellegrino 2022). The separation of collected microparticles primarily relies on density separation methods using high-density chemicals as wells as physical separation under microscopes (Hidalgo-Ruz et al. 2012). However, electrostatic separation methods have also been proposed in recent studies as an alternative (Felsing et al. 2018).

The sampling techniques and requirements to collect the sediment as well as water samples are listed in the table below (Table 11.3).

Table 11.3 Sampling techniques and requirements for environmental forensic investigations

Sampling method/technique	Description	Tools/materials used	Key references
Bulk sampling	Collection of large volumes of water for analysis	Grab methods, pump methods, pre-cleaned motor pumps	Hidalgo-Ruz et al. (2012), Selvam et al. (2021a, b)
Volume reduction sampling	Collection of large volumes followed by reduction before analysis	Manta trawl, plankton net, other trawling systems	Hidalgo-Ruz et al. (2012), Chauhan et al. (2021), Sathish et al. (2020)
Sediment sampling	Collection of sediment samples from seabeds, riverbeds, or lakes	Van Veen grab, sediment samplers, stainless steel scoops, and containers	Bharath et al. (2021), Chauhan et al. (2021), Karthik et al. (2018)
Beach sediment collection	Sampling beach sediments for plastic or particle analysis	Stainless steel scoop, stainless steel containers	Karthik et al. (2018)
Mesh size selection	Filtration of microplastics using defined mesh sizes	Mesh nets (typically 300–390 μm); 333 μm recommended by MSFD guidelines	Marrone and Pellegrino (2022)
Microplastic separation	Separation of microplastics from environmental matrices	Density separation with high-density solutions (e.g., NaCl, $ZnCl_2$); electrostatic methods	Hidalgo-Ruz et al. (2012), Felsing et al. (2018)
Sampling guidelines	Ensuring quality control and standardization during sampling	NOAA standard protocols including site selection, handling techniques, QA/QC measures	Opfer et al. (2012), Lippiatt et al. (2013)

11.7 Analysis of Microplastics

The collected samples are analyzed by examining their morphological characteristics using microscopes, polymer composition through spectroscopic techniques, aging properties, metal contamination, and finally the chemical composition.

The preliminary examination of morphological characters is done using stereomicroscopes (Patterson et al. 2021), polarizing microscopes (Woodall et al. 2015), fluorescent microscopes (Tiwari et al. 2019), and scanning electron microscopes (SEM) (Patel et al. 2020). Analyzing and categorizing the various polymer types of microplastics is critical as each polymer has distinct environmental impacts. Fourier-transform infrared (FTIR) spectroscopy has been widely utilized for the characterization and identification of microplastics (Kumar and Varghese 2021a, b), with attenuated total reflectance FTIR (ATR-FTIR) also being frequently applied (Veerasingam et al. 2021; Ashwini and Varghese 2019). The carbonyl index calculated from FTIR data provides insights into the aging of microplastics (Veerasingam et al. 2016a, b).

Raman spectroscopy is another powerful tool used to classify microplastics based on polymer type, making it valuable in forensic analysis (Devi et al. 2020; Sruthy and Ramasamy 2017; Anu Pavithran 2021). For detecting metals on the surface of microplastics, inductively coupled plasma-mass spectroscopy (ICP-MS) and inductively coupled plasma optical emission spectrometry (ICP-OES) are commonly used (Selvam et al. 2021a; Sathish et al. 2020). Micro-Raman spectrometry is employed for the chemical characterization of microplastics, providing detailed insights into their composition (Prata et al. 2021).

11.8 Conclusion

India's marine ecosystems are significantly affected by the contamination of microplastics. As proper evidence matters for taking necessary legal actions and regulations, characterizing the types and origins of microplastics enables authorities and organizations to continuously monitor specific pollution sources and implement stringent regulations upon them. Although there exist several limitations in this field, with extended collaborations and support from various organizations as well as future advancements in technology, the field of environmental forensics is poised to become more precise and efficient.

References

Amrutha K, Warrier AK (2020) The first report on the source-to-sink characterization of microplastic pollution from a riverine environment in tropical India. Sci Total Environ 739:140377. https://doi.org/10.1016/j.scitotenv.2020.140377

Amrutha K, Shajikumar S, Warrier AK, Sebastian JG, Sali YA, Chandran T, Sivadas S, Naik R, Amrish VN, Kumar A, Unnikrishnan V (2023) Assessment of pollution and risks associated with microplastics in the riverine sediments of the Western Ghats: a heritage site in southern India. Environ Sci Pollut Res 30(12):32301–32319. https://doi.org/10.1007/s11356-022-24437-z

Anu Pavithran V (2021) Study on microplastic pollution in the coastal seawaters of selected regions along the northern coast of Kerala, southwest coast of India. J Sea Res 173:102060. https://doi.org/10.1016/j.seares.2021.102060

Ashwini SK, Varghese GK (2019) Environmental forensic analysis of the microplastic pollution at "Nattika" Beach, Kerala Coast, India. Environ Forensics 21(1):21–36. https://doi.org/10.1080/15275922.2019.1693442

Bakir A, Rowland SJ, Thompson RC (2014) Enhanced desorption of persistent organic pollutants from microplastics under simulated physiological conditions. Environ Pollut 185:16–23. https://doi.org/10.1016/j.envpol.2013.10.007

Balasubramaniam M, Phillott AD (2016) Preliminary observations of microplastics from beaches in the Indian Ocean. 23. www.iotn.org

Barnes DKA, Galgani F, Thompson RC, Barlaz M (2009) Accumulation and fragmentation of plastic debris in global environments. Philos Trans R Soc Lond B Biol Sci 364(1526):1985–1998. https://doi.org/10.1098/rstb.2008.0205

Bharath KM, Srinivasalu S, Natesan U, Ayyamperumal R, Kalam SN, Anbalagan S, Sujatha K, Alagarasan C (2021) Microplastics as an emerging threat to the freshwater ecosystems of Veeranam lake in South India: a multidimensional approach. Chemosphere 264:128502. https://doi.org/10.1016/j.chemosphere.2020.128502

Chauhan JS, Semwal D, Nainwal M, Badola N, Thapliyal P (2021) Investigation of microplastic pollution in river Alaknanda stretch of Uttarakhand. Environ Dev Sustain 23(11):16819–16833. https://doi.org/10.1007/s10668-021-01388-y

Derraik JGB (2002) The pollution of the marine environment by plastic debris: a review. Mar Pollut Bull 44(9):842–852. https://doi.org/10.1016/S0025-326X(02)00220-5

Devi SS, Sreedevi AV, Kumar AB (2020) First report of microplastic ingestion by the alien fish Pirapitinga (Piaractus brachypomus) in the Ramsar site Vembanad Lake, South India. Mar Pollut Bull 160:111637. https://doi.org/10.1016/j.marpolbul.2020.111637

Felsing S, Kochleus C, Buchinger S, Brennholt N, Stock F, Reifferscheid G (2018) A new approach in separating microplastics from environmental samples based on their electrostatic behavior. Environ Pollut 234:20–28. https://doi.org/10.1016/j.envpol.2017.11.013

Gopinath K, Seshachalam S, Neelavannan K, Anburaj V, Rachel M, Ravi S, Bharath M, Achyuthan H (2020) Quantification of microplastic in Red Hills Lake of Chennai city, Tamil Nadu, India. Environ Sci Pollut Res 27(26):33297–33306. https://doi.org/10.1007/s11356-020-09622-2

Gwinnett C, Harrison E, Osborne A, Pivato A, Varghese G (2021) Sampling microplastics for environmental forensic applications. In: Detritus, vol 14. Cisa Publisher – Eurowaste Srl, pp I–III. https://doi.org/10.31025/2611-4135/2021.14080

Hidalgo-Ruz V, Gutow L, Thompson RC, Thiel M (2012) Microplastics in the marine environment: a review of the methods used for identification and quantification. Environ Sci Technol 46(6):3060–3075. https://doi.org/10.1021/es2031505

Jambeck JR, Geyer R, Wilcox C, Siegler TR, Perryman M, Andrady A, Narayan R, Law KL (2015) Plastic waste inputs from land into the ocean. Science 347(6223):768–771. https://doi.org/10.1126/science.1260352

James K, Vasant K, Padua S, Gopinath V, Abilash KS, Jeyabaskaran R, Babu A, John S (2020) An assessment of microplastics in the ecosystem and selected commercially important fishes off Kochi, south eastern Arabian Sea, India. Mar Pollut Bull 154:111027. https://doi.org/10.1016/j.marpolbul.2020.111027

James K, Vasant K, Sikkander Batcha SM, Padua S, Jeyabaskaran R, Thirumalaiselvan S, Vineetha G, Benjamin LV (2021) Seasonal variability in the distribution of microplastics in the coastal ecosystems and in some commercially important fishes of the Gulf of Mannar and Palk Bay, Southeast coast of India. Reg Stud Mar Sci 41:101558. https://doi.org/10.1016/j.rsma.2020.101558

Janardhanam M, Sivakumar P, Srinivasan G, Sivakumar R, Marcus PN, Balasubramaniam S, Rajamanickam K, Raman T, Singaram G, Harikrishnan T (2022) Microplastics in demersal sharks from the Southeast Indian Coastal Region. Front Mar Sci 9. https://doi.org/10.3389/fmars.2022.914391

Jovanović B (2017) Ingestion of microplastics by fish and its potential consequences from a physical perspective. In: Integrated environmental assessment and management, vol 13(3). Wiley-Blackwell, pp 510–515. https://doi.org/10.1002/ieam.1913

Karthik R, Robin RS, Purvaja R, Ganguly D, Anandavelu I, Raghuraman R, Hariharan G, Ramakrishna A, Ramesh R (2018) Microplastics along the beaches of southeast coast of India. Sci Total Environ 645:1388–1399. https://doi.org/10.1016/j.scitotenv.2018.07.242

Koelmans AA, Besseling E, Shim WJ (2015) Nanoplastics in the aquatic environment. Critical review. In: Marine anthropogenic litter. Springer International Publishing, pp 325–340. https://doi.org/10.1007/978-3-319-16510-3_12

Kumar AS, Varghese GK (2021a) Microplastic pollution of Calicut beach – contributing factors and possible impacts. Mar Pollut Bull 169:112492. https://doi.org/10.1016/j.marpolbul.2021.112492

Kumar AS, Varghese GK (2021b) Source apportionment of marine microplastics: first step towards managing microplastic pollution. Chem Eng Technol 44(5):906–912. https://doi.org/10.1002/ceat.202000482

Kumar VE, Ravikumar G, Jeyasanta KI (2018) Occurrence of microplastics in fishes from two landing sites in Tuticorin, South east coast of India. Mar Pollut Bull 135:889–894. https://doi.org/10.1016/j.marpolbul.2018.08.023

Kumar R, Sinha R, Refat Jahan Rakib M, Padha S, Ivy N, Bhattacharya S, Dhar A, Sharma P (2022) Microplastics pollution load in Sundarban delta of Bay of Bengal. J Hazard Mater Adv 7:100099. https://doi.org/10.1016/j.hazadv.2022.100099

Leistenschneider C, Burkhardt-Holm P, Mani T, Primpke S, Taubner H, Gerdts G (2021) Microplastics in the Weddell Sea (Antarctica): a forensic approach for discrimination between environmental and vessel-induced microplastics. Environ Sci Technol 55(23):15900–15911. https://doi.org/10.1021/acs.est.1c05207

Lippiatt S, Opfer S, Arthur C (2013) Marine debris monitoring and assessment. NOAA Technical Memorandum NOS-OR&R-46

Maharana D, Saha M, Dar JY, Rathore C, Sreepada RA, Xu XR, Koongolla JB, Li HX (2020) Assessment of micro and macroplastics along the west coast of India: abundance, distribution, polymer type and toxicity. Chemosphere 246:125708. https://doi.org/10.1016/j.chemosphere.2019.125708

Marrone A, Pellegrino D (2022) A plethora of microplastic pollution studies: the need for a forensic approach. Detritus 18:50–57. https://doi.org/10.31025/2611-4135/2022.15164

Mato Y, Isobe T, Takada H, Kanehiro H, Ohtake C, Kaminuma T (2001) Plastic resin pellets as a transport medium for toxic chemicals in the marine environment. Environ Sci Technol 35(2):318–324. https://doi.org/10.1021/es0010498

Mudge SM (2008) Environmental forensics and the importance of source identification. In: Environmental forensics. The Royal Society of Chemistry, pp 1–16. https://doi.org/10.1039/9781847558343-00001

Mugilarasan M, Venkatachalapathy R, Sharmila N, Gurumoorthi K (2017) Occurrence of microplastic resin pellets from Chennai and Tinnakkara Island: towards the establishment of background level for plastic pollution. Indian J Geo Mar Sci 46(06). www.pelletwatch.org

Naik A, Desai A, Nanajkar M, Rathore C (2021) Microplastics in water, sediment and seafood as an emerging threat to the marine environment: a case study in Goa, west coast of India. Chemosphere 270:129359

Nanninga GB, Horswill C, Lane SM, Manica A, Briffa M (2020) Microplastic exposure increases predictability of predator avoidance strategies in hermit crabs. J Hazard Mater Lett 1:100005. https://doi.org/10.1016/j.hazl.2020.100005

Napper IE, Baroth A, Barrett AC, Bhola S, Chowdhury GW, Davies BFR, Duncan EM, Kumar S, Nelms SE, Hasan Niloy MN, Nishat B, Maddalene T, Thompson RC, Koldewey H (2021) The abundance and characteristics of microplastics in surface water in the transboundary Ganges River. Environ Pollut 274:116348. https://doi.org/10.1016/j.envpol.2020.116348

Opfer S, Arthur C, Lippiatt S (2012) NOAA Marine Debris Shoreline Survey Field Guide. www.MarineDebris.noaa.gov

Patel AK, Bhagat C, Taki K, Kumar M (2020) Microplastic vulnerability in the sediments of the Sabarmati River of India, pp 127–138. https://doi.org/10.1007/978-981-15-4668-6_7

Patterson J, Jeyasanta KI, Laju RL, Edward JKP (2021) Microplastic contamination in Indian edible mussels (Perna perna and Perna viridis) and their environs. Mar Pollut Bull 171:112678. https://doi.org/10.1016/j.marpolbul.2021.112678

Prata JC, da Costa JP, Fernandes AJS, da Costa FM, Duarte AC, Rocha-Santos T (2021) Selection of microplastics by Nile Red staining increases environmental sample throughput by micro-Raman spectroscopy. Sci Total Environ 783:146979. https://doi.org/10.1016/j.scitotenv.2021.146979

Rashid CP, Arunpandi N, Abhijith VT, Vidhya V, Gupta GVM, Ramanamurty MV (2021) Microplastics in zooplankton in the eastern Arabian Sea: the threats they pose to fish and corals favoured by coastal currents. Mar Pollut Bull 173:113042

Razeghi N, Hamidian AH, Wu C, Zhang Y, Yang M (2021) Microplastic sampling techniques in freshwaters and sediments: a review. Environ Chem Lett 19(6):4225–4252. https://doi.org/10.1007/s10311-021-01227-6

Robin RS, Karthik R, Purvaja R, Ganguly D, Anandavelu I, Mugilarasan M, Ramesh R (2020) Holistic assessment of microplastics in various coastal environmental matrices, southwest coast of India. Sci Total Environ 703:134947. https://doi.org/10.1016/j.scitotenv.2019.134947

Rochman CM, Hoh E, Kurobe T, Teh SJ (2013) Ingested plastic transfers hazardous chemicals to fish and induces hepatic stress. Sci Rep 3. https://doi.org/10.1038/srep03263

Rochman CM, Kurobe T, Flores I, Teh SJ (2014) Early warning signs of endocrine disruption in adult fish from the ingestion of polyethylene with and without sorbed chemical pollutants from the marine environment. Sci Total Environ 493:656–661. https://doi.org/10.1016/j.scitotenv.2014.06.051

Ryan PG (1987) The incidence and characteristics of plastic particles ingested by seabirds. Mar Environ Res 23(3):175–206. https://doi.org/10.1016/0141-1136(87)90028-6

Sarkar DJ, Das Sarkar S, Das BK, Manna RK, Behera BK, Samanta S (2019) Spatial distribution of meso and microplastics in the sediments of river Ganga at eastern India. Sci Total Environ 694:133712. https://doi.org/10.1016/j.scitotenv.2019.133712

Sarker S, Huda AS, Niloy MNH, Chowdhury GW (2022) Trophic transfer of microplastics in the aquatic ecosystem of Sundarbans mangrove forest, Bangladesh. Sci Total Environ 838:155896. https://doi.org/10.1016/j.scitotenv.2022.155896

Sathish N, Jeyasanta I, Patterson J (2020) Microplastic contamination in river Kosasthalaiyar, Associated Creek and Lake of Chennai, Tamil Nadu, India. J Aquatic Biol Fish 68

Selvam S, Jesuraja K, Venkatramanan S, Roy PD, Jeyanthi Kumari V (2021a) Hazardous microplastic characteristics and its role as a vector of heavy metal in groundwater and surface water of coastal South India. J Hazard Mater 402:123786. https://doi.org/10.1016/j.jhazmat.2020.123786

Selvam S, Manisha A, Roy PD, Venkatramanan S, Chung SY, Muthukumar P, Jesuraja K, Elgorban AM, Ahmed B, Elzain HE (2021b) Microplastics and trace metals in fish species of the Gulf of Mannar (Indian Ocean) and evaluation of human health. Environ Pollut 291:118089. https://doi.org/10.1016/j.envpol.2021.118089

Sruthy S, Ramasamy EV (2017) Microplastic pollution in Vembanad Lake, Kerala, India: The first report of microplastics in lake and estuarine sediments in India. Environ Pollut 222:315–322. https://doi.org/10.1016/j.envpol.2016.12.038

Sundar S, Chokkalingam L, Roy PD, Usha T (2021) Estimation of microplastics in sediments at the southernmost coast of India (Kanyakumari). Environ Sci Pollut Res 28(15):18495–18500. https://doi.org/10.1007/s11356-020-10333-x

Suresh A, Vijayaraghavan G, Saranya KS, Neethu KV, Aneesh B, Bijoy Nandan S (2020) Microplastics distribution and contamination from the Cochin coastal zone, India. Reg Stud Mar Sci 40:101533. https://doi.org/10.1016/j.rsma.2020.101533

Tiwari M, Rathod TD, Ajmal PY, Bhangare RC, Sahu SK (2019) Distribution and characterization of microplastics in beach sand from three different Indian coastal environments. Mar Pollut Bull 140:262–273. https://doi.org/10.1016/j.marpolbul.2019.01.055

Tourinho PS, Ivar do Sul JA, Fillmann G (2010) Is marine debris ingestion still a problem for the coastal marine biota of southern Brazil? Mar Pollut Bull 60(3):396–401. https://doi.org/10.1016/j.marpolbul.2009.10.013

Tsering T, Sillanpää M, Sillanpää M, Viitala M, Reinikainen SP (2021) Microplastics pollution in the Brahmaputra River and the Indus River of the Indian Himalaya. Sci Total Environ 789:147968. https://doi.org/10.1016/j.scitotenv.2021.147968

Veerasingam S, Mugilarasan M, Venkatachalapathy R, Vethamony P (2016a) Influence of 2015 flood on the distribution and occurrence of microplastic pellets along the Chennai coast, India. Mar Pollut Bull 109(1). www.pelletwatch.org

Veerasingam S, Saha M, Suneel V, Vethamony P, Rodrigues AC, Bhattacharyya S, Naik BG (2016b) Characteristics, seasonal distribution and surface degradation features of microplastic pellets along the Goa coast, India. Chemosphere 159:496

Veerasingam S, Ranjani M, Venkatachalapathy R, Bagaev A, Mukhanov V, Litvinyuk D, Mugilarasan M, Gurumoorthi K, Guganathan L, Aboobacker VM, Vethamony P (2021) Contributions of Fourier transform infrared spectroscopy in microplastic pollution research: a review. Crit Rev Environ Sci Technol 51(22):2681–2743. https://doi.org/10.1080/10643389.2020.1807450

Vidyasakar A, Krishnakumar S, Kasilingam K, Neelavannan K, Bharathi VA, Godson PS, Prabha K, Magesh NS (2020) Characterization and distribution of microplastics and plastic debris along Silver Beach, Southern India. Mar Pollut Bull 158. https://doi.org/10.1016/j.marpolbul.2020.111421

Watts AJR, Urbina MA, Goodhead R, Moger J, Lewis C, Galloway TS (2016) Effect of microplastic on the gills of the Shore Crab Carcinus maenas. Environ Sci Technol 50(10):5364–5369. https://doi.org/10.1021/acs.est.6b01187

Woodall LC, Gwinnett C, Packer M, Thompson RC, Robinson LF, Paterson GLJ (2015) Using a forensic science approach to minimize environmental contamination and to identify microfibres in marine sediments. Mar Pollut Bull 95(1):40–46. https://doi.org/10.1016/j.marpolbul.2015.04.044

Forensic Analysis of Soil and Sediments 12

Varsha Rani Patel, Shivam Chourasiya, Anjali Dubey, and Baljeet Yadav

Abstract

The scientific and analytical value of soil and sediment evidence in criminal and civil procedures is examined in this chapter. Often disregarded, soil and sediments are rather special molecules with important associative evidence connecting a suspect or object to a specific location. Starting with a definition of soil and sediment, this chapter addresses its composition, methods of formation, and classification depending on physical, chemical, biological, and geological criteria. It notes their use in forensic settings—including buried remains, aquatic environments, and terrestrial crime scenes—as well as the basic distinctions between soil and sediment. Essential to maintaining the integrity of soil and sediment samples, the chapter provides a complete review of the chain of custody, preservation, and collection techniques. Priority one is reducing contamination, highlighting stratigraphy and collecting control soil samples for a comparison study. Microscopic inspection, mineralogical profiling, chemical characterisation (pH, elemental analysis, ICP-MS, AAS), biological markers (pollen, spores, microorganisms), and experimental techniques like FTIR and SEM-EDS are among the advanced analytical techniques that are then covered. Additionally, introduced is the use of geographic information systems (GIS) and geospatial techniques in

V. R. Patel (✉)
Faculty of Forensic Science, MATS University, Raipur, Chhattisgarh, India
e-mail: varshap@matsuniversity.ac.in

S. Chourasiya
Faculty of Forensic Science, Galgotias University, Greater Noida, Uttar Pradesh, India

A. Dubey
Faculty of Forensic Science, Medi-Caps University, Indore, Madhya Pradesh, India

B. Yadav
School of Basic and Applied Sciences, K. R. Mangalam University, Gurugram, Haryana, India

M. S. Sankhla et al. (eds.), *Introducing the Synergy Between Forensic Science and Environmental Analysis*, https://doi.org/10.1007/978-981-95-0199-1_12

forensic soil analysis. Using case studies and pragmatic applications, the chapter shows how soil and sediment evidence has proved crucial in settling a range of crimes, from sexual assault and murder to environmental breaches and illegal dumping. The subject addresses the reconstruction of crime scene dates depending on sediment deposition patterns and the evidential value of dirt transfer on shoes, clothes, tools, and automobiles. The chapter also evaluates critically the challenges and restrictions in forensic soil and sediment analysis, including the need of standardised processes, sample heterogeneity, and environmental changes. It also highlights fresh advances using metagenomics, artificial intelligence, and data-driven soil databases that are increasing the accuracy and dependability of forensic interpretations.

Keywords

Significance of soil · Examination of soil · Forensic analysis · Soil analysis · Biological markers

12.1 Introduction

Forensic soil science uses soil morphology, particularly investigations using pre-existing soil maps and spatially held soil data, minerals, chemistry, geophysics, biology, and molecular biology to answer legal questions, issues, or assumptions (Fitzpatrick 2009). One kind of trace evidence that can be found in a number of criminal cases, such as rape, murder, and other cases of a similar nature, is soil. Soil can maintain its integrity when applied to a wide range of items due to its remarkable adhesive properties, including but not limited to shoes, fingernails, toenails, tires, hair, and clothing materials. Soil is a crucial piece of evidence in criminal investigations due to the qualities mentioned above. It can be used to demonstrate a link between a number of people who have engaged in criminal activity and an individual who was there at a specific crime scene, whether they were a suspect or a criminal. This trace evidence evaluation is based on Locard's Exchange Principle, which states that there will always be a material exchange between two objects anytime they come into contact with one another (Chisum and Turvey 2000). An association between the person and the scene, as well as his possessions, can be established. Soil as trace evidence plays an important role in forensic investigation. When these trace hints provide clear evidence linking one or more persons to the crime, they serve as important evidence or corroborative evidence (Marumo et al. 1995).

The analysis of soil evidence in criminal examination began almost a century ago in a fictional story called "Sherlock Holmes" written by Sir Arthur Conan Doyle (Saferstein 2004). In 1904, soil was used for the first time as evidence to solve a criminal case by Georg Popp (Ruffell and McKinley 2005). Being a physically transferable substance, soil can be moved from one person's feet or shoes to another who has visited that place. Ultimately, by assessing the type and point of touch, it helps the person or object be linked to the crime scene (Fitzpatrick et al. 2009).

12.2 Nature and Composition of Soil and Sediments

Ground and subsurface conditions of the Earth mostly rely on soil and sediment. Their formation processes, compositions, locations, and scientific uses are quite varied even if they all follow the weathering and breakdown of rocks and organic components. Precise study and interpretation in disciplines including forensic science, geology, hydrology, and environmental science depend on the capacity to separate soil from silt. Comprising organic components, water, air, minerals, and living entities, soil is the highest layer of the crust of the Earth. Rock weathering and biological activity help plants survive and grow throughout time (Soil Science Society of America, 2008a, b, c). Climate, organisms, geography, parent material, and time are the five characteristics defining soil. Thus, soil develops in place and shows several horizons or layers: O, A, B, and C (Osman 2013).

Conversely, silt is described as solid particles carried and deposited by wind, water, ice, or gravity. It might comprise mineral particles, organic matter, or both.

12.2.1 Soil Horizons and Profile Interactions

Comprising organic materials, water, air, and mineral particles forming different horizontal layers known as soil horizons, soil is a layered natural organism. These horizons arise throughout time via leaching, organic breakdown, weathering, and other paedogenic processes. Usually, a soil profile consists of five primary master horizons: O, A, E, B, and C as well as the unweather R horizon below. Every horizon has unique composition, texture, colour, and purpose that can all provide significant forensic cues (Soil Survey Staff 2014).

Comprising microbial biomass, plant waste, and decomposing leaves, the topmost layer—known as the O horizon—is rich in organic matter. The great humus concentration causes it to typically be black. In agricultural fields, this horizon is non-existent; it is common in wooded areas. In forensics, the O horizon can provide DNA from pollen grains, decomposing trash, or plant or microbial sources that could help identify the location or compute the period of time following deposition (Bullock et al. 1985).

Located beneath it, the A horizon—sometimes known as topsoil—is composed of a range of organic elements together with mineral particles including clay, silt, and sand. It is the most biologically active zone having microbial life and root activity. Usually sticking to shoes, tools, or clothing and used as trace evidence to link a person or object to a specific location, this horizon is vital in forensic investigations (Kirk et al. 2021). It also routinely displays human-generated disturbances, which might assist in the search for graveyards.

Because of the leaching (eluviation) of minerals like clay, iron, or aluminium oxides, some soil profiles show a lighter colour. Though less common, the E horizon can be quite important for separating layers of soil and determining whether a picture is naturally or artificially stratified.

By a process known as illuviation, materials that have leached from the top layers—including iron, aluminium oxides, clay, and occasionally calcium carbonate—accumulate in the B horizon, or subsoil. Usually showing a more constant mineral content and a denser structure, this layer examining deeper soil samples or cemeteries in forensic research requires the B horizon since it can help to keep evidence less disturbed by surface activity (Tibble et al. 2008).

C horizon consists of somewhat worn-out parent material. Its composition is mainly silt, sand, or broken rock; thus, biological activity has less effect. Although the mineral profile of this layer can be used to determine the geologic origin of the soil sample and link it to various terrain, its forensic relevance in terms of biological composition is lessened (Soil Science Society of America, 2008a, b, c).

The R horizon, which stands for unweathered geological material, is bedrock. For geological crime scene mapping or buried body recovery, it can be crucial even if it is rarely sampled in forensic contexts and is usually inaccessible without mechanical aid.

Using knowledge of the vertical variance in horizons, forensic soil analysis experts can compare questioned and known soil samples, determine the depth of evidence burial, and recreate events at a site. Every horizon in criminal investigations has special chemical, physical, and biological characteristics that can support location discrimination, time estimate, and trace evidence comparison.

12.2.2 Major Components of Soil in Forensic Context

Soil consists of a complex, dynamic mixture of organic matter, mineral material—usually from in situ or transported weathered rocks—and a permeable phase of air and water (Fig. 12.1). Whereas gaseous emissions from the soil can change the composition of soil water, solutes can change the composition of soil air. Moreover, the composition of organic and mineral compounds allows a lot of possibility (Dawson and Mayes 2015).

The bulk of inorganic, or mineral, soils are composed of minerals—many of which come from the geological parent materials (Dixon and Schultze 2002). Usually formed on transported or residual geological waste, soil can be found in direct relation to the source rocks. Apart from the minerals produced by processes of soil forming, all the minerals present in rocks are also present in soils. Furthermore, different specimens of the same material can be distinguished from one another and consequently have traits that help in sample discrimination. The most important of the several soil-forming elements is parent material and degree of weathering; however, the common set of soil-forming minerals is usually further confined over great geographical areas. Usually found in a normal soil are 6–10 main minerals including quartz, feldspar, carbonates, and clay minerals (Deng and Dixon 2002).

Soil organic matter consists in dead and decaying plant, animal, and bacterial remains as well as living bacteria and plant roots. Usually, the most common sort of soil organic matter is humus, a rather stable compound produced from very broken-down biological material. Comprising a spectrum of organic components including

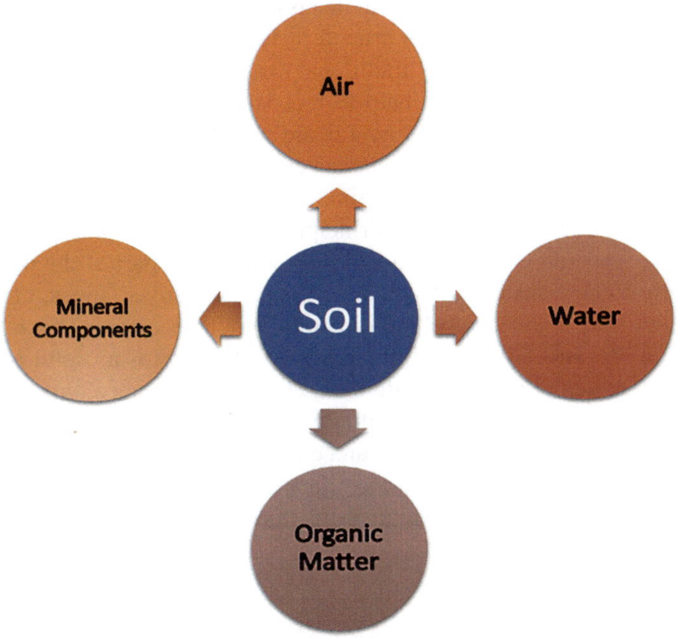

Fig. 12.1 Soil components

complex polysaccharides, lipid and wax compounds, humic acids, suberin, cutin, etc., the humic fraction is often the largest pool of organic matter. While managed soils—agricultural, municipal, or residential—may also gain organic inputs from the application of manures, slurry, composts, or mulches, the bulk of the organic matter in most soils originates from litter from surrounding vegetation. Besides, the ground might include a great range of living entities. The species and relative abundance of organisms that inhabit and are connected with soil depend much on plant species composition, soil chemistry (such as pH and elemental composition), physical structure (such as particle size), and environmental factors (such as temperature and precipitation). Apart from the physical remains of certain discovered species, the decaying imported inputs provide a broad spectrum of biochemical traces alongside the natural inputs (Dawson and Mayes 2015).

Forensic Significance of Soil Evidence**:** Soil is a varied amalgamation of organic substances, elements, air, water, and organisms. Owing to its distinctive composition and spatial variations, soil can function as a potent kind of trace evidence in forensic analyses (Pye and Croft 2004a, b; Ruffell and Wiltshire 2004). Analysing soil taken from a suspect's car, clothes, or footwear against soil from a crime scene helps one to determine whether the suspect was present at or close by that site (Ruffell and Dawson 2009). The evidential value of soil samples can be greatly improved by the presence of rare minerals, distinctive colours, textures, plant remains, pollen, and man-made materials including glass shards or paint fragments (Bull et al. 2006a, b, c, d). When alternative forms of evidence, such fingerprints or

DNA, are either absent or inconclusive, this is especially important (Morgan 2022). Often overlooked, soil evidence is crucial in forensic investigations especially in exterior events or in cases lacking traditional evidence types. When combined with other types of evidence, soil can help to reconstruct events and either support or refute investigative ideas. Inspired by Locard's Principle and the ideas of persistence and transferability, forensic scientists can examine and interpret soil to forge important links among the suspect, victim, and crime scene (Fig. 12.2).

According to Locard's Exchange Principle, "every contact leaves a trace". This means that if a person comes into touch with a location, part of the soil from that area is likely to be transferred to them or their possessions (shoes, clothes, tires, etc.), and they may also leave behind dirt from another location. This is relevant to soil evidence. Because of this reciprocal relationship, soil is a useful type of trace evidence (Gaensslen et al. 2008; Saferstein 2015). For instance, dirt particles on a suspect's shoes may be found when they walk through a muddy field at a crime scene. These particles can be gathered and examined to determine a link to the incident. This idea emphasises the basic justification for looking for and keeping soil evidence when conducting criminal investigations (Houck and Siegel 2010).

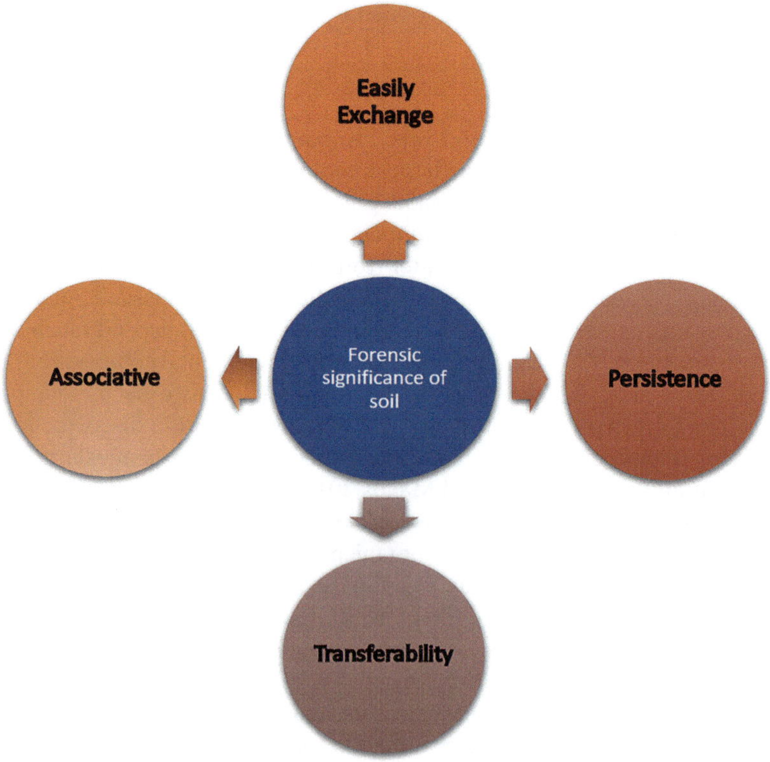

Fig. 12.2 Forensic significance of soil

The term "soil persistence" describes how long dirt particles stay on an item or person after first coming into contact. Because it can provide information on the type and timing of touch between a suspect or object and a specific environment, this idea is crucial to forensic science. A number of interconnected elements control soil persistence. One of the main factors is adhesion; particles with finer grains, particularly those that are rich in clay or organic materials, have a higher propensity to stick to surfaces. These particles are more likely to last over time because they adhere better to rough, fibrous, or porous materials, such shoe treads or clothing textiles (Pye and Croft 2004a, b). Persistence is significantly impacted by subsequent action following the original contact. Walking on different types of ground, coming into contact with different surfaces, or deliberate cleaning (e.g., brushing or washing shoes) can remove or reduce the amount of soil and hence affect its evidentiary value (Grieve and Wiggins 2001). Precipitation, humidity, wind, and temperature are examples of environmental factors that affect soil dynamics. Water can dissolve or erode soil residues, wind can spread weakly bonded particles, and extreme temperatures can alter the soil's adhesion characteristics (Bull et al. 2006a, b, c, d). When assessing whether soil on a person or object is fresh or the result of previous, unrelated contacts, persistence is essential. For instance, moist, fresh soil tends to adhere better and last longer than loose, dry dirt, which is more likely to be moved during transportation or exposure to the elements (Morgan 2019a, b), Therefore, by better knowledge of soil persistence, forensic investigators can more precisely ascertain the date and relevance of soil traces in the framework of a crime scene.

Soil transferability is the capacity of soil particles to migrate between one surface upon contact. This attribute is absolutely vital in forensic investigations, especially when deciding whether an object or suspect has come into touch with a specific environment (Pye 2007). Soil transferability is influenced by several interacting elements. Important considerations are the soil's texture and composition. Fine-grained soils high in clay or organic matter are more transferable than coarse, sandy, or gravel soils because of their smaller particle size and cohesive qualities (Bull et al. 2006a, b, c, d). These kinds of soils stick more to surfaces and remain buried in shoe tread patterns or fabric weaves. Whereas dry, loose soil may not adhere as firmly and is prone to loss or displacement, moist or wet soil shows greater adhesion and transfers more readily upon contact (Grim 1968). The kind of surface that comes into touch with the ground is also quite crucial. Rough or porous surfaces with enhanced retention qualities—such as denim, carpet fibres, and unfinished wood—allow dirt particles to become caught inside the material (Ruffell and Wiltshire 2004). Conversely, dirt is less likely to gather and stay on smooth, nonporous surfaces like metal or plastic. Additionally, the amount of soil transported is influenced by the pressure applied and the length of contact. Standing, kneeling, and pulling a heavy object across dirt are examples of prolonged, forceful contact that will result in a more substantial transfer than light or quick touch (Pye and Blott 2009). Determining whether soil discovered on a suspect, car, or object is the consequence of secondary transfer—such as walking on previously contaminated surfaces—or direct contact with a crime scene requires an understanding of these dynamics. To ascertain the soil evidence's probative value and the likelihood of transmission channels, a forensic scientist must carefully examine its context and condition.

In forensic investigations, soil can provide vital associative evidence that establishes connections between the crime scene, the victim, and the offender. The usefulness stems from the significant regional variations in soil that can be attributed to differences in organic matter, mineral composition, texture, moisture content, and human-caused contaminants. It might be evident that a suspect was present at or close to a crime scene when soil particles found on their belongings—such as shoes, clothing, or car tires—match the soil composition at the scene. The likelihood of direct contact and presence may be increased if atypical industrial soil with synthetic chemicals or unique mineral profiles is found on a suspect's vehicle and at the crime scene (Pye 2007). In a similar way, soil on a victim's clothing, under their fingernails, or on their body may help investigators determine the victim's location before death or assault, particularly in cases of outdoor attacks or secret burials (Bull et al. 2006a, b, c, d). In some cases, dirt can prove a link between a suspect and a victim if they both have the same unusual soil type from a same environment, which would otherwise be impossible to explain. This strengthens the hypothesis stating that the victim and suspect were at a particular place. Careful and exact study is necessary to enable soil to create associations. Generally speaking, forensic soil analysis combines macroscopic assessments covering colour, texture, and stratification with microscopic methods looking at pollen grains, spores, diatoms, and mineral content. Sophisticated instrumental techniques including scanning electron microscopy (SEM), X-ray diffraction (XRD), and Fourier Transform Infrared (FTIR) spectroscopy (Ruffell and Wiltshire 2004) greatly help to identify and analyse soil samples at the molecular and structural levels. By means of these analytical methods, forensic investigators can determine whether two soil samples originate from the same source, enabling the reconstruction of movements and the identification of relationships between the principal individuals and sites engaged in a crime.

12.2.3 For Forensic Science, the Importance of Soil Sediment Collection and Preservation

The collection and preservation of soil and sediment samples in forensic investigations must be given great thought and adherence to scientific processes since they could offer vital associative or circumstantial evidence linking suspects, victims, and crime scenes. The first step in good collecting is selecting pertinent sampling sites. These sites could comprise the crime scene and the suspect's clothes, shoes, car, and other objects that might have come into touch with the ground (Pye 2007). Control samples must be gathered from the area surrounding the evidence location as well as from several points across the crime scene if one is to understand the inherent variety of soil in that area. Usually using clean plastic or metal tools, researchers extract soil into sterile, sealed containers such as glass jars or plastic vials. The containers are meticulously tagged with the date, time, location, and particular reference points (Griff Griffin et al. 2015), guaranteeing correct documentation. Sending the whole item to the forensic laboratory when soil is taken from clothes or shoes rather than trying removal in the field is usually advised to help to avoid loss or contamination (Bull et al. 2006a, b, c, d).

Safeguarding is maintaining the integrity and background of the sample. Damp soil has to air-dry first before sealing to prevent chemical changes or microbial degradation (Morgan 2019a, b). Moreover, it is imperative to prevent cross-contamination between soil samples; every sample should be transported separately and instruments should be cleaned in between collections. Samples must be kept dry and refrigerated and accompanied by the suitable chain-of- custody documentation if they are legally admissible in court. These techniques are absolutely vital since soil is quite heterogeneous—even small spatial changes can generate distinctive mineral, chemical, or biological profiles. Forensic comparisons use microscopy, X-ray diffraction, SEM, FTIR spectroscopy, and chemical analysis among other techniques to identify unique indicators in the composition of the soil (Ruffell and Wiltshire 2004). These assessments may be jeopardised by improper collection or preservation, leading to the determination that the evidence is untrustworthy or inadmissible. As a result, careful management guarantees that evidence from soil and sediment can consistently substantiate theories about the presence, movement, or interaction of people or items at particular sites during criminal incidents.

12.2.4 Sampling and Collection Procedure of Soil at Crime Scene

Due to the potential of soil to provide significant associative evidence, forensic investigations are required to sample and collect soil at crime scenes. The initial phase involves identifying all potential sources and locations where soil may have been disseminated, including footwear, garments, equipment, vehicles, and sites where individuals may have interred or deposited remains. To address natural variability in soil composition, researchers should collect control (reference) samples from the crime scene and surrounding areas, alongside questioned (trace) samples from objects or suspects (Pye and Croft 2004a, b). Additional samples should be taken at regular intervals throughout the area (e.g., every 1–2 m) and at different depths, particularly if digging or burial is suspected. Control samples should be taken from the precise location where the suspect or object was thought to have come into contact with the ground (Bull et al. 2006a, b, c, d). Clean, non-contaminating implements like plastic scoops, trowels, or stainless-steel spatulas are commonly used for collecting. To avoid contamination, investigators should refrain from using their hands.

Maintaining the integrity of the soil sample depends on the container selection. The most popular containers are glass jars, polyethylene bags, or clean, airtight plastic vials. To stop mould growth or microbial degradation, wet or moist soil should be sealed in paper bags or allowed to air-dry before being sealed (Griffin et al. 2015). To maintain the original condition and enable laboratory-based recovery and analysis, soil that has stuck to clothing, tools, or footwear should not be scraped off at the scene. Instead, the entire item should be placed in cardboard boxes or paper evidence bags (Ruffell and McKinley 2005). For analytical comparison, samples weighing between 20 and 50 grammes are advised for loose soil. The date, time, precise position (ideally with GPS), depth of collection, and collector's name must all be appropriately marked on each container. Each sample should also come with a chain-of-custody form to provide legal traceability. Following these guidelines to the letter guarantees that soil samples are clean, representative, and admissible in court. It also gives forensic labs the confidence to conduct microscopic, chemical, mineralogical, and biological analysis (Morgan 2019a, b).

12.3 Analysis of Soil

12.3.1 Physical Analysis

Examining and measuring the physical characteristics of soil, such as its texture, structure, porosity, density, colour, and moisture content, is known as physical analysis. In the fields of agriculture, environmental science, civil engineering, and forensics, these characteristics are crucial for understanding soil behaviour (Hillel 2008; Brady and Weil 2016). Aristotle and other philosophers started studying soil in the fourth century BCE, but it wasn't until the late nineteenth and early twentieth centuries that systematic physical research gained traction, thanks to the efforts of soil scientists like Vasily Dokuchaev, who is regarded as the father of soil science. He underlined the importance of soil as a naturally occurring entity with distinct physical properties derived from its production processes (Brevik and Hartemink 2010). As research has developed, modern physical soil analysis techniques have evolved significantly. Traditional methods including sieve analysis for texture and soil core sampling for bulk density (Baveye et al. 2010; Dexter 2004) have been supplemented by digital imaging for colour and aggregation studies, X-ray computed tomography (CT) for visualising soil structure, and laser diffraction particle size analysis. These modern tools enable three-dimensional non-destructive analysis of soil samples and higher precision.

12.3.1.1 Particle Size Distribution

Common analysis techniques include thermogravimetric methods for estimating soil moisture, core and clod methods for ascertaining bulk density, and hydrometer and pipette methods for particle size distribution (Blake and Hartge 1986; Gee and Bauder 1986). These techniques help evaluate the soil's mechanical stress resistance, capacity to retain water, or ability to support plant development. By means of comparative soil profiling, physical soil analysis is indispensable in forensic science in connecting suspects or objects to crime sites. Because of texture, colour, and structure, every site has a different soil signature that makes soil a consistent trace evidence source. In environmental science and agriculture as well, knowledge of physical characteristics helps one ascertain soil fertility, water retention capacity, and suitability for use in construction or farming (Gee and Or 2002). Physical soil analysis is crucial since it affects land use planning, crime scene investigation, and environmental management decisions including major ones. Enhanced agricultural yields, better conservation techniques, and evidence-based court rulings in forensic settings all depend on a comprehensive awareness of these physical traits.

12.3.1.2 Soil Colour Analysis

In forensic science, geotechnical engineering, pedagogy, and environmental research, soil colour analysis is a key criterion. This has bearing on the methodical measurement and observation of soil colour, which offers vital information regarding the minerals, organic matter, moisture content, and composition of the soil. While the first methodical approach to characterise soil colour originated in the early 1900s, the field changed drastically in the 1910s when Albert H. Munsell created the Munsell Soil Colour System. Standardising soil colour descriptions depending on hue, value (lightness), and chroma (intensity) helps to simplify cross-disciplinary and cross-location comparisons (Munsell Colour Company, 2000). At first, subjective interpretation made eye observation in natural light the only way to gauge soil colour quite difficult. Eventually, forensic and agricultural sciences adopted homogeneous charts, such the Munsell Soil Colour Book, in order to lower bias and boost reliability. The advancement of digital technology has helped to better analyse soil colour. Modern methods including spectrophotometry, digital image processing, and hyperspectral imaging help to quantify soil colour and spatial variability (Viscarra Rossel et al. 2006). These methods provide more objective and higher-resolution data fit for machine learning algorithms and geographic information systems (GIS).

In forensic science, particularly with regard to trace evidence analysis, the colour of the soil is crucial for reconstructing crime scenes. Using colour evaluation in conjunction with further physicochemical testing, soil taken from a crime scene to a suspect's shoes or vehicle can be investigated to link the sample with its source and support or refute alibis. Important considerations in agronomic and environmental evaluations are also soil colour, which reveals drainage issues, organic matter (dark brown to black), and iron content (yellowish or reddish tones). Consequently, soil colour analysis has evolved from a subjective visual assessment to a scientifically sound and technologically advanced technique, greatly advancing the disciplines of agriculture, forensics, and the environment.

12.3.1.3 Soil Texture

The relative distribution of mineral particle sizes—especially those of sand, silt, and clay—impacts both the physical and chemical properties of soil. In soil science and geotechnical engineering, particle size distribution (PSD) research is absolutely essential for establishing the size range and percentage distribution of particles. Early in the 1900s, the US Department of Agriculture (USDA) and the International Society of Soil Science (ISSS) created well-accepted classification systems for soil texture. Thanks to advances in soil analysis including automated sedimentation systems, image analysis, and laser diffraction techniques (Gee and Or 2002), particle size distribution measurements are today far more accurate and efficient.

The texture and PSD of soil define its permeability, porosity, aeration, and nutrient retention, affecting agriculture, environmental research, civil engineering, and forensic investigation. For smaller particles (silt and clay), traditional particle size analysis methods include sedimentation techniques (e.g., hydrometer and pipette techniques) and mechanical screening for bigger particles (sand and gravel). The basis of these techniques is Stokes' Law, which links the sedimentation rate of particles in a fluid with their diameter and density. Modern laboratories use laser diffraction analysers to offer high-resolution particle size distribution data by measuring the light scattering pattern of soil particles, enabling faster and more complete assessments free from the need of chemical dispersants (Beuselinck et al. 1998). Understanding soil texture and particle size distribution helps one evaluate soil fertility, erosion susceptibility, drainage characteristics, and load-bearing capacity—all of which greatly influence land use planning, agricultural management, and building techniques. In forensic science, comparisons of particle size and soil texture help to recreate accident scenes or link suspects to crime scenes (Murray 2004a, b). In forensic geology and sustainable agriculture as well as in many other fields, informed decision-making depends on an awareness of soil texture and particle size distribution.

12.3.1.4 Moisture Content Analysis

It addresses about determining soil water content, usually stated as a percentage of the dry mass of the soil. For many disciplines, including hydrology, forensic science, agriculture, and geotechnical engineering, this property is vital since it influences soil behaviour, fertility, stability, and how it interacts with mechanical or environmental forces. First measured by crude, mostly qualitative techniques based on tactile evaluation and physical observation, was soil moisture. In the nineteenth century, standardising moisture measurement was accomplished using scientific techniques including gravimetric drying. The gravimetric method, which involves weighing the soil both before and after it is oven-dried at 105 °C, is still the gold standard today due to its accuracy. Offering non-destructive, in situ, real-time technology, the evolution of temporal domain reflectometry (TDR), electrical resistance sensors, and neutron probe techniques has revolutionised moisture monitoring (Gardner 1986).

Modern techniques for measuring soil moisture comprise of highly accurate gravimetric (oven-drying) methods; time domain reflectometry (TDR), which measures dielectric properties using electromagnetic pulses; and inexpensive alternative often used in farming capacitance sensors. Although neutron scattering techniques are restricted and require the use of radioactive isotopes, they offer exceptional accuracy in detecting hydrogen atoms. Electrical resistance sensors and gypsum blocks are used for less expensive field applications, although their precision is diminished. These techniques have improved soil moisture assessment's precision and speed, which is advantageous for forensic investigations, agriculture, and climate modelling (Gardner 1986; Topp and Ferre 2002; Robinson et al. 2008; Vereecken et al. 2008).

12.3.1.5 Bulk Density and Porosity

Important physical characteristics of soil significantly influence plant growth, water transport, nutrient availability, and soil vitality. Bulk density, which is typically expressed in g/cm^3, is the mass of dehydrated soil per unit volume, including air spaces and organic matter, whereas porosity is the percentage of soil volume that is made up of voids (Hillel 1998). These studies have their roots in agricultural research conducted in the early nineteenth century, when researchers looked into how soil structure affected crop output and water retention. The first techniques were straightforward using oven-drying techniques and volume displacement. These strategies evolved with changes in soil physics and the manufacturing of ever more precise sensors. While the following formula is used to indirectly calculate porosity, core samples and subsequent oven-drying at 105 °C are often used today to ascertain bulk density:

$$\text{Porosity}(\%) = \left[1 - (\text{Bulk Density} / \text{Particle Density})\right] \times 100$$

Although the particle density of mineral soils is usually regarded to be 2.65 g/cm^1 (Brady and Weil 2016), recent developments in non-destructive evaluation of soil structure and porosity include the application of geophysical techniques including ground-penetrating radar and 3D imaging technologies including X-ray computed tomography. Although porosity determines the capacity of the soil to retain and move air and water, which directly influences microbial activity and plant viability, bulk density is absolutely essential for knowledge of soil compaction and root penetration (Reynolds et al. 2002).

12.3.2 Chemical Analysis of Soil

The methodical assessment of soil's chemical characteristics, such as pH, organic matter, cation exchange capacity (CEC), nutritional content (nitrogen, phosphorus, and potassium), and the existence of contaminants, is known as chemical analysis. When agricultural scientists like Justus von Liebig emphasised the importance of minerals and chemical nutrients in plant development in the nineteenth century, chemical soil analysis was born. Because of improvements in instrumentation and a better understanding of soil-plant interactions, soil chemistry has evolved over time (Sparks 2003). Because of advancements in technology, modern soil analysis combines wet chemistry and instrumental techniques. Modern techniques include atomic absorption spectroscopy (AAS), inductively coupled plasma mass spectrometry (ICP-MS), X-ray fluorescence (XRF), and ion chromatography (IC). Earlier techniques included manual titration and colorimetry. These offer accurate measurements of the soil's heavy metals, trace elements, and micro- and macronutrients (Tan 2010). Measurements of pH and electrical conductivity (EC) aid in determining the acidity or salinity of the soil, which has a direct impact on plant health and nutrient availability. Among the disciplines depending on chemical analysis of soil are environmental science, forensic science, geology, and agriculture. It directs methods of fertilisation and soil amendment in agriculture. In forensic science, soil chemistry can assist in crime scene investigations by means of recognisable soil fingerprints connecting suspects to particular sites (Murray et al. 2009). Furthermore, used in environmental assessments are chemical data tracking of pollution levels and soil degradation, vital for sustainable land management.

12.3.2.1 Identification of pH and Conductivity of Soil

Soil analysis is an essential tool in forensic investigations since it can link objects or people to particular geographical areas. pH and electrical conductivity (EC) are two basic soil indicators used to assess the chemical and environmental traits of soil samples. While conductivity gauges the capacity of the soil to conduct electricity, which is mostly controlled by soluble salts and ion concentration, pH indicates how acidic or alkaline the soil is, influencing microbial activity and nutrient availability.

Early in the twentieth century, soil first found application in forensic investigations. Dirt and other trace evidence were first used in crime scene investigations by Austrian criminalist Hans Gross. Later, in the 1930s, Dr. Edmond Locard, sometimes called the "father of forensic science", identified soil as a form of trace evidence, thereby validating the idea that "every contact leaves a trace" (Murray and Tedrow 1975).

Sensitivity and specificity of soil analysis have been much raised by recent advances. Portable pH and EC metres, ion-selective electrodes, and automated titration systems allow on-site and lab-based assessments these days. Geospatial mapping technologies and geochemical databases also help to increase location matching and evidentiary value (Carter and Gregorich 2008).

Analytical Techniques: Usually, the pH of soil is measured with a pH metre including a glass electrode. The mixture is let to equilibrate before the measurement after mixing soil with distilled water in a 1:1 or 1:2.5 ratio. The pH of the soil helps one ascertain its acidity, neutrality, or alkalinity (Thomas 1996; Rayment and Higginson 1992).

Electrical conductivity (EC) analysis—which involves suspending soil in deionised water and measuring the conductivity of the solution—is a typical technique for spotting soluble salts like sodium, potassium, and chloride ions (Rhoades et al. 1990; Hendershot et al. 2008).

More recently, advanced methods for large-scale soil characterisation apply remote sensing technologies, capacitance sensors, and electrochemical impedance spectroscopy (Smith and Doran 1996).

When paired with mineralogical and biological study, pH and EC enhance the evidentiary link between a scene and a suspect. In forensic settings, particularly when other properties (colour, texture) are comparable, pH and conductivity can separate soil samples from various sources. These criteria could show whether a suspect has been in a given environment (such as acidic forest soil against alkaline desert soil). They also assist in identifying disturbed ground, as in illicit excavation or covert grave search (Pye and Blott 2004).

12.4 Elemental Analysis of Soil in Forensic Science

Elemental assessment of soil is the technique for ascertaining and evaluating the elemental composition of soil samples. Forensic science is vital since it allows detectives to trace soil evidence to a specific place or suspect, therefore enabling them to connect crime scenes, vehicles, or individuals. This method is predicated on the theory that different mineral, chemical, and environmental factors produce different elemental fingerprints in soil from different geographic areas (Bull et al. 2006a, b, c, d; Ruffell and Wiltshire 2004).

Early in the 1900s, Dr. Hans Gross and Edmond Locard identified the clear evidentiary value of trace elements such as soil, which helped forensic science to evolve from soil analysis. Based on Locard's Exchange Principle, which holds that "every contact leaves a trace", soil forensics was born (Locard 1930; Murray 2004a, b). Early techniques consisted mostly in visual comparison, microscopy, and basic chemical tests. Over the years, nevertheless, the development of instrumental techniques such X-ray fluorescence (XRF), atomic absorption spectroscopy (AAS), and inductively coupled plasma mass spectrometry (ICP-MS) has changed the sensitivity, precision, and reach of soil analysis (Koons and Buscaglia 2002; Rawlins et al. 2006).

Among the most powerful modern tools for trace element detection in the parts per trillion (ppt) level is ICP-MS. The material is ionised using a high-temperature plasma. Then mass-to-charge ratio of the ions is determined. Often used for profiling trace elements, particularly rare earth metals, this approach is rather successful in forensic geolocation studies (Rodushkin and Ruth 1997). ICP-MS is perfect for differentiating between identical soil types since it is sensitive, quick, and able to analyse numerous elements at once.

Another often used technique that finds the concentration of specific metal components by use of light absorption at different wavelengths is AAS. Although AAS analyses one element at a time and is less sensitive than ICP-MS, it is nevertheless somewhat cost-effective and useful for forensic circumstances involving the measurement of lead, arsenic, cadmium, and other heavy metals (Welz and Sperling 1999). Especially relevant is managing toxicological studies involving soil pollution.

X-ray fluorescence, or XRF, is a non-destructive method used to track the secondary X-rays emitted by atoms upon main X-ray source excitation. It is routinely used in field studies for elemental mapping of soil and sediment as a fast screening tool. Portable XRF (pXRF) devices are particularly helpful for on-site examination at crime scenes, therefore enabling quick decisions (Sharma et al. 2017). Although it can analyse a large number of components, XRF typically has greater detection limits than ICP-MS.

Other analytical techniques include Neutron Activation Analysis (NAA), Fourier Transform Infrared Spectroscopy (FTIR) for mineral identification, and scanning electron microscopy in combination with Energy Dispersive X-ray Spectroscopy (SEM-EDS) for microscopic elemental imaging and particle morphology.

In forensic science, soil elemental analysis is important because it helps to support environmental forensics, create an evidentiary link between the suspect and the scene, and separate soils from different geographical sites. For example, matching the soil on a suspect's shoe to that at a crime scene can be quite crucial in investigations on homicide or kidnapping. Moreover, soil composition is a reliable kind of physical evidence since it stays basically consistent across time. Table 12.1 shows the various soil elemental analysis and destructive or non-destructive techniques used in forensic science.

Apart from providing a habitat for plants, soil is a necessary component of the terrestrial ecology since it naturally filters pollutants. But quick industrialisation, intensive farming, and poor waste management have turned soil into a major storehouse of environmental contaminants like heavy metals, persistent chemicals, and surplus fertilisers. These poisons gravely endanger human health, biodiversity, and the sustainability of agriculture as well as its respective foundations. Deposition of heavy metals such as lead (Pb), cadmium (Cd), arsenic (As), chromium (Cr), and mercury (Hg) in soils due to anthropogenic activities like mining, industrial emissions, and wastewater irrigation. As stated by Alloway (2013), pharmaceuticals generally infiltrate the food chain through plant absorption; they are non-biodegradable and can persist in the environment for decades. Moreover, aggravating environmental damage is use of chemical fertilisers, especially those based on nitrogen and phosphate, which causes nutritional imbalance, soil acidification, and nitrate leaking into groundwater (Chen et al. 2018).

Strong analytical methods are used to detect and measure contaminants in soil, evaluating and reducing these risks. Inductively coupled plasma mass spectrometry (ICP-MS) has great sensitivity and multi-elemental capabilities, which allows one to routinely detect trace and ultra-trace levels of heavy metals. For routine metal analysis including copper, iron, and zinc (Welz and Sperling 1999), atomic absorption spectroscopy (AAS) presents a reasonably cost-effective method. Rapid and non-destructive field screening of elemental content uses X-ray fluorescence (XRF). Fourier Transform Infrared Spectroscopy (FTIR) is useful for spotting organic pollutants and fertiliser residues depending on their chemical structures (Griffiths and de Haseth 2007). Moreover, Neutron Activation Analysis (NAA) offers high-precision information for many elements without the need of chemical digestion; yet, its availability is limited since it depends on a nuclear reactor (Greenberg et al. 2011). These techniques guarantee exact assessment of soil pollution, enabling environmental monitoring, forensic investigations, and regulatory compliance. These methods ensure precise evaluation of soil pollution; hence, they help to support environmental monitoring, forensic investigations, and regulatory compliance. Table 12.2 provides a comprehensive overview of soil pollutants, including contamination types, common substances, their environmental and health impacts, and the analytical methods used for detection.

Table 12.1 Soil elemental analysis techniques in forensic science

Technique	Procedure	Target elements	Destructive/ non-destructive	Advantages	Disadvantages
ICP-MS (inductively coupled plasma mass spectrometry) (Rodushkin and Ruth 1997)	Soil is digested with strong acids (HNO_3/HCl) and nebulized into plasma. Ions are detected by mass spectrometer.	Wide range including trace and rare earth elements (e.g., Pb, Cd, As, U, rare earth elements)	Destructive	Extremely sensitive (ppt level)Multi-element detectionHigh precision and speed	Expensive equipmentRequires skilled operatorsSample digestion needed (hazardous acids)
AAS (atomic absorption spectroscopy) (Welz and Sperling 1999)	Soil digested in acid; solution aspirated into flame or furnace, absorption of light at element-specific wavelength measured	Primarily metals: Pb, Cd, Cu, Zn, Fe, Mg, Ca, Ni, etc.	Destructive	Cost-effectiveReliable for known metalsSimple operation	One element at a timeLess sensitive than ICP-MSSample prep needed
NAA (Neutron Activation Analysis) (Greenberg et al. 2011)	Soil is irradiated with neutrons in a reactor, forming radioactive isotopes; emitted gamma rays analysed	Over 60 elements including rare earths, halogens, and heavy metals	Non-destructive (but makes sample radioactive temporarily)	Highly sensitiveNo chemical prep neededMulti-element detection	Requires access to nuclear reactorExpensive and time-consumingRadioactive handling issues
FTIR (Fourier Transform Infrared Spectroscopy) (Griffiths and de Haseth 2007)	Soil sample (powdered or KBr pellet) analysed by IR beam; molecular vibrations produce spectra	Mineral composition: quartz, kaolinite, calcite, clays, organic matter	Non-destructive	Rapid and simpleNo sample digestionIdentifies minerals and organics	Cannot detect metalsOverlapping spectra can cause interferenceLess quantitative

(continued)

Table 12.1 (continued)

Technique	Procedure	Target elements	Destructive/ non-destructive	Advantages	Disadvantages
SEM-EDS (Scanning Electron Microscopy + Energy Dispersive X-ray Spectroscopy) (Goldstein et al. 2003)	Soil particles coated (usually with gold), imaged under SEM; EDS detects X-rays emitted from elements	Surface elements: Si, Al, Fe, Mg, Ca, K, Ti, etc.	Partially destructive (coating alters sample)	High-resolution imagingMicro-scale element mappingMorphology + composition analysis	Limited to surface elementsSample prep neededLess suitable for trace elements
XRF (X-ray Fluorescence Spectroscopy) (Jenkins et al. 1995; Sharma et al. 2017)	Soil sample (pressed pellet or loose powder) irradiated with X-rays; secondary X-rays measured	Wide range: Na to U (e.g., Fe, Mn, Zn, Cr, Ti, Pb)	Non-destructive	Field-portable versions availableNo chemical prepFast multi-element analysis	Higher detection limitsLight elements (e.g., Na, Mg) harder to detectMatrix effects possible

Table 12.2 Study of soil's contents' pollutants

Contaminant type	Common substances	Sources	Environmental and health impact	Analytical methods
Heavy metals (Kabata-Pendias and Pendias 2001; Rodushkin and Ruth 1997)	Lead (Pb), cadmium (Cd), arsenic (As), mercury (Hg), chromium (Cr), nickel (Ni), zinc (Zn), copper (Cu)	Industrial emissions, sewage sludge, battery waste, mining, paint, pesticides	Toxic to plants and animals, bioaccumulation in food chain, neurological and renal damage in humans	ICP-MS, AAS, XRF, SEM-EDS
Persistent organic pollutants (POPs) (Jones and de Voogt 1999)	PAHs, PCBs, dioxins	Petrochemical spills, incineration, pesticide use	Carcinogenic, endocrine disruptors, long-term soil persistence	GC-MS, HPLC, FTIR (for functional group analysis)
Fertilizer residues (Bockman et al. 1990; Brady and Weil 2010)	Nitrates (NO_3^-), phosphates (PO_4^{3-}), potassium (K), urea	Agricultural runoff, over-fertilisation	Eutrophication of water bodies, nitrate poisoning in humans, soil nutrient imbalance	UV-vis spectrophotometry, ion chromatography, colorimetry
Pesticides and herbicides (Racke 1992; Fenner et al. 2013)	Organochlorines, organophosphates, glyphosate, atrazine	Agricultural practices, fumigation, pest control	Toxicity to soil microbiota, groundwater contamination, cancer risk	GC-MS, LC-MS/MS, ELISA
Salts and soil acidity (pH-related) (Richards 1954; Brady and Weil 2010)	Sodium (Na^+), chloride (Cl^-), aluminium ions (Al^{3+})	Irrigation with saline water, acid rain, industrial waste	Decreased soil fertility, plant toxicity, nutrient leaching	Flame photometry, titration, EC, and pH metre
Microplastics and plastics (Rillig 2012; Zhang and Liu 2018)	Polyethylene (PE), polypropylene (PP), polystyrene (PS)	Plastic waste degradation, sewage sludge	Alters soil structure, affects water retention, harms microorganisms	FTIR, Raman spectroscopy, SEM

12.5 Importance of Microbial DNA Profiling (Metagenomics, 16SrRNA Sequencing)

Microbial DNA profiling is a useful tool especially for the analysis of soil samples in forensic investigation to ascertain the provenance, chronology, or evidence transfer. Unlike conventional soil analysis, which emphasises physical and chemical properties, microbial profiling studies the particular microbial populations present in soil, acting as a biological fingerprint of a given site (Benschop et al. 2020). Metagenomics and 16S rRNA gene sequencing define two main methods that rule forensic microbial research. By sequencing the entire DNA acquired straight from soil samples without the need of growing microorganisms, metagenomics enables a thorough research of the variety of microorganisms, including viruses, fungi, bacteria, and archaea (Thompson et al. 2017). By including common and rare species, this method presents an objective picture of the composition and operation of the microbial population. Finding location-specific microbial markers facilitates especially the connection of a suspect or object to specific surroundings.

But 16S rRNA gene sequencing amplifies and sequences the highly conserved 16S ribosomal RNA gene and concentrates especially on bacteria. Since its different sections let bacterial species vary from one another, this gene is a useful tool for taxonomic classification and microbial community comparison in soil (Janda and Abbott 2007). Because of its dependability and cost—especially when sample quantity is limited—16S rRNA profiling is often used in forensic investigations for microbial trace analysis.

Even in cases when their physical and chemical properties are similar, microbial soil profiling has forensic relevance since it can differentiate between soils from different locations depending on the make-up of their microbial populations. Soil can so adhere to equipment, clothing, shoes, or vehicles and leave behind microbial DNA that can be gathered and matched to reference samples from known locations. By considering microbial succession, this tool is quite useful for matching suspects to crime scenes or graveyards and for computing the post-mortem interval (PMI) in decomposition studies (Metcalf et al. 2015).

Finally, by use of metagenomics and 16S rRNA sequencing for microbial DNA profiling, forensic soil analysis has advanced noticeably. By allowing forensic professionals to use the invisible microbial population as trace evidence, these techniques provide them a fresh and delicate approach to relate the surroundings and attribution of sources in criminal investigations. Comparison between 16S rRNA gene sequencing and metagenomic sequencing is shown in Table 12.3.

12.6 Case Studies and Applications

12.6.1 Case I

First reported in December 2023, the SHO of Jhotwara in Jaipur said that criminals had excavated a tunnel close to a vegetable store and subterranean store housed by a bank. The suspects depart the scene. They move the earth that has been excavated using a tractor and undercover clothing. Police track for those bits of evidence using the suspect's clothes and dirt taken from one place to another (Bansal et al. 2024).

Table 12.3 A comparison between 16S rRNA gene sequencing and metagenomic sequencing

Parameter	Metagenomic sequencing (Ranjan et al. 2016; Thomas et al. 2012)	16S rRNA gene sequencing (Janda and Abbott 2007; Yarza et al. 2014)
Definition	Whole-genome shotgun sequencing of all microbial DNA in a sample, capturing both taxonomic and functional information	Amplification and sequencing of the 16S ribosomal RNA gene specific to bacteria and archaea for taxonomic identification
Analysis process	1. DNA is extracted from soil. 2. DNA is randomly fragmented and sequenced using platforms (e.g., Illumina, PacBio). 3. Reads are quality-checked, assembled, and annotated using bioinformatics tools (e.g., MG-RAST, MEGAHIT). 4. Functional gene analysis is performed (e.g., antibiotic resistance, metabolic pathways).	1. DNA is extracted from soil. 2. PCR amplifies conserved 16S rRNA regions (e.g., V3–V4). 3. Amplicons are sequenced (commonly via Illumina MiSeq). 4. Sequences are aligned and compared against databases (e.g., SILVA, Greengenes) for taxonomic classification.
Microbial targets	Bacteria, archaea, fungi, viruses, protists—all genetic material present.	Bacteria and archaea only—based on 16S rRNA gene
Key applications	Functional gene analysis (e.g., biodegradation genes) Forensic soil microbiome comparison Pathogen identification Biogeochemical process study	Microbial community composition Forensic geolocation Environmental monitoring Human or animal activity tracing via soil microbiota
Advantages	High resolution (species/strain level) Detects both known and novel microbes Functional and taxonomic insights More comprehensive than amplicon-based methods	Cost-effective Well-established databases Highly reproducible Suitable for large-scale microbial diversity comparisons
Limitations	Expensive and computationally intensive Requires high-quality DNA Complex bioinformatics required	Only provides taxonomic information Limited resolution (typically genus level) Biases due to primer selection and PCR
Destructive/ non-destructive	Destructive (requires soil DNA extraction)	Destructive (requires soil DNA extraction)

12.6.2 Case II

In a murder case, a guy gets stabbed with a knife on a highway and strikes in the skull with a stone. His dead body was found next to the highway track. Once his brother found the dead body of the victim, he contacted the investigating agency. Following an inquiry at the crime scene, the investigative team collects some evidence there: the victim's watch, blood-stained stone, blood-stained clothes, blood-stained ground, control soil near the body, and hair sample. Ten to twelve days later, the investigating officer followed the suspects. After noticing another crime site, the IO asked forensic material to be sent to the forensic science lab for further examination. The soil samples were delivered to the physics section to match the control soil and validate the location of the main and secondary crime sites by means of blood-smeared dirt. Blood on every other piece of evidence exposes specifics on the suspect's and the victim's relationship. The soil from the crime scene and the victim's clothes served as the sole hints regarding the main and secondary crime scene sites (Bansal et al. 2024).

12.6.3 Case Study III: The Murder of Tony Parsons (Scotland, 2017)

Charity rider Tony Parsons, 63, disappeared from view in 2017. His body turned up in a remote area of Scotland in 2020. Forensic soil analysis was fundamental for the investigation. Professor Lorna Dawson examined gravesite soil samples, tying them to the workplace farm of the offenders. Strong proof that helped to convict the perpetrators came from the presence of specific plant elements and soil composition (The Scottish Sun 2024, December 1).

12.6.4 Case Study IV: Murder Case in Paraná State (Brazil, 2019)

A murder case in Paraná State, Brazil, benefited much from soil evidence. Matching soil samples from the crime site with those from a suspect's vehicle was accomplished using mineralogical and chemical analysis. After the results confirmed the soil on the automobile matched the crime scene, the suspect was found close to the place where the victim's body was discovered. The conviction of the suspect was mostly attributed to this proof (de Carvalh and de Souza 2019)..

12.7 Forensic Soil Analysis: Challenges, Significance, and Legal Admissibility in India

Soil analysis has developed into a vital tool in forensic science, allowing investigators to use trace evidence to link objects, suspects, or victims to particular crime sites. Adopted in 2024, the Bharatiya Nyaya Sanhita (BNS), Bharatiya Nagarik Suraksha Sanhita (BNSS), and Bharatiya Sakshya Adhiniyam (BSA) modernise India's criminal justice system and stress the need of forensic evidence. Using both international and Indian points of view, this chapter explores the difficulties and limits of forensic soil analysis, the importance of soil matching in investigations, and the BNS and BSA frameworks' criteria for expert testimony and legal admissibility.

12.7.1 Challenges and Limitations in Forensic Soil Analysis

Since soil is naturally complex and unpredictable, forensic soil analysis offers several difficulties. Clearly pointing a sample to a specific location could prove challenging since soil composition—which is influenced by geology, climate, vegetation, and human activity—can vary significantly over small distances (Fitzpatrick 2008). The precision of associative evidence suffers when variables including microbial profiles, organic content, or mineral composition vary within metres (Bhattacharya et al. 2024). One big drawback is the lack of approved methods for compiling and evaluating soil. The findings differ depending on jurisdiction since forensic labs apply a range of techniques including spectrometry, microscopy, and microbial DNA profiling (Pye 2007). This irregularity limits cross-case comparisons and also calls for less reliability of the testimony in court. These issues are considerably exacerbated by contamination risks. Inappropriate handling, storage, or transportation might introduce extraneous particles that cause false positives or equivocal results (Murray and Tedrow 1992). Cross-contamination from shared tools or packing, for example, could compromise sample integrity. To understand soil data, one must understand many disciplines including geology, chemistry, and biology. Misinterpretation can lead to erroneous conclusions; for example, stressing too much similarities while neglecting inherent variation (Ruffell and McKinley 2008). The small sample sizes employed in forensic cases—often just a few grammes and may not fairly represent the soil profile of a site—limit thorough examination (Bhattacharya et al. 2024). Another problem is temporal changes. Human activity (including building) can alter the composition of soil over time; hence, samples collected weeks later would not match the crime scene (Fitzpatrick 2008). Finally, the dearth of large-scale soil data limits contextual analysis. Unlike DNA or fingerprint databases, which enable one to ascertain the uniqueness of a sample or estimate match probabilities, there are few international or regional soil databases (Pye 2007). India, whose forensic infrastructure is still developing, aggravates these challenges by underlining the need of consistent practices and better training.

12.7.2 Significance of Soil Matches in Forensic Investigations

Essential for forensic investigations, soil matching provides associative evidence linking objects or individuals to crime scenes. Soil is so ubiquitous that it can be found on equipment, clothes, shoes, and vehicles; it is therefore a popular type of trace evidence in crimes ranging from environmental breaches to murders (Murray and Tedrow 1992). Soil matches their evidential value because of the particular mix of physical, chemical, and biological traits at a given site—such as mineral composition, particle size, or microbial DNA (Ruffell and McKinley 2008). In prominent Indian trials, soil evidence has proved absolutely vital. For rural homicide investigations, for example, soil matches have been used to link suspect vehicles to crime scenes, supporting other evidence or providing recommendations for additional investigation (Bhattacharya et al. 2024). In outdoor crime scenes, when fibres and other hints could fade, soil comes really handy. In cases involving hidden graves, soil analysis can help with body retrieval and suspect identification by spotting disturbances or alien elements (Fitzpatrick 2008). Comprehensive information from small samples allows advanced techniques including X-ray diffraction, scanning electron microscopy, and microbial profiling to increase the accuracy of soil matching (Ruffell and McKinley 2008). Soil evidence may also expose secondary transfer, including soil carried by a vehicle, thereby helping to rebuild the dynamics of a crime scene. Furthermore, soil matches can refute alibis; if a suspect's clothes match a crime site, it raises doubts on their claimed location (Pye 2007). Determining the relevance of soil matching calls both careful collecting, analysis, and interpretation. Professionals who correctly explain the probative relevance of soil evidence raise the likelihood that the BNSS's demand that forensic teams participate in major Indian cases will result in convictions or direct enquiries (Verma 2024).

12.7.3 Legal Admissibility and Expert Testimony under BNS and BSA

Giving forensic evidence first importance, the BNS, BNSS, and BSA have transformed India's criminal justice system with effect from July 1, 2024. The BSA, which controls admissibility of evidence and names soil as "real evidence" to prove facts, replaced the Indian Evidence Act of 1872 (Section 2, BSA, 2023). Admissibility calls for relevance, dependability, and a recorded chain of custody to help prevent contamination or manipulation (Verma 2024). The BNSS mandates forensic team visits for offences carrying a 7-year or greater penalty, therefore ensuring the scientific collection of soil samples (Section 176, BNSS, 2023). Expert witness is absolutely vital when presenting soil evidence before courts. Under Section 39–45 of the BSA, the views of forensic soil scientists are admissible as relevant facts provided they show competency and use scientific principles (Forensics Digest 2024). Experts should clarify the need of a soil match by adding mentioning spectrometry or microscopy and pointing out potential disadvantages such as soil variability or secondary transfer concerns (Fitzpatrick 2008). Precedents like Kishore Chand v. State of Himachal Pradesh say that stressing the certainty of a match is immoral and could lead to the evidence being rejected (1991) (Forensics Digest 2024). Like the Daubert criteria, the BSA follows worldwide guidelines and requires that methods be widely approved, tested, and subject to peer review (Faigman et al. 2015). But India's poor forensic system and lack of skilled personnel compromise admissibility (Forensics Digest 2024). In cases involving different soils, the lack of standardised soil analysis methods could make judges doubt the validity of the findings (Bhattacharya et al. 2024). When combined with other data, soil evidence usually supports the arguments presented in the case. The focus on forensic science by the BNS and BNSS emphasises the need of strong training and infrastructure to guarantee that soil evidence fits legal criteria, enabling professionals to properly handle the complexity of science and judicial requirements.

12.8 Future Prospects in Forensic Soil Analysis: Novelties and Worldwide Viewpoints

Forensic soil analysis is a fundamental technique for tying suspects, victims, or objects to crime scenes using trace evidence. Though its potential is great, problems include soil heterogeneity, a lack of standards, and a lack of databases compromise its dependability. Emerging technologies such soil metagenomics, geographic information systems (GIS), remote sensing, artificial intelligence (AI), and worldwide soil databases seem promising in order to overcome these limits and enhance accuracy and applicability. These changes primarily relate to India, where forensic evidence is highly valued by the Bharatiya Nyaya Sanhita (BNS) and Bharatiya Sakshya Adhiniyam (BSA), which go into effect in 2024. This chapter investigates these future opportunities utilising case studies and global research to show their revolutionary prospects.

12.9 Advances in Soil Metagenomics

Soil metagenomics is the study of microbial DNA from environmental samples and is changing forensic soil analysis by providing biological markers unique to a certain location. Metagenomics creates a "biological fingerprint" from the distinct microbial populations in soil, in contrast to traditional techniques that concentrate on physical or chemical characteristics (Young et al. 2017). Thanks to high-throughput sequencing methods such as 16S rRNA sequencing, forensic investigators can separate soil samples from various geographical areas even in trace levels (Habtom et al. 2017). The forensic value of metagenomics is demonstrated through case studies. In a fictitious rural murder case in India, microbial profiles from soil on a suspect's shoes could connect them to a particular field, supporting tangible evidence like mineral content (Bhattacharya et al. 2024). By distinguishing between urban and rural soils, metagenomics has aided research worldwide, according to studies with sample differentiation accuracy of over 90% (Young et al. 2017). Complex data interpretation, exorbitant sequencing costs, and environmental-induced temporal fluctuations in microbial populations are some of the difficulties (Habtom et al. 2017). In addition to microbiological reference databases specifically designed for forensic applications and standardised procedures, future plans call for the development of low-cost sequencing platforms. India will be able to match the BNS's emphasis on scientific evidence and improve case resolution in a variety of settings by integrating metagenomics into forensic labs.

12.10 Integration of GIS and Remote Sensing

By providing contextual and spatial insights, remote sensing and geographic information systems (GIS) are revolutionising forensic soil analysis by mapping soil characteristics like mineralogy or organic content. GIS helps law enforcement to recreate crime scenes and track down suspects over long distances (Ruffell and McKinley 2008). Drones or satellite photos used for remote sensing can identify changes in vegetation or disturbances in the soil that could indicate illegal activity, like hidden graves (Fitzpatrick 2008). These tools have proven to be very helpful in forensic work.

Investigators were directed to the victim's remains in a 2019 Brazilian homicide case after remote sensing detected soil anomalies in a cemetery (de Carvalho and de Souza 2019). GIS mapping of the kinds of soil discovered in rural crime sites could help to support the BNSS mandate for forensic participation in major crimes in India (Verma 2024). In a bank burglary case in Jaipur, for example, GIS may have followed dirt carried by a tractor back to its source, therefore supporting the evidence (Bansal and Sharma 2024).

Future advances include building real-time remote sensing systems and merging GIS with hyperspectral imagery for more exact soil classification. Obstacles are high costs, the difficulty of data processing, and the need for specific knowledge. Broad acceptance made possible by investments in forensic infrastructure could help spatial forensics under the BSA framework to be further developed in India.

12.11 AI and Machine Learning in Soil Pattern Recognition

Forensics soil analysis is being transformed by artificial intelligence (AI) and machine learning (ML) automating pattern recognition and improving data interpretation. AI systems search complex soil records looking for minute patterns and limit human error by combining physical, chemical, and microbiological traits (Pye 2007). Machine learning models can be trained using big datasets, which will then enable precise classification of soil samples or prediction of their geographical source. Recent research emphasises the promise of artificial intelligence. Using deep learning and spectral data, Zhang et al. (2023) categorised soil samples such that they could find origins with above 90% accuracy. AI-assisted analysis could have sped up soil comparisons between the suspects' farm and the funeral site in Scotland's 2017 Tony Parsons murder case, therefore strengthening the case (The Scottish Sun 2024). Artificial intelligence (AI) could automate soil comparisons in India in events such as the Jaipur highway murder, where soil on the victim's clothes confirmed the whereabouts of crime scenes (Bansal and Sharma 2024). Obstacles include the need for large, diverse training datasets and the danger of overfitting models to certain geographic locations (Bhattacharya et al. 2024). For forensic soil analysis, metagenomic and GIS data will be combined with open-source artificial intelligence systems going forward. Adoption of artificial intelligence in India could enhance the necessity for scientific rigour of the BSA if labs are equipped with the required instruments and guidance.

12.12 International Soil Databases for Forensic Use

Lack of thorough soil databases limits forensic soil analysis and makes it challenging to assess the uniqueness or match likelihood of a sample (Pye 2007). The development of worldwide soil databases combining microbiological, chemical, and physical traits from numerous geographical locations marks a vital future route. Estimating match likelihoods allows forensic experts to substantiate their conclusions in court using these databases. Global efforts are under way. The European Soil Database provides thorough soil maps (Ruffell and McKinley 2008), therefore facilitating forensic comparisons. To improve casework, the FBI's forensic soil database for the United States combines mineralogical and chemical information (Young et al. 2017). India's lack of a national forensic soil database (Bhattacharya et al. 2024) could find roots in regional microbial research. Future initiatives aim to create interoperable worldwide databases, maybe under the administration of INTERPOL, standardising data and enabling cross-border investigations. Obstacles include funding, data protection, and international collaboration considering different soil types. Particularly with regard to interstate crimes, a national database compliant with BNS and BSA criteria could help to enhance forensic skills in India.

12.13 Conclusion

12.13.1 Future of Soil Forensics

Forensic soil analysis will need to combine soil metagenomics, GIS, remote sensing, artificial intelligence, and multinational databases in order to get above current limitations. Cases from Brazil, Scotland, and Jaipur demonstrate how these advances provide better dependability, speed, and precision. Once infrastructure and training gaps are resolved, using these technologies in India inside the BNS and BSA systems would enhance forensic investigations. By following these advances, forensic soil analysis could become a basic element of both Indian and international criminal justice systems. Though forensic soil analysis has great potential to link suspects to crime sites, its flaws—homogeneity, lack of standardisation, and database scarcity—demand ongoing research and development. The BNS and BSA frameworks' efficacy in India, which increase the value of forensic evidence, depends on filling in infrastructural gaps and ensuring that expert opinion conforms with legal and scientific criteria. By means of overcoming these challenges, soil analysis will be able to prove itself as a basic component of forensic investigations, therefore strengthening the administration of justice in India's revised criminal justice system.

12.13.2 Summary of Forensic Utility of Soil and Sediments

Often engaged in crimes ranging from environmental breaches to murders, soil and sediments can attach to clothes, shoes, vehicles, and tools. Their forensic relevance comes from their variable composition—minerals, organic compounds, and microbial communities—that varies with location and creates a "geological fingerprint" (Murray and Tedrow 1992). Given its uniqueness, soil can be used as associative evidence to link objects or persons to specific sites. In outdoor crime scenes, soil evidence is quite helpful since other traces, including fibres, might degrade. In a 2017 murder case, soil from a burial site featuring specific plant ingredients linked suspects to a farm, therefore supporting their conviction (The Scottish Sun 2024). Like this, soil hauled by a tractor generated investigative leads in a 2023 bank theft in Jaipur, highlighting the need of soil in tracing down movement (Bansal and Sharma 2024). Sediments, such those found in aquatic habitats, which preserve evidence like pollen or diatoms (Pye 2007) help with drowning or burial instances, therefore augmenting this benefit. Analytical techniques raise the obvious worth of soil. Using metagenomics, spectrometry, and microscopy, soil samples—even in minute quantities—can be extensively characterised (Young et al. 2017). Unlike hair or blood evidence, soil found on the victim's clothes in a highway murder in Jaipur confirmed the main and secondary crime scene sites (Bansal and Sharma 2024). As shown in cases when a suspect's claimed whereabouts runs counter to their dirt on their shoes, soil can likewise refute alibis (Ruffell and McKinley 2008). Notwithstanding its benefits, soil evidence has some challenges including homogeneity, contamination risks, and lack of thorough databases (Fitzpatrick 2008). Given the BSA's focus on scientific evidence, these limitations demand for rigorous criteria for data collecting and analysis to ensure reliability in court (Verma 2024).

12.13.3 Importance of Multidisciplinary Approaches

The intricacy of soil and sediment evidence calls for multidisciplinary methods including knowledge from geology, chemistry, biology, and data science. One field cannot sufficiently characterise the several elements of soil, including organic compounds, minerals, and microbial DNA. Forensic geologists look at mineral content, chemists review chemical profiles, and microbiologists look at microbial populations (Bhattacharya et al. 2024) to offer a complete analysis. Emerging technologies underline the need of collaboration. Bioinformatics understanding is essential to grasp sequencing data for soil metagenomics, which detects microbial profiles unique to a certain area (Habtom et al. 2017). In a 2019 Brazilian murder case, a suspect was linked to the crime site using geological knowledge in concert with mineralogical and chemical examination of car soil (de Carvalho and de Souza 2019). Working using geographic information systems (GIS) and remote sensing, geographers and data analysts map soil features or spot anomalies (Ruffell and McKinley 2008). Multidisciplinary approaches also help the courtroom to look better. To combine complicated facts into understandable, legally admissible evidence, expert witness calls both scientific knowledge and communication abilities (Fitzpatrick 2008). Cases like Kishore Chand v. State of Himachal Pradesh (1991) show that, although courts seek accuracy and scientific rigour (Forensics Digest 2024), the BSA in India allows expert opinions as relevant facts. Working together, forensic scientists and solicitors ensure that soil evidence meets these criteria. Particularly in developing countries like India (Bhattacharya et al. 2024), coordination between disciplines, different analytical standards, and resource constraints creates challenges to multidisciplinary approaches. Future projects should concentrate on creating multidisciplinary training programmes and standardising methods if we are to enable cooperation and keep up with international forensic discoveries.

12.13.4 Role in Strengthening Criminal Investigations

Soil and sediment evidence considerably improve criminal investigations by providing strong, scientifically based links between suspects and crime sites. Its purpose transcends mere association; it also supports other sources of evidence, guides searches, and helps to reinterpret crime scene dynamics. Soil on clothes and hauled by a tractor in the Jaipur bank break-in helped police track the perpetrators and concentrate their activities (Bansal and Sharma 2024). Using soil study, detectives in the Tony Parsons case found a burial site, leading to convictions (The Scottish Sun 2024). When there is no direct evidence in cases such as environmental crimes or hidden burials, soil evidence is particularly crucial. In Brazil, the lack of eyewitnesses was overcome when dirt from a suspect's automobile turned up at a murder scene (de Carvalho and de Souza 2019). The BNSS strengthens the investigative power of soil in India by means of forensic team involvement for serious crimes (Verma 2024). As the Jaipur highway murder shows, where soil matching helped to clear the chronology of events, soil can also differentiate between major and secondary crime scenes (Bansal and Sharma 2024). Innovations like artificial intelligence and machine learning raise the relevance of soil. Using AI models that can categorise soil samples with above 90% accuracy (Zhang et al. 2023) can help to speed investigations. Notwithstanding their constraints, international soil databases could improve court presentations by providing statistical background for events (Pye 2007). Building a national database in India would help with cross-state investigations in line with BNS goals. Legal admissibility of soil determines its investigative value. The BSA requires a recorded chain of custody and reliable methods for soil evidence to pass legal review (Forensics Digest 2024). Expert testimony can close the discrepancy between scientific complexity and legal standards, as shown in the Brazilian case where mineralogical evidence met exact criteria (de Carvalho and de Souza 2019). Obstacles including India's poor forensic system need for expenses if we are to fully use the potential of soil.

References

Alloway BJ (2013) Heavy metals in soils: Trace metals and metalloids in soils and their bioavailability, 3rd edn. Springer

Bansal K, Sharma M (2024) Forensic soil analysis in criminal investigations: case studies from Jaipur, India. [Publication details unavailable]

Bansal K, Sharma M, Daga SS (2024) Soil: forensic interpretation – some case studies. Indian J Forensic Community Med 11(3):126–129. https://doi.org/10.18231/j.ijfcm.2024.027

Baveye PC, Laba M, Otten W, Bouckaert L, Sterpaio PD, Goswami RR, Grinev D (2010) Observer-dependent variability of soil structure visualization using X-ray computed tomography. Geoderma 157(1–2):51–63. https://doi.org/10.1016/j.geoderma.2010.03.009

Benschop CCG, Boonstra N, Meiland HC, van der Beek CP (2020) Microbiome analysis in forensic soil comparisons: A pilot study. Forensic Sci Int Genet 47:102298. https://doi.org/10.1016/j.fsigen.2020.102298

Beuselinck L, Govers G, Poesen J, Degraer G, Froyen L (1998) Grain-size analysis by laser diffractometry: comparison with the sieve-pipette method. Catena 32(3–4):193–208. https://doi.org/10.1016/S0341-8162(98)00051-4

Bhattacharya S, Dutta S, Das S (2024) Soils in forensic analysis: a review. Int J Adv Res 12(7):928–936. https://doi.org/10.21474/IJAR01/17192

Blake GR, Hartge KH (1986) Bulk density. In: Klute A (ed) Methods of soil analysis: part 1—physical and mineralogical methods, 2nd edn. ASA and SSSA, pp 363–375

Bockman OC, Kaarstad O, Lie OH, Richards I (1990) Agriculture and fertilizers. Norsk Hydro

Brady NC, Weil RR (2010) Elements of the nature and properties of soils, 3rd edn. Pearson Education

Brady NC, Weil RR (2016) The nature and properties of soils, 15th edn. Pearson Education

Brevik EC, Hartemink AE (2010) Early soil knowledge and the birth and development of soil science. Catena 83(1):23–33. https://doi.org/10.1016/j.catena.2010.06.011

Bull PA, Morgan RM, Dunkerley S (2006a) Forensic geoscience: applications of geology, geomorphology and geophysics to criminal investigations. Earth Sci Rev 69(3–4):245–261. https://doi.org/10.1016/j.earscirev.2004.08.003

Bull PA, Morgan RM, Dunkerley S (2006b) The transfer and persistence of trace particulates: experimental studies using clothing fabrics. Sci Justice 46(3):141–150. https://doi.org/10.1016/S1355-0306(06)71585-9

Bull PA, Morgan RM, Dunkerley S (2006c) The transfer and persistence of trace particulates: Forensic geoscience and crime reconstruction. Sci Justice 46(2):107–124. https://doi.org/10.1016/S1355-0306(06)71596-7

Bull PA, Morgan RM, Dunstan D (2006d) Forensic geoscience and crime detection. Minerva Med 126(1):29–42

Bullock P, Fedoroff N, Jongerius A, Stoops G, Tursina T (1985) Handbook for soil thin section description. Waine Research Publications

Carter MR, Gregorich EG (2008) Soil sampling and methods of analysis, 2nd edn. CRC Press

Chen N, Lu Y, Ding X, Li L, Fu C (2018) Effects of nitrogen and phosphorus fertilizers on soil nutrient and heavy metal concentrations in suburban vegetable fields in Nanjing, China. Sustainability 10(11):3786. https://doi.org/10.3390/su10113786

Chisum W, Turvey B (2000) Evidence dynamics: Locard's exchange principle and crime reconstruction. J Behav Profil 1:1–15

Dawson LA, Mayes RW (2015) Criminal and environmental soil forensics: soil as physical evidence in forensic investigations. In: Introduction to environmental forensics. Academic Press, pp 457–486

de Carvalho MA, de Souza AB (2019) Forensic soil evidence in a murder investigation in Paraná State, Brazil. [Journal article, publication details unavailable]. PubMed

Deng Y, Dixon JB (2002) Soil organic matter and organic-mineral interactions. Soil Mineral Environ Appl 7:69–107

Dexter AR (2004) Soil physical quality: Part I. Theory, effects of soil texture, density, and organic matter, and effects on root growth. Geoderma 120(3–4):201–214. https://doi.org/10.1016/j.geoderma.2003.09.004

Faigman DL, Slobogin C, Monahan J (2015) Gatekeeping science: Using the structure of scientific research to distinguish between admissibility and weight in expert testimony. Northwest Univ Law Rev 110(1):859–904. https://ssrn.com/abstract=2665847

Fenner K, Canonica S, Wackett LP, Elsner M (2013) Evaluating pesticide degradation in the environment: Blind spots and emerging opportunities. Science 341(6147):752–758. https://doi.org/10.1126/science.1236281

Fitzpatrick RW (2008) Soil: forensic analysis. In: Encyclopedia of forensic sciences. Elsevier, pp 1737–1745. https://doi.org/10.1016/B978-012382165-2/00263-7

Fitzpatrick RW (2009) Soil: forensic analysis. Wiley encyclopedia of forensic science. John Wiley & Sons, Chichester, pp 2377–2388

Fitzpatrick RW, Raven MD, Forrester ST (2009) A systematic approach to soil forensics: criminal case studies involving transference from crime scene to forensic evidence. In: Criminal and environmental soil forensics. Springer Netherlands, Dordrecht, pp 105–127
Forensics Digest (2024, Oct 7) Key case laws and rules of evidence in forensic science: an Indian perspective. Forensics Digest. https://forensicsdigest.com/key-case-laws-and-rules-of-evidence-in-forensic-science-an-indian
Gaensslen RE, Harris HA, Lee HC (2008) Introduction to forensic science and criminalistics. McGraw-Hill
Gardner WH (1986) Water content. In: Klute A (ed) Methods of soil analysis: Part 1—physical and mineralogical methods, 2nd edn. ASA and SSSA, Madison, pp 493–544
Gee GW, Bauder JW (1986) Particle-size analysis. In: Klute A (ed) Methods of soil analysis: Part 1—physical and mineralogical methods, 2nd edn. ASA and SSSA, pp 383–411
Gee GW, Or D (2002) Particle-size analysis. In: Dane JH, Topp GC (eds) Methods of soil analysis: Part 4 physical methods. Soil Science Society of America, pp 255–293
Goldstein J, Newbury D, Joy D, Lyman C, Echlin P, Lifshin E, Sawyer L, Michael J (2003) Scanning electron microscopy and x-ray microanalysis, 3rd edn. Springer
Greenberg RR, Bode P, De Nadai Fernandes EA (2011) Neutron activation analysis: a primary method of measurement. Spectrochim Acta B At Spectrosc 66(3–4):193–241. https://doi.org/10.1016/j.sab.2011.03.010
Grieve MC, Wiggins KG (2001) Fibres and their forensic significance. Sci Justice 41(1):71–75. https://doi.org/10.1016/S1355-0306(01)71778-1
Griffin R, Briley DS, Ley D (2015) Forensic soil analysis: standards and protocols for evidence collection. JFI 65(3):245–259
Griffiths PR, de Haseth JA (2007) Fourier transform infrared spectrometry, 2nd edn. Wiley
Grim RE (1968) Clay mineralogy, 2nd edn. McGraw-Hill
Habtom H, Pasternak Z, Matan O, Blau K, Jurkevitch E (2017) Applying microbial biogeography in soil forensics. Forensic Sci Int Genet 34:195–203. https://doi.org/10.1016/j.fsigen.2018.02.014
Hendershot WH, Lalande H, Duquette M (2008) Soil reaction and exchangeable acidity. In: Carter MR, Gregorich EG (eds) Soil sampling and methods of analysis, 2nd edn. CRC Press, pp 173–178
Hillel D (1998) Environmental soil physics. Academic Press, San Diego
Hillel D (2008) Soil in the environment: crucible of terrestrial life. Academic Press
Houck MM, Siegel JA (2010) Fundamentals of forensic science, 2nd edn. Academic Press
Janda JM, Abbott SL (2007) 16S rRNA gene sequencing for bacterial identification in the diagnostic laboratory: Pluses, perils, and pitfalls. J Clin Microbiol 45(9):2761–2764. https://doi.org/10.1128/JCM.01228-07
Jenkins R, Gould RW, Gedcke D (1995) Quantitative X-ray spectrometry. CRC Press
Jones KC, de Voogt P (1999) Persistent organic pollutants (POPs): state of the science. Environ Pollut 100(1–3):209–221
Kabata-Pendias A, Pendias H (2001) Trace elements in soils and plants, 3rd edn. CRC Press
Kirk C, Dawson L, Mountain BW (2021) Forensic soil analysis: applications and recent developments. Forensic Sci Int 321:110730. https://doi.org/10.1016/j.forsciint.2021.110730
Koons RD, Buscaglia J (2002) Forensic significance of elemental composition and morphology of glass fragments. J Forensic Sci 47(3):524–530
Locard E (1930) The analysis of dust traces. Am J Pol Science 1:276
Marumo Y, Sugita R, Seta S (1995) Soil as evidence in criminal investigation. Identification Reference Centre, National Research Institute of Police Science, National Police Academy, Japan
Metcalf JL, Xu ZZ, Weiss S, Lax S, Van Treuren W, Hyde ER et al (2015) Microbial community assembly and metabolic function during mammalian corpse decomposition. Science 351(6269):158–162. https://doi.org/10.1126/science.aad2646
Morgan RM (2019a) Conceptualising forensic science and forensic reconstruction. Sci Justice 59(6):515–522. https://doi.org/10.1016/j.scijus.2019.05.003
Morgan RM (2019b) Conceptualising forensic science and forensic reconstruction. Sci Justice 59(3):218–224. https://doi.org/10.1016/j.scijus.2019.01.003
Morgan RM (2022) Forensic science: a very short introduction. Oxford University Press

Munsell Color Company (2000) Munsell soil color charts. GretagMacbeth, New Windsor, NY
Murray RC (2004a) Evidence from the earth: forensic geology and criminal investigation. Mountain Press Publishing Company
Murray RC (2004b) Evidence from the ground: The role of soil science in criminal investigations. Geotimes 49(10):24–27
Murray RC, Tedrow JCF (1975) Forensic geology: earth sciences and criminal investigation. Rutgers University Press
Murray RC, Tedrow JCF (1992) Forensic geology. Prentice Hall
Murray RC, Tedrow JCF, Foulk GA (2009) Forensic geoscience: principles, techniques and applications. Geological Society of London
Osman KT, Osman KT (2013) Factors and processes of soil formation. Soils: Principles, Properties and Management, pp 17–30
Pye K (2007) Geological and soil evidence: Forensic applications. CRC Press. https://doi.org/10.1201/9781420004755
Pye, K., & Blott, S. J. (2004). Development of a forensic methodology for soil analysis. Forensic Sci Int, 145(2–3), 195–203. doi:https://doi.org/10.1016/j.forsciint.2004.04.070
Pye K, Blott SJ (2009) Development of a forensic database of soil properties and characteristics. Sci Justice 49(4):204–212. https://doi.org/10.1016/j.scijus.2009.07.001
Pye K, Croft DJ (2004a) Forensic geoscience: applications of geology, geomorphology and geochemistry to criminal investigations. Earth Sci Rev 69(3–4):193–226. https://doi.org/10.1016/j.earscirev.2004.08.002
Pye K, Croft DJ (2004b) Forensic geoscience: applications of geology, geomorphology and geochemistry to criminal investigations. Earth Sci Rev 69(3–4):271–291. https://doi.org/10.1016/j.earscirev.2004.08.002
Racke KD (1992) Degradation of organophosphorus insecticides in environmental matrices. ACS Symp Ser 505:1–18
Ranjan R, Rani A, Metwally A, McGee HS, Perkins DL (2016) Analysis of the microbiome: advantages of whole genome shotgun versus 16S amplicon sequencing. Biochem Biophys Res Commun 469(4):967–977
Rawlins BG, Lark RM, O'Donnell KE, Tye AM (2006) The use of soil geochemical and mineralogical data to help source artefacts. Forensic Sci Int 162(2–3):139–150. https://doi.org/10.1016/j.forsciint.2005.10.015
Rayment GE, Higginson FR (1992) Australian laboratory handbook of soil and water chemical methods. Inkata Press
Reynolds WD, Drury CF, Yang XM, Tan CS (2002) Use of indicators and pore volume-function characteristics to quantify soil physical quality. Geoderma 110(1–2):131–146. https://doi.org/10.1016/S0016-7061(02)00242-8
Rhoades JD, Kandiah A, Mashali AM (1990) The use of saline waters for crop production (FAO Irrigation and Drainage Paper No. 48). Food and Agriculture Organization of the United Nations
Richards, L. A. (Ed.). (1954). Diagnosis and improvement of Saline and Alkali soils. USDA Agricultural Handbook No 60, 154
Rillig MC (2012) Microplastic in terrestrial ecosystems and the soil? Environ Sci Technol 46(12):6453–6454
Robinson DA, Campbell CS, Hopmans JW, Hornbuckle BK, Jones SB, Knight R et al (2008) Soil moisture measurement for ecological and hydrological watershed-scale observatories: a review. Vadose Zone J 7(1):358–389. https://doi.org/10.2136/vzj2007.0143
Rodushkin I, Ruth T (1997) Determination of trace metals in estuarine and seawater reference materials by high resolution inductively coupled plasma mass spectrometry. J Anal At Spectrom 12(10):1181–1185
Ruffell A, Dawson L (2009) Forensic soil analysis. In: Ritz K, Dawson L, Miller D (eds) Criminal and environmental soil forensics. Springer, pp 113–128. https://doi.org/10.1007/978-1-4020-9204-6_8
Ruffell A, McKinley J (2005) Forensic geoscience: applications of geology, geomorphology and geophysics to criminal investigations. Earth Sci Rev 69(3–4):235–247
Ruffell A, McKinley J (2008) Geoforensics. Wiley-Blackwell. https://doi.org/10.1002/9780470985083

Ruffell A, Wiltshire P (2004) Conjunctive use of quantitative and qualitative X-ray diffraction analysis of soils and rocks for forensic analysis. Forensic Sci Int 145(1):13–23. https://doi.org/10.1016/j.forsciint.2004.02.023, https://doi.org/10.1016/j.forsciint.2004.03.018, https://doi.org/10.1016/j.forsciint.2004.03.013
Saferstein R (2004) Criminalistics: An introduction to forensic science
Saferstein R (2015) Criminalistics: An introduction to forensic science, 11th edn. Pearson
Sharma R, Mehra R, Walia V, Singh M (2017) Application of portable XRF and GIS to map heavy metals in agricultural soils. Environ Monit Assess 189(3):115. https://doi.org/10.1007/s10661-017-5826-z
Smith JL, Doran JW (1996) Measurement and use of pH and electrical conductivity for soil quality analysis. In: Doran JW, Jones AJ (eds) Methods for Assessing Soil Quality. Soil Science Society of America, pp 169–185
Soil Science Society of America (2008a) Glossary of soil science terms 2008. SSSA
Soil Science Society of America (2008b) Glossary of soil science terms. https://www.soils.org/publications/soils-glossary
Soil Science Society of America (2008c) Glossary of soil science terms 2008. ASA-CSSA-SSSA
Soil Survey Staff (2014) Keys to soil taxonomy, 12th edn. USDA-NRCS
Sparks DL (2003) Environmental soil chemistry. Academic Press
Tan KH (2010) Principles of soil chemistry, 4th edn. CRC Press
The Scottish Sun (2024, Dec 1) Forensic expert: how soil analysis helped solve the Tony Parsons murder case. https://www.thescottishsun.co.uk/news/13830860/forensic-expert-expert-witness-scots-murder/
Thomas GW (1996) Soil pH and soil acidity. In: Sparks DL (ed) Methods of soil analysis: part 3—chemical methods. Soil Science Society of America, pp 475–490
Thomas T, Gilbert J, Meyer F (2012) Metagenomics-a guide from sampling to data analysis. Microb Inform Exp 2(1):3
Thompson LR, Sanders JG, McDonald D, Amir A, Ladau J, Locey KJ et al (2017) A communal catalogue reveals Earth's multiscale microbial diversity. Nature 551(7681):457–463. https://doi.org/10.1038/nature24621
Tibble J, Graham E, Dawson LA (2008) Forensic implications of soil horizon variation in body recovery and burial depth determination. Sci Justice 48(3):132–137. https://doi.org/10.1016/j.scijus.2007.12.002
Topp GC, Ferre PA (2002) Time domain reflectometry: a seminal technique for measuring mass and energy in soil. Soil Sci Soc Am J 66(6):1442–1454. https://doi.org/10.2136/sssaj2002.1442
Vereecken H, Huisman JA, Bogena H, Vanderborght J, Vrugt JA, Hopmans JW (2008) On the value of soil moisture measurements in vadose zone hydrology: a review. Water Resour Res 44(W00D06). https://doi.org/10.1029/2008WR006829
Verma A (2024, Sept 8) Types of evidence under Indian evidence law (Bhartiya Sakshya Adhiniyam, 2023). Law Article. https://lawarticle.in/types-of-evidence-under-indian-evidence-law-bhartiya-sakshya-adhiniyam-2023
Viscarra Rossel RA, Walvoort DJJ, McBratney AB, Janik LJ, Skjemstad JO (2006) Visible, near infrared, mid infrared or combined diffuse reflectance spectroscopy for simultaneous assessment of various soil properties. Geoderma 131(1–2):59–75
Welz B, Sperling M (1999) Atomic absorption spectrometry, 3rd edn. Wiley-VCH
Yarza P, Yilmaz P, Pruesse E, Glöckner FO, Ludwig W, Schleifer KH et al (2014) Uniting the classification of cultured and uncultured bacteria and archaea using 16S rRNA gene sequences. Nat Rev Microbiol 12(9):635–645
Young JM, Weyrich LS, Cooper A (2017) Forensic soil DNA analysis using high-throughput sequencing: a review. Forensic Sci Int Genet 26:76–85. https://doi.org/10.1016/j.fsigen.2016.10.013
Zhang GS, Liu YF (2018) The distribution of microplastics in soil aggregate fractions in southwestern China. Sci Total Environ 642:12–20
Zhang Y, Li X, Wang J (2023) Machine learning applications in forensic soil analysis: a review. J Forensic Sci 68(4):1123–1135. https://doi.org/10.1111/1556-4029.15234

Future Directions in Environmental Forensics

13

Baljeet Yadav, Anu Singla, Saba Rashid, Girraj Sharma, Vaibhav Saran, and Chandra Kumar

Abstract

Environmental forensics is a developing field of study applying science, law, and inquiry to identify pollution sources, timing, and accountability. Environmental forensics has been developed from pollution tracking and regulatory enforcement into a complex discipline that examines difficult environmental crimes. Pollution and environmental degradation are examined via time analysis, source attribution, and pollution fingerprinting. Modern environmental forensic investigations apply both traditional and innovative approaches. Geospatial technology, biological markers, genetic and microbiological forensics provide ecological and spatial background for contamination events. Predictive modeling, pattern discovery, and data harmonization across environmental datasets are enabled by artificial intelligence, machine learning, and big data. Research is advanced by 3D environmental reconstruction, eDNA applications, satellite, and drone observation. Despite scientific developments, environmental forensics faces legal challenges, data access, and confidentiality of environmental crimes. The future

B. Yadav (✉) · S. Rashid
School of Basic and Applied Sciences, K. R. Mangalam University, Gurugram, Haryana, India
e-mail: baljeet.yadav@krmangalam.edu.in

A. Singla
APJ Abdul Kalam Institute of Forensic Science and Criminology, Bundelkhand University, Jhansi, India

G. Sharma
Jindal Institute of Behavioral Sciences, O.P. Jindal Global University, Sonipat, Haryana, India

V. Saran
Department of Forensic Science, Amity University, Mohali, Punjab, India

C. Kumar
School of Engineering, Faculty of Science, Engineering and Technology, Universidad Mayor, Santiago, Chile

M. S. Sankhla et al. (eds.), *Introducing the Synergy Between Forensic Science and Environmental Analysis*, https://doi.org/10.1007/978-981-95-0199-1_13

of environmental forensics rests on its capacity to welcome innovation, adapt to new challenges, and link research and policy. As environmental crimes get increasingly complex and commercially motivated, environmental forensics will be more crucial than ever in ensuring responsibility and safeguarding of ecosystems.

Keywords

Environment · Contaminants · Forensics · Predictive model · Geospatial

13.1 Introduction to Environmental Forensics

Environmental forensics has always focused on tracking the origins as well as timing and responsible parties of environmental pollution. The discipline matches the expansion of public anxiety about environmental safety and industrial responsibility along with regulatory procedures. This field practiced historical pollution management until it evolved into preventive operations for monitoring environmental risks to minimize future damage. The future of environmental forensics will see expanded development because of technological, computational, analytical, and international cooperation along with multidisciplinary interaction. For successful progress, the field requires both technological progress along with considered attention to moral regulations and societal and legal standards. The upcoming decades will see environmental forensics guided by the main developments and directions that this field will follow.

Environmental forensics is a subdiscipline of analytical chemistry that involves the use of chemical analysis to help determine the source of contamination events that take place within the environment (Stephen 2008). A forensic investigation usually involves the establishment of the source of a contamination event, which can occur as a result of a myriad of factors—be it from a direct chemical spill of waste chemical or via a variety of fugitive emissions of chemicals. Such investigations usually take place when it has been established that there is contamination of a particular matrix above a regulatory threshold level or as a result of public, NGO, or media pressure. As more commonly used compounds become regulated and their unrestricted uses become banned, so forensic investigation into more recalcitrant or emerging compounds becomes more common.

Regardless of the nature of the incident, a successful forensic investigation relies on a well-designed sampling program as an integral part of any environmental forensics' investigation. There are many factors that can potentially bias the data collected and lead to false or erroneous data and subsequent results. Therefore, such studies must be conducted meticulously, such that the chemical data can withstand possible legal scrutiny. To acquire the necessary chemical data to answer the most asked questions in an environmental forensics investigation, it is first essential to understand the environmental chemistry of the compound in question. These questions may include but are not limited to what the potential sources of the contamination are, what is the likely extent of the contamination, and how did the contamination occur.

13.2 Historical Context of Environmental Forensics

Environmental forensics is the application of scientific methods and techniques to help resolve questions of environmental contamination and poisoning. It is a relatively new and rapidly developing evidence-based discipline, with roots in numerous scientific fields. The turn of the twenty-first century heralded a growing awareness of its potential value to regulatory and legal agencies. Additionally, as the range of contaminants capable of being analyzed has expanded, so has the capability of understanding how such compounds entered the environment and their subsequent behavior. Proposals are presented for future developments in several key areas: cross-discipline collaboration, training within the disciplines involved, legislation to facilitate and support the development of environmental forensics, development of criteria against which the validity of environmental forensic studies can be judged, and development of international networks to share developments and best practice in each area (Megson et al. 2016). Despite wide-ranging legislation designed to control chemicals in the environment, accidental releases and illegal dumping still occur. In addition to having detrimental effects on ecosystems and human health, the resulting environmental contamination can lead to contentious legal disputes between governments, industry, and communities. The capacity to predict the mobility of organic contaminants in porous media is paramount for the establishment of effective disposal and isolation procedures for chemical waste (McCarthy and Zachara 1989). Establishing liability for such events can be problematic as many chemicals are either not detected in the environment or have widely varying post-release behavior, making source differentiation difficult. Networking forensic chemists' lawyers and stakeholders involved with contaminated land helps to reduce the barriers to the crossover of knowledge and technology and to enhance the application and understanding of forensic techniques.

13.3 Current Techniques in Environmental Forensics

Forensic science is the application of science to answer questions in the judicial system (Yadav et al. 2022). Environmental forensics is a relatively new discipline that focuses on the concerns of environmental crime. It is a multidisciplinary subject with applications in many different areas such as water pollution, soil contamination, and climate change. Environmental contaminants can be volatile organic compounds, petroleum hydrocarbons, halogenated solvents, or other specific compounds. Volatile organic compounds can be used as forensic markers for contamination (Ashworth and Yates 2007). Chemical fingerprinting is a key technique in environmental forensics. Forensic scientists on occasion act more as detectives than chemists, piecing together clues in much the same way a detective would do. The forensic scientist must be all too aware of human intervention, searching for signs of it at every turn and being familiar with possible methods of cover-up. Each case is unique and not every piece of evidence is applicable to the next case. In the past, this has led to a somewhat disjointed approach to the science resulting in different laboratories developing their own methods of investigation and others being neglected altogether. Chemometrics is a statistical technique that can be applied to chemical data in order to extract relevant information or knowledge. It is commonly used in environmental forensics for the assessment of general oil types and also for the determination of source. The chemical fingerprints generated from any given analytical technique can be subjected to a range of statistical or mathematical manipulations facilitating easier interpretation and aiding the pursuit of the truth. Gas chromatography mass spectrometry (GCMS) is a technique ideally suited to the determination of volatile organic compounds because it is able to detect and separate them from interference compounds very easily. This technique can identify and quantify a number of different classes of compounds from alcohols to phenols; however, only 16 US EPA priority pollutants are subject to quantitative analysis. Nontarget screening includes all possible analytes apart from the specified analytes, and these can include hydrocarbon and solvents. Overall, whether the specific class of compounds is nontarget or target, if there is an indication of a contamination event, it may warrant further investigation. Environmental forensics is a relatively new discipline that includes many aspects which would not normally be considered to be forensic. There are a great number of potential contaminants and all samples with a wide variety of chemical compounds and properties. Determination of those contaminants is not straightforward.

13.3.1 Biological Indicators

A biological indicator is a population of organisms or types of biota that are used to ascertain the environmental quality of a habitat or ecosystem (i.e., upland, wetland, marine, etc.). Factors that may influence biota and its behavior are geographic variables such as latitude, altitude, and proximity to the equator as well as environmental variables such as temperature, salinity, exposure to sunlight, humidity, rates of evaporation, substrate characteristics, and storminess (Megson et al. 2016). The association of biota with an environment is binary in that a certain indicator species for a specific habitat can be clearly identified. Since certain biota are far more sensitive to habitat disturbances than others, they can be given more weight in evaluating an environment for forensic purposes. Alteration of biological indicators in an environment can indicate a change in environmental conditions. Monitoring these changes over time can indicate a trend. Monitoring biological indicators can also support observations made using physical and chemical indicators.

13.3.2 Geospatial Technologies

Spatial data is becoming more and more common in all kinds of investigations. This kind of data is often about the occurrences of events, movements, and positions of entities in space and time and can be extracted from different sources. Everything is now equipped with sensors and communicates with other devices, usually via the internet. Therefore, many digital evidences for crime investigation include a geospatial component. GPS coordinates of captured images, information from satellite imagery, or access-log records from websites show locations involved in criminal investigations (Zichar 2016). So the geospatial data plays an important role in crime investigation. It not only helps to track suspects but also can be used for profiling serial offenders, recognizing trends in criminal activities, and so on. In criminal crimes and national security investigations, geospatial data is often available in the seized pool of data. This pool can contain thousands or millions of images, documents, and other types of files, and many of those seized files need to be examined. To help investigators in finding and focusing on "the right files," different automatic and semiautomatic analysis tools and applications for rendering spatial data are developed.

13.4 Advances in Analytical Techniques

Environmental forensics traditionally used a limited set of analytical techniques, LC-HRMS, GC-HRMS, IR spectroscopy, and elemental analysis. However, due to the enormous increase in the number of potential environmental contaminants, new technologies are developing at an unprecedented pace. Environmental scientists are becoming more interested in testing and applying these new techniques, and forensic laboratories are now competing for environmental assessments, expanding the use of a wider array of analytical techniques. "Niche" techniques like SPME, electron microprobe, ASV, IR microscopy, and MALDI are being used to study environmental contaminants. Environmental forensics will benefit from much of this new methodology; however, it is evident that many of these techniques are still in their infancy and more research and method development is needed. In recent years, there has been increasing interest in the identification of the type of explosives, and infrared spectroscopy appears to be one of the most promising techniques for this purpose (Banas et al. 2023).

Environmental forensics researchers and practitioners need to keep abreast with advancements in analytical technique and methodology. There are many promising new techniques and instrumental developments on the market, in the market pipeline, or used in the research context that may benefit environmental forensic investigations. Key needs are the pursuit of automated and efficient field techniques, in situ sampling analyses, miniaturization of laboratory instrumentation, and continued technological development and innovation for mass spectrometry.

Environmental forensics has a core need for quantitative analyses of organic and inorganic constituents in natural materials. New on-/at-line approaches of high temperature or evaporative analyses of particulates in the sub-mg range will help to address this need. New thermal desorption and pre-concentration technology allow greater spread of technology used in laboratories for laboratory and field and mobile applications to provide very sensitive analysis at lower capital cost to laboratories. Addressing analytical needs here would broaden the potential "toolkit" for environmental forensics. Analytical technologies are developing rapidly, warranting keeping abreast of developments and where appropriate the application of advanced methodologies.

13.4.1 High-Resolution Mass Spectrometry

The potential existence of unknown compounds is what makes forensic chemistry unique within the field of analytical chemistry. Forensic chemists try to assess whether a material is an illicit drug or pollutant, for example, or to provide other chemical information to a criminal investigation. Yet most of those assessments are done on substances which have been invoked by previous knowledge and literature for a suspect. On occasions, however, an environmental sample is obtained for analysis which does not match any known patterns or materials. These so-called "unknowns" are notoriously difficult to assess using mass spectrometric methods common in environmental chemistry. In a criminal investigation, the consequences of such unknowns can be dire. If standards cannot be produced for analysis, and no library hits occur by NIST or custom or in-house library searches, analytically inclined forensic chemists become hampered in providing chemical information. For many standard mass spectrometric methods, hit or no hit is a binary result, and relying on library matches where a "mystery" sample does not match one of known suspected compounds is often unhelpful. To date, few methods can quickly and accurately determine the chemical identity of an unknown compound in an environmental sample.

On occasions, however, high-resolution mass spectrometric methods in particular Fourier transform mass spectrometry (FTMS) have revolutionized how environmental analyses are performed. With its unsurpassed mass accuracy, high resolution, and sensitivity, FTMS is the first mass spectrometer in the world to give chemists a tool for assessing such difficult and complex environmental analyses. Two features of FTMS are responsible for changing the face of how complex, multitiered environmental analyses are accomplished. FTMS analyses can be tried as a "wide net" approach. That is, without any chemical change to the sample, it is brought into the FTMS adequately dried/solubilized. Once in the mass spectrometer, thousands of scans are recorded for the high-resolution mass-to-charge ratio (m/z) dataset. Subsequently, the data is deconvoluted to reflect a "fingerprint" of chemical families with which to assess the data. The other complementary method for assessing unknown compounds in environmental samples that have or could have societal consequences is to assist isomers in their assessment.

13.4.2 Isotope Ratio Analysis

Isotope ratios are defined as the relative abundances of isotopes of a given atomic element that differ in mass. It is a measure of the differences in size, shape, and behavior of the isotope(s) upon formation of new bonds. Forensic scientists can measure isotope ratios of most elements in a variety of materials such as bones, teeth, faeces, soil, or body fluids. Isotropic ratios reflect the local geology or climate experienced by the individual and its ecology (Bartelink and Chesson 2019). Thus, given suitable conditions, isotope analysis on human remains can offer information on the most likely geographical origin of individuals. It can be used as an investigative tool to focus inquiries as well as for evidential purposes in court. Theory and methodology on isotope analysis and forensic interpretation of results have been developed to an impressive degree, highlighting the application of isotope techniques in hydrological forensics (Hsieh and Bugna 2008).

Thus, isotope analysis is not widely used, nor is it the centerpiece of evidence in most countries and laboratories. In these counties, an effort is being made to drive the field to be more efficient and reliable as a forensic technique. In doing so, attempts are being made to increase its use in social and environmental justice cases involving either human remains, wildlife parts, or for assessment of environmental exposure or authenticity control. Isotope ratio analysis effectively distinguishes anthropogenic from natural contamination sources (Alberts et al. 2019).

This section focuses on isotope ratio analysis in terms of the theory behind isotope analysis, the isotopes commonly used to provenance food and human remain, the type of materials suited for analysis, cautionary notes on sample selection and handling, and the logistics of specimen transportation and storage. Applications of isotope analysis in forensic anthropology and forensic science contexts are also described along with current problems in the field and suggested future directions.

13.4.3 Integration of Big Data and Machine Learning

In environmental forensics, the integration of big data and machine learning has become increasingly important. Machine learning, a subset of artificial intelligence (AI), allows computers to learn from experience without being explicitly programmed to capture information. Rather than relying on a set of rules, machine learning analyzes massive amounts of data to detect patterns and make predictions. Algorithmic and computational considerations render predictable behavior in dynamic systems, useful in forensics (Comess et al. 2020). Today, practitioners of machine learning in forensics can reflect on over 25 years of experience, from the first applications of artificial neural networks to large, national databases of compiled forensic image signatures used for criminal fingerprint matching. The term "big data," which refers to massive amounts of computerized data, is used in tandem with "machine learning" to describe the field of AI that has revolutionized understanding in multiple disciplines.

A successful example of integrating big data with AI and machine learning to build high-performing predictive models is given. Select researchers and government agencies are developing modeling approaches to leverage data and build computer models that better predict the outcomes of exposures to chemicals with unknown toxicity. The scanning of scholarly literature for environmental health-related chemical exposure and biological outcome data is ongoing, and results are processed through an automated pipeline to aggregate data into structured computational models. These projects demonstrate the efficacy of pipelines, built with collaboration between software developers, health scientists, and data scientists, which turn decentralized data into centralized models. Keeping models in siloed communities is counterproductive to research as the fundamental methods of model creation rely on open data. Collaboration between foundational researchers and software engineers is essential to expanding the breadth of modeling approaches leveraged by data, improving the transferability and robustness of these approaches. As more effort is put forth to develop models, better strategies for model organization and dissemination that aid community engagement should be investigated.

13.4.4 Predictive Modeling and Pattern Recognition

Predictive observer and pattern recognition methodologies developed in other scientific spheres, like meteorology and biology, should progress to become a more pressing part of the environmental forensics' toolkit as their respective usage in forensic work is enhanced and refined. There remains a temptation to focus on questions concerning exact compounds or sources, which are often, necessarily, the first focus of investigations. However, these are often better dealt with via moiety, multi-compound time-span compliance, and broader-brush studies of these class-a-source matters. Several groups across the globe are working on this, and through publication of algorithms and improvements in recognitional databases, great progress is anticipated in the next decade. Meanwhile, phenomenological modeling should be pursued with renewed vigor. Studies have shown that often, even well-characterized sources give rise to the formation of very complex mixtures when released to the environment. It is vital to fully understand these processes to ascertain source terms. This applies equally to concerns over organic contaminants and nutrient surpluses, both of which topics are likely to broaden in scope and deepen in quality and number as computational power increases and the predictive capacities of models continue to improve. In a similar spirit, strictly ab initio modeling efforts into the fate and transport of organic contaminants should be continued as the enormous capabilities of quantum chemistry in chasing down fates are just beginning to be explored in these contexts. Establishment of major international collaborative groups or centers is necessary to define, focus, and distribute skills and training in the fields where the highest impact is anticipated over the next decade. Thus, for example, a center focusing on work to codevelop or standardize algorithms, modeling frameworks, and databases for use across networks in predictive observer efforts would have far-reaching consequences. The efforts could be worldwide; extensive use of the internet through both conventional visualization facilities and avatars would ensure competence and understanding of the methods would be disseminated globally.

13.4.5 Data Harmonization and Sharing

To promote data integration, interoperability, and reuse, investment, infrastructure, and changes in research culture are needed. While funding and organizing tools are essential, they must meet community needs and change the way research is done. More broadly, transdisciplinary conversation and deliberation are needed to close the gap between the scientific and public policy communities. It is critical to develop and document needs, standards, and models to promote data accessibility and reuse while protecting proprietary information. Logistics of data documentation, access, use, and follow-up citation must be developed through deep conversation. Policies and governance structures must be developed to enable the activities. Initiatives should address a range of important issues, including, but not limited to (Heacock et al. 2022), the following: research efficiency, data use, and citizen engagement; addressing the policy relevance of data; learning and experiential opportunities; developing and disseminating resources, tools, and methods; promoting data citation, acknowledgment, and credit; investigating ethical use, protection, and privacy across data types; defining the research domain and collective identity; developing metadata, controlled vocabularies, and interoperability standards; developing incentives and a collective sense of ownership; and scaling up to address grand challenges. Pattern recognition helps interpret complex environmental datasets (Alimi and Hassine 2018).

A series of working groups, convened in person or virtually, and sustained by collaborative platforms, are likely to be productive. While it is early to articulate next steps, the suggested groundwork for transdisciplinary conversation and deliberation aligns with essential issues likely to arise next.

13.4.6 Remote Sensing and Spatial Forensics

The future of environmental forensics is likely to benefit from ongoing developments in the fields of remote sensing and spatial analytics, which, combined with traditional forensics, can enhance the verification of allegations of environmental crimes. Remote sensing has provided significant advances in the assessment of environmental problems at large spatial scales across various areas, including the identification of sources of large spills, quality and change assessments before and after large environmental incidents, land cover and pollution impact assessments of oil spills, and the monitoring of attributes of oil sources (van der Linden et al. 2019). Satellite-based remote sensing approaches will enhance the monitoring of oil spills and assist forensic investigations. Examined oil fingerprinting and microbiological tracking within the realm of environmental forensics (Eiserbeck et al. 2012). Remote sensing approaches for local case investigations are also becoming available to detect changes in land cover and resulting pollution, allowing for an assessment of land use pre- and post-incident. Integrating remote sensing, routine pollutant sampling, and spatial assessment approaches will enhance the closure of cases. Future outcomes of these advancements will allow combining traditional sampling and monitoring networks with change detection and verification in the same model framework at various time intervals and for a range of pollutants. Provided a global historical emissions inventory for persistent organic pollutants (POPs) (Breivik et al. 2002). Historical spatial data are projected to become more readily available, providing large datasets of various variables and measures in a time and spatially granular manner across regions. The efforts to develop new spatial indicators for environmental evaluation along with related hazard assessment techniques will close the gap between measured pollutants and input datasets for mitigation and management models to capture environmental risks. Nonindependent sources of evidence for expansion-based failures and recently emerging evidence-sharing methods can broaden the applicability of spatial forensics to various types of cases (Zichar 2016).

13.4.7 Satellite and Drone Technologies

Astrobiology is a relatively young science, still in its infancy compared with molecular biology and comparative planetology. As such, it is likely that in the coming decades, new tech-sub-disciplines will emerge as it has in the existing planetary sciences. It showed how drones aid in environmental monitoring and data collection (Kaplan and Haenlein 2019). Here, "tech-sub-disciplines" refer to focused areas of study employing specific technology or methodologies. For example, Earth-bound life was able to evolve and become complex because the Earth cooled sufficiently for liquid water to exist, and 'stable" plate tectonics enables long-term climate control over Gigaannums. The specific modern methods of temporal analysis of subsurface and textural materials employed in the fields of geology and geochemistry were not necessary or relevant pre- or early civilization, when people had little impact and no understanding of their planet on such scales (Stuart et al. 2019). The likelihood is that astro-exploration will get ever cheaper and easier analogously to how one does now with planetary Earth. Likewise, new methods of detection will evolve to keep pace, involving new tools and techs to cope with these ever-growing sets of data.

While all ages of human endeavor are colored by the zeitgeist and the brand of knowledge systems available, considering future directions with reference to humanity's earliest astro-intuition seems appropriate. Early in the exploration of worlds other than Earth, concerns were raised about littering extraterrestrial environments with earthlings and Earth-sentient microbes with, as inevitabilities, infallible caution debates in any post-1950 CE ETI extension endeavors undertaking observation via powerful instruments. Given limits of currently available methods, the ground-based and orbiting instruments now and near-term employed were developed and deployed to face, unprepared, exoplanets at the 1879 blossom of Goorang (Ustin and Middleton 2021). The capability following these earlier efforts is daunting, to a degree unproven. Some exoplanetary environments were expected to not weather, filter, or cover up the original biosignature probes, while states of Martian habitable wetness and Venusian cloud/drum storms were expected and embraced to not end life and/or fossilizability prospects too quickly.

13.4.8 3D Environmental Reconstruction

The rapid development of network technologies, 3D imaging and printing technologies, are widely used to detect and reconstruct explosions, murder crimes, architectural destruction, and other crime scenes (Ren et al. 2018). However, there are many corresponding factors affecting the reconstruction result. Therefore, it is significant to make a better presentation for a more convincing result. In this paper, a simple model of crime scene is created and a sketch-based rapid modeling method is proposed. It shows that again, there is interested in using sketch as the tool to input the modeling information of subjective sense about crime scene, and then a series of relevant geometrical model is constructed using basic modeling primitives (Carew et al. 2021). After that, this scene model is exported and rendered in the visualization engine, where lighting and material can be freely adjusted to get optimal scene presentation. Finally, the application of the reconstruction crime scene is investigated. The important aspect in criminal investigations, criminal scene modeling and presentation, is addressed.

With the development of network technologies, more and more related forensic methods and tools are employed in the forensic field. One of the key points is to present the detection result. Conventionally, crime scene images taken by digital cameras are used as the primary data source of the presentation. Consequently, the subjective explanation or judgment is usually made because that the original 2D images cannot display the 3D information of the crime scene. But 3D images can make for clearer guidance, it is found that the required reconstruction time exponentially increases with the complexity of the scene. Therefore, a better representation of the current reconstruction can be made to ensure a more convincing 3D visualization.

With the increasing complexity of scene modeling, there are only a few methods raising the requirement of rapid modeling on dealing with large point cloud data including sketches as the input. In the interactive modeling, it is hard to get a sketch that is free from occlusions. However, there are some limitations in using sketch-based method. The sketch of the input model should be cleaned in advance before using it. In addition, a changing reconstruction window is needed during the reconstruction process, which makes it difficult to preserve the perspective consistency.

13.4.9 Molecular and Microbial Forensics

The future of microbial forensics is very exciting as the development of new techniques, especially in sequencing and bioinformatics, is rapidly expanding the number of potential forensic targets that can be used to tell us information about a sample and its sources. Nonetheless, there are many unanswered questions on the constraints, limits, and boundaries on the power of these methods (Robinson et al. 2021). It is currently unclear whether microbiomic methods are truly elegant, high-sensitivity tools that can outperform or supersede existing forensic methods. Without rigorous scientific vetting and validation like that of existing forensic techniques, adequacy of preparation for the potential forensic use of microbiomic methods is currently low. Research must be conducted independently of law enforcement to establish solely scientific bases for the potential uses of microbiomic evidence.

Microbial forensics is an emerging field that combines microbiology and forensic science. It is defined as the application of microbiology to legal investigations. Its scope is like traditional forensics in that it encompasses scientific investigations on settled or suspected nonnatural deaths, issues of personal identification, or the analysis of evidence related to a crime. Unlike traditional forensics, microbial forensics focuses on the study of "outbreak" events (Castro and Ungria 2022). The study of these events aims to understand the transmission of an individual microbe across the affected population. Outbreak events are usually studied by epidemiologists using a number of classical and modern investigative methods. The target pathogen's genetic, chemical, and physical information as well as the analysis of epidemiological data are relied upon in implementing tests to identify and characterize the microbes of interest.

The reliable characterization of molecular variations through the application of phylogenetic analysis or genetic tests provides investigators many tools that can help identify microorganisms down to the substrain level. Generally, molecular surveillance has been shown to be preventative and proactive, with most studies undertaken to assess the effects of specific interventions, such as vaccination, and to monitor the ongoing effects of biocontrol on pathogen populations. In microbial forensics, the accurate identification of the microbial agent to the substrain level has the potential to identify the person responsible for the initial dispersal of the contaminant (Biswas et al. 2007). For constituents of evidence samples to be considered source indicators, expert groups carrying out microbial forensic analyses should be in formal or informal networks with the relevant expertise available to them.

13.4.10 Environmental DNA (eDNA) Applications

A need is identified for the inclusion of novel techniques into environmental forensic investigations, which introduce new ways to study environments and communities, particularly at a more empirical level. Examples include denaturing gradient gel electrophoresis for studying eDNA by identifying and quantifying DNA fingerprint curves or amplicon length distributions, the use of ion torrent and mitogenome sequencing for characterizing eDNA, eDNA metabarcoding techniques to analyze a community's genetic identities, and eDNA environmental tracking, barcoding, and air analysis. These techniques are expected to lead to exciting new developments in environmental forensics and increase its relevance to real-world issues.

Novel forensics techniques are presented with the potential to change how environments are studied, that have seen limited application to date in environmental forensics, and which can be used with existing site investigation data as part of an integrative package of methods. eDNA applications are expected to reveal substantial hidden biodiversity in environments that have previously been inferred too poor in species and to identify new putative eDNA sources, for example, from organisms not yet detected in a habitat. Until now, the eDNA focus has been on identifying and characterizing only eDNA in water; however, a related approach looking for DNA in air is expected to be employed increasingly in forensics. A proof of concept is presented here that animal DNA can be collected from airborne eDNA by filtration of wind such as of a fan blowing air for several minutes onto a membrane. Primer sets for PCR amplification designed to target high molecular weight or low molecular weight vertebrate-specific mitochondrial 16S rRNA sequences are preferred. An animal community is reflected in the detection of DNA from wild mice and bush rats in the air, and successful modeling and characterization of species detection probabilities and estimates is also demonstrated.

Technical improvements have been made to DNA extraction, storage, and transport methods for air-collected samples to maximize DNA yields. This is expected to make exhausting sub-malignant sources more tractable and resolve more difficult cases. It is hoped that air sampling will be further adapted to target particulates such as soil and vegetation for study with existing DNA characterization techniques. With ongoing improvements in sampling methods and amplification targeting, air eDNA would seem likely to become the next avenue of bioinformation gaining investigators' interests.

13.4.11 Artificial Intelligence and Automation

The future of environmental forensics lies in a deepening integration of artificial intelligence (AI) and automation in investigation workflows. Today, investigators are drowning in ever-expanding treasuries of data ranging from hard drives, cloud services, and live systems as well as Web, social media, and IoT services. To manage these new complexities, tools are needed that use AI to assist and/or automate elements of the forensic workflow (Du et al. 2020). On the automation side, ancillary tasks like collecting data, controlling acquisition devices, running analysis, assembling reports, and interfacing with multiple tools and data sources will persist into the foreseeable future. These tasks create opportunities for technologies like robotic process automation (RPA) and no-code/low-code solutions that migrate routine processes from manual to automated execution (James and Gladyshev 2013).

Forensic case management and triage can potentially be automated as well to reduce the lead times of investigations and analysis, but they will require cognitive systems with a much higher maturity level. Facilitating the query and investigation of results from different tools can also enrich the investigative ecosystem; however, this is a challenging data integration problem that researchers have struggled with for decades. Other AI technologies such as predictive recommendation and generative design can assist investigators in planning and conducting their analysis. However, these tasks call for explainability capabilities so that any suspicion of being influenced can be mitigated. In addition, mechanisms must be implemented for accountability.

Still, much of the discussion on AI in forensics focuses on inserting machine learning models into detective workflows to automate routine techniques. While the automation of analysis will lower the burden on investigators, it must come with safeguards. An analysis result that highlights criminal activity must also highlight why those activities were flagged over many benign possibilities. The very nature of AI exacerbates this problem since the internal workings of neural networks are notoriously opaque. Any system that places reliance on an AI model must also account for how those models arrived at their conclusions.

13.5 Emerging Technologies

This article discusses some emerging technologies that have potential applications in environmental forensics. Machine learning approaches are being integrated into GIS-based environmental modeling and monitoring of pollution, and smartphone apps are being developed for citizen science in urban pollution monitoring (Davis et al. 2023). New genomic approaches can rapidly fingerprint the sources of water pollution. New DNA-based technology can rapidly fingerprint soil and feces in pollution forensics. These new approaches have already been adopted in some instances, and hybrid approaches involving both conventional and emerging technologies are promising as early implementations of the new technologies (Megson et al. 2016).

Establishment of Machine Learning (ML) Within GIS-Based Environmental Modeling: In the Earth and environmental sciences, the integration of machine learning approaches into GIS-based modeling of water pollution on a catchment scale is emerging in response to the volume and complexity of available data. Such ML-enhanced integrated modeling has the potential to enhance predictions of pollution by E. coli and nitrate from diffuse sources, which is of interest to environmental forensics (specifically the use of modeling predictions and field surveys to link pollution events to on-the-ground activities by people) (Morrison et al. 1970). There is a need for modeling frameworks able to assimilate all available data on catchments and pollution events, and ML methods may assist the establishment and application of such frameworks. Enhanced modeling may help identify pollution sources, which may in turn assist the management of contaminated catchments.

Smartphone Applications for Citizen Science Environmental Monitoring: A number of smartphone-based citizen science applications are emerging for the monitoring of urban water pollution, including Water Smart and Waterkeeper in Canada, Water Watch in Australia, and MyCoast in New Zealand. The former two apps can collect both geo-referenced data on pollution observations and time-series water-quality measurements, which can then be visualized and gleaned for historical and seasonal patterns, respectively (whether citizen science observations correspond with wet-weather plumes and the seasonal distribution of such pollution catastrophes). There is a need to investigate the applicability and utility of such smartphone-based environmental monitoring citizen science applications within a UK context, and an approach is presented using Plymouth Sound as a pilot study site for future work.

13.5.1 Remote Sensing Applications

Remote Sensing Applications in the Study of Major Environmental Changes.

Several studies dealing with various subject matters on the impedance of printers and the effect of humidity on the response of inks were conducted on multiple modern designs of ink jet printers. The scientific methods were applied, results were tabulated, and an estimation of the numerous factors in action was had. Natural change detection with remote sensing relies heavily on the accurateness of the earth ground truths because change in product futility offset detector performance on the planet Earth, which is usually difficult or even impossible to acquire. Current remote sensing study in this field is a major flood disaster in China, which has been analyzed with other geospatial data including DEMs for building and evaluation of flood model/river flows and meteorological data for rainfall derivation. Pasture and farm irrigation maps were used for building and evaluation of economical damage assessment model. Verdicts: Precise calibration factor for early flooding indicator was retrieved; AG-HSI model was mainly used for evaluation of vegetation flooding damage.

The study on anthropogenic change detection with RS faces difficulties in the acquisition of high-resolution ground truth data, which is expensive/time consumed. Multi-scaled or multi-date Global Land Cover change products retrievals/replacement/consistency is imperative for better international environmental policy. RS study on other advanced sensors will focus on a project supported by National Science Foundation. UAS-based crimes detection, police assistance, and public safety monitoring are another research direction. In a self-developed UAS system with an effective aerial detection range of more than 5400 m, a detection limit of 2 cm of building change was spent with a 6000 m ground length. Collaboration with a 4D urban change monitoring project is another current direction.

13.5.2 Nanotechnology in Forensics

Nanotechnology is poised to play an increasingly important role in forensic monitoring of environmental contaminant threats in the future. There are, at present, a number of important considerations related to the environmental safety evaluation, risk governance, and potential toxicological implications of nanomaterials in the context of nanotechnology, particularly with respect to developments in nanotechnology in forensics (Linkov et al. 2013). Risks posed by the release of nanoparticles are inseparable from release predictions. With a high degree of uncertainty regarding the release of nanoparticles and their properties during use and transformation, there is great interest in undertaking a release prediction for individual products.

Each product is characterized with respect to nDPs size, primary particle size, shape, constituent material, and surface coating, all of which affect its physicochemical properties and subsequent environmental fate and behavior. By assigning a product category group to each product, a broad range of important modeling studies can be undertaken, linking release to exposure scenarios in which estimates of important nanoform characteristics (size, state of agglomeration, coating, durability, etc.) can be developed. An assessment framework is needed for improving nanoecotoxicology. Difficulty in environmental monitoring of nanoparticles stems from several science and technology challenges related to material characterization, methods development, model validation, and data assimilative modeling.

Nanocomposite materials combining gold-coated silica spheres with a carbon conducting polymer were synthesized to be used for electrochemical sensing of environmental pollutants (Rahmati et al. 2021). As a proof of concept, the analytical capabilities of this material were tested toward electroactive species of heavy metals such as lead ions and hydrogen peroxide. The sensors exhibited impressive sensitivity and selectivity toward the dangerous heavy metals by applying a linear sweep voltammetry protocol. Nanocomposite materials have great potential as effective electrochemical sensors, especially for environmental monitoring applications.

13.5.3 Artificial Intelligence and Machine Learning

The application of artificial intelligence (AI) and machine learning (ML) is shaping a new way of detecting anomalies and improving efficiency in industrial environments. Current algorithms provide companies with specific data of a vast range of usual patterns, helping to better manage resources such as energy. This, in turn, fosters sustainability. However, complicated deep learning models and big data tools demand for computing resources that, on one hand, can be dangerous to the industry's finances and, on the other hand, question the environmental sustainability of the AI implementation in Industry 4.0 in general. Future ML models will need to be both high performance and environmentally friendly, a relevant issue that is initiating the conversation around Green AI. When applying the newest deep learning models to detect anomalies, more resources in hardware expense and energy consumption are to be expected. The trend tends to follow the curve created by other industries. However, traditional algorithms do an excellent job accurately detecting anomalies with low resource consumption. The majority of industries only use a small amount of data, and the obsolescence of assets is a slow process. There are still many options to explore unknown large anomaly scenarios with simpler models. Hence, the study concluded that regardless of the approach to novel models, resource consumption should always be measured (Huertas-García et al. 2023).

The study focuses on what many overlook in current technologies, which is reducing as much as possible not only the monetary costs of possible innovations but also the environmental impact of the technology. It deliberates on using a simple tree model to explain what companies can do with only 0.005% of the data that most companies do not even use for traditional algorithms. Simple models are still highly efficient, competitive with the latest research, and have a higher performance/cost ratio. A multi-objective optimization approach was included to assess the trade-off between a model's performance and the environment. The rationale of keeping models simple and measuring their impact on the ecosystem should be the guiding rules in changing a company's fabric to a more data-driven one (Miller et al. 2018).

13.6 Challenges in Environmental Forensics

Over the past few decades, environmental forensics has developed into a broad and mature discipline with many applications. Theory is established and analytical chemists regularly obtain very high-quality data that helps solve complex cases. Several "success stories" highlight how to uncover the truth. On the other hand, every scientist has also seen failure—pollution not sourced, methods not working, lawyers wishing they would better understand the science and, suggestively, no go for smelling a rat. Some warnings should always be kept in mind (Megson et al. 2016).

Environmental forensics is complicated as source characterization is multidisciplinary and requires input from various stakeholders. Pollution is a core issue but questions are highly interdisciplinary; for the lawyer, geologist, and hydrologist, this is the "kickoff." Environmental forensic chemistry answers pollution-specific questions and generates data of a widely varying quality. Where there are "tricks of the trade" in analytical chemistry, there are certainly many where scientists have decided on a personal approach. Understanding how to interpret and use these often-proprietary data is a challenge for less experienced lawyers. The lawyer says what she/he wants to know scientifically and should trust the scientist but has seen many cases fail due to inappropriate or low-quality data. To successfully apply the acquired insight, lawyers must be mere chemists for a day. For the chemist, understanding the level of proficiency of his/her own methods is vital to avoid unreasonably rigid expectations.

Every case is unique. Generally, the lawyer has experience on the legal side while the analyst has expertise in the data generation process. From experience, scientific methods may differ widely in their applicability, outcome, and interpretability. Understanding the choice will aid greatly in the interpretation of the data. The most important options to characterize the input of a case should be discussed. At the early stage, general chemistry and data should be available as they guide lawyers on the level of interpretable detail. Motivations for analytical choices should be backed by data and accessible to all stakeholders, as the choice of pollutants to analyze, sample preparation, extraction, and analytical technique. The common choices and potential pitfalls on both the analytical and interpretative level should be addressed.

13.7 Future Research Directions

Future research directions in the field of environmental forensics could include the application of somatic mutations as molecular markers for investigating environmental incidents. The DNA of a vertebrate individual can be damaged by the environmental presence of genotoxic compounds in ecosystems. Some of them generate mutations in regions of DNA that could function as proper markers in forensics studies. The identification of either known/natural (mutated) alleles or potentially new mutations in a pool of "wild-type" alleles would require a dedicated research effort. Nevertheless, by knowing the sequence of the target markers in prior victim species, it appears possible to identify some of them using biochemistry methods. This approach is a promising area of research for addressing environmental issues of diverse types.

The lack of empirical successes in analyzing isotopes of earth-abundant elements in environmental forensics represents a focus of research. Specifically, it would be interesting to know whether the empirical successes in leveraging isotopic anomalies of certain elements could be repeated with another environmental concern. Mass-independent isotope fractionation processes affecting silicon isotopes have ambiguous origins, suggesting that its abiogenic sources could facilitate the linkage of a cause to a pollution incident. Similarly, phosphate pollution could also be traced to its sources based on its anomalous isotopes. Certain isotopes might also allow inferring anthropogenic contamination through analyzing environmental samples.

The future of some of the topics proposed should clarify whether they fall within the realm of environmental forensics. Focus on using simple mass spectrometric methods to analyze toxic compounds or their metabolites or environmental indicators would be enlightening. For instance, determining chlorinated hydrocarbons in an unknown pollution incident involving a vulcanization plant might not be regarded as environmental forensics in most forensics study contexts, while utilizing certain spectrometry techniques to detect toxic aromatic amine spills would be clearly regarded as environmental during incident or risk assessment.

13.7.1 Innovative Analytical Techniques

Increasingly complex, multicomponent, and ever-improving technologies provide exciting challenges for the analytical chemist. When legacy products enter into new countries via other products, other companies, or as illegal disposal, new analytical techniques will be required to determine which new products are contaminating the environment. Forensic scientists will provide the new-product information to the essential sponsors (Richter et al. 2020). Characteristic fragmentation and ionization in mass spectrometry can provide information on an unknown composition, sometimes providing a perfect description of its chemical structure. If the unknown does not fit with known components, not only does one's world appear bleaker, but the opportunity also exists to develop new chemistry, which represents a challenge and an intellectual goal. Eventually environmental geochemistry was introduced in forensic science applications (Hursthouse 2001). Forensic science is a community of many interrelated sciences using the same materials and is therefore typically separate from environmental analysis communities, which average substances differently. Environmental investigations typically want to identify as many components as possible in an entire sample, while forensic investigations typically want as near as possible to a "perfect match" against reference materials. Consequently, forensic investigations very quickly restrict the timing and sophistication of the search, illustrated in the first analysis of eight sets of four "grains of dust" on a filter. Environmental forensics challenges raise many intellectual issues and provide a mechanism for more forensic scientists to enter environmental science with a new perspective. Environmental forensics sessions may be held at two forensic meetings due to the close connection with the GAC, or a separate session may be held in conjunction with an environmental science meeting. Scientists from the environmental science community should be invited to describe environmental forensic perspectives at these meetings. Environmental analytical techniques successfully and cost-effectively developed for one problem need to be sought for similar environmental forensics' cases. In parallel, forensic results that are misused as a publicity tool need to be better disclosed and challenged. High-profile cases often attract much media attention.

13.7.2 Longitudinal Studies

Longitudinal studies can be defined as the examination of a location over an extended period of time (vis-a-vis a cross-sectional study which may look at the same location at a single point in time). Environmental forensics investigations mainly use cross-sectional studies to provide a snapshot in time although some analyses may benefit from a time-lapse or longitudinal component (Lee et al. 2019). An area may have been polluted at various stages over past decades. A source may operate for a decade or more before it is either shut down or redirected, with the lake then receiving a more modern source. Such scenarios could be exhibited by looking at recent to very old sediment cores. Or early research may compare epiphytic lichens which have thrived for some time and responses on the bark of trees planted more recently. National Environmental Agencies may have long-term monitoring programs in place that could offer a time-lapse angle to an inquiry. Longitudinal approaches can be useful with faded, erratic, intermittent, or expired emissions sources. If chemicals persist, they can be detected at a later date after once being prevalent. If chemicals are detected at soil, sediment, or biota at a water body, this excludes all other inputs. Alternatively, if they are detected at a sediment core, this could exclude biogenic sources of metals. Longitudinal studies can be used for planning future studies, with previous studies establishing correct methodologies for more recent studies. Longitudinal studies have the potential to shed still more light on pollution events that are several decades old and even conclude events of new interest.

13.8 Conclusion

The word "future" can mean different things: good and bad, short and long term, and imaginable and unimaginable. This statement applies generally to environmental science, and especially to environmental forensics. With this in mind, the following text will consist of several topics relating to future directions in environmental forensics, in order of increasing timescale. Futures of environmental forensics will be defined as the timeframe of next several decades. This is approximately the lifespan of analytical techniques. Here, futures will be discussed regarding the increasing use of chemical fingerprints, the input of molecular biology, and the future of chemistry itself. On a shorter timescale, for example, the next decade, there might be remarkable improvements in existing approved methods, but these tend to be trivial compared to the possible qualitative changes mentioned above.

While environmental forensics comprises chemical, biological, and narrative investigations, much of the discussion here will focus on chemical investigations as these form the basis of most environmental forensics. It should be acknowledged that biological investigations are important and are in fact a key to the success of chemical investigations. Environmental forensics were once just an extension of conventional forensic science, and thus there was concern about confidentiality of environmental issues. However, with recent advances in analytical techniques and better understanding of chemicals, some chemicals are now recognized to be globally important or critical, making it difficult to keep these chemicals secret. Environmental crime was, and still is, difficult to detect but is considered relatively simpler and thus is not expected to be more sophisticated in the future. Environmental crime was initially mostly about dumping, which remains the most common crime today. However, many environmental issues are on the rise, such as the damage from overexploitation of resources, climate change, and biopiracy. Such issues have a huge economic impact, thereby attracting the interest of criminals. With increasing amounts of money at stake, the sophistication of environmental crime is expected to increase.

References

Alberts J, Lollar BS, Lacrampe-Couloume G (2019) Isotopic tracing of environmental contaminants. Environ Sci Technol 53(2):475–489. https://doi.org/10.1021/acs.est.8b04235

Alimi H, Hassine A (2018) Pattern recognition techniques in environmental data analysis. Environ Monit Assess 190(7):387. https://doi.org/10.1007/s10661-018-6793-z

Ashworth DJ, Yates SR (2007) Environmental fate of volatile organic compounds: Implications for forensic investigations. J Environ Qual 36(5):1211–1220. https://doi.org/10.2134/jeq2007.0064

Banas AM, Banas K, Breese MBH (2023) Classification of the residues after high and low order explosions using machine learning techniques on fourier transform infrared (FTIR) spectra. Mol 28(5):2233. https://doi.org/10.3390/molecules28052233
Bartelink, E. J., & Chesson, L. A. (2019). Recent applications of isotope analysis to forensic anthropology. ncbi.nlm.nih.gov
Biswas N, Bewtra JK, Hubberstey AV, Cimenti M (2007) Alternative methods in tracking sources of microbial contamination in waters. PDF
Breivik K, Sweetman A, Pacyna JM, Jones KC (2002) Towards a global historical emission inventory for selected POPs. Environ Pollut 120(1):1–24. https://doi.org/10.1016/S0269-7491(02)00139-X
Carew, R. M., French, J., & Morgan, R. M. (2021). 3D forensic science: a new field integrating 3D imaging and 3D printing in crime reconstruction. ncbi.nlm.nih.gov
Castro, A. E., & De Ungria, M. C. A. (2022). Methods used in microbial forensics and epidemiological investigations for stronger health systems. ncbi.nlm.nih.gov
Comess S, Akbay A, Vasiliou M, Hines RN, Joppa L, Vasiliou V, Kleinstreuer N (2020) Bringing big data to bear in environmental public health: challenges and recommendations. ncbi.nlm.nih.gov
Davis GB, Rayner JL, Donn MJ (2023) Advancing "Autonomous" sensing and prediction of the subsurface environment: a review and exploration of the challenges for soil and groundwater contamination. ncbi.nlm.nih.gov
Du X, Hargreaves C, Sheppard J, Anda F, Sayakkara A, Le-Khac NA, Scanlon M (2020) SoK. In: Exploring the state of the art and the future potential of artificial intelligence in digital forensic investigation. PDF
Eiserbeck C, Bennett B, Larter S (2012) Environmental forensics: from oil fingerprinting to microbial source tracking. Sci Total Environ 414:701–714. https://doi.org/10.1016/j.scitotenv.2011.11.007
Heacock ML, Lopez AR, Amolegbe SM, Carlin DJ, Henry HF, Trottier BA, Velasco ML, Suk WA (2022) Enhancing data integration, interoperability, and reuse to address complex and emerging environmental health problems. ncbi.nlm.nih.gov
Hsieh YP, Bugna GC (2008) Environmental isotope techniques in hydrology. Hydrol Process 22(3):473–487. https://doi.org/10.1002/hyp.6637
Huertas-García Á, Martí-González C, Maezo RG, Rey AE (2023) A comparative study of machine learning algorithms for anomaly detection in industrial environments: performance and environmental impact. In: International conference on trends in sustainable computing and machine intelligence (pp. 373–389). Singapore: Springer Nature Singapore
Hursthouse AS (2001) Application of environmental geochemistry in forensic science. Environ Geochem Health 23(4):209–213. https://doi.org/10.1023/A:1012201311101
James J, Gladyshev P (2013) Challenges with automation in digital forensic investigations. PDF
Kaplan A, Haenlein M (2019) Siri, Siri, in my hand: Who's the fairest in the land? On the interpretations, illustrations, and implications of artificial intelligence. Bus Horiz 62(1):15–25. ISSN 0007-6813. https://doi.org/10.1016/j.bushor.2018.08.004. (https://www.sciencedirect.com/science/article/pii/S0007681318301393). (n.d.)
Lee SB, Mills DEK, Morse SA, Schutzer SE, Budowle B, Keim P (2019) Education and training in microbial forensics. ncbi.nlm.nih.gov
Linkov I, Steevens J, Adlakha-Hutcheon G, Bennett E, Chappell M, Colvin V, Davis JM, Davis T, Elder A, Foss Hansen S, Hakkinen PB, Hussain SM, Karkan D, Korenstein R, Lynch I, Metcalfe C, Ramadan AB, Satterstrom FKK (2013) Emerging methods and tools for environmental risk assessment, decision-making, and policy for nanomaterials: summary of NATO Advanced Research Workshop. PDF
McCarthy JE, Zachara JM (1989) Subsurface transport of contaminants
Megson DP, Reiner EJ, Jobst KJ, Dorman FL, Robson M, Focant J-F (2016) A review of the determination of persistent organic pollutants for environmental forensics investigations, vol 941. PDF, p 10

Miller T, Gallidabino M, Macrae J, Hogstrand C, Bury N, Barron L, Snape J, Owen S (2018) Machine learning for environmental toxicology: a call for integration and innovation, vol 52. PDF, p 12953

Morrison J, Watts G, Hobbs G, Dawnay N (1970) Field-based detection of biological samples for forensic analysis: Established techniques, novel tools, and future innovations. PDF

Rahmati S, Doherty W, Amani Babadi A, Che Mansor MSA, Muhd Julkapli N, Hessel V Ostrikov K (Ken) (2021) Gold–carbon nanocomposites for environmental contaminant sensing. ncbi.nlm.nih.gov

Ren P, Shui W, Liu J, Fan Y, Zhao W, Zhou M (2018) A sketch-based rapid modeling method for crime scene presentation. PDF

Richter CF, Lortie CJ, Kelly, T. L., Filazzola, A., Nunes, K. A., Sarkar, R. (2020). Online but not remote: Adapting field-based ecology laboratories for online learning. ncbi.nlm.nih.gov

Robinson JM, Pasternak, Z., Mason, C. E., Elhaik, E. (2021). Forensic applications of microbiomics: a review. ncbi.nlm.nih.gov

Stephen M (2008) Environmental forensics and the importance of source identification. https://doi.org/10.1039/9781847558343-00001

Stuart MB, McGonigle AJS, Willmott JR (2019) Hyperspectral imaging in environmental monitoring: a review of recent developments and technological advances in compact field deployable systems, vol 19. PDF

Ustin SL, Middleton EM (2021) Current and near-term advances in Earth observation for ecological applications. ncbi.nlm.nih.gov

Van der Linden S, Okujeni A, Canters F, Degerickx J, Heiden U, Hostert P, Priem F, Somers B, Thiel F (2019) Imaging spectroscopy of urban environments. PDF

Yadav B, Singla A, Gupta P, Rashid S (2022) Study of toxin 'Ricinine' present in Ricinus communis by TLC and GC-MS A forensic perspective. Res J Pharm Technol 1018–1022:1018. https://doi.org/10.52711/0974-360X.2022.00170

Zichar MARIANNA (2016) Guidelines for cost-effective geovisualization in digital forensics, vol 10. PDF, p 117

GPSR Compliance

The European Union's (EU) General Product Safety Regulation (GPSR) is a set of rules that requires consumer products to be safe and our obligations to ensure this.

If you have any concerns about our products, you can contact us on ProductSafety@springernature.com

In case Publisher is established outside the EU, the EU authorized representative is:

Springer Nature Customer Service Center GmbH
Europaplatz 3
69115 Heidelberg, Germany

Batch number: 10755329

Printed by Printforce, the Netherlands